CELLULAR NEURAL NETWORKS

CELLULAR NEURAL NETWORKS

Edited by

T. Roska
Computer and Automation Institute,
Hungarian Academy of Sciences
Hungary

J. Vandewalle
Katholieke Universiteit Leuven
Belgium

JOHN WILEY & SONS
Chichester · New York · Brisbane · Toronto · Singapore

Other Wiley Editorial Offices

John Wiley & Sons, Inc., 605 Third Avenue,
New York, NY 10158-0012, USA

Jacaranda Wiley Ltd, G.P.O. Box 859, Brisbane,
Queensland 4001, Australia

John Wiley & Sons (Canada) Ltd, 22 Worcester Road,
Rexdale, Ontario M9W 1L1, Canada

John Wiley & Sons (SEA) Pte Ltd, 37 Jalan Pemimpin #05-04,
Block B, Union Industrial Building, Singapore 2057

British Library Cataloguing in Publication Data

A catalogue record for this book is available from the British Library

ISBN 0 471 93836 X

Typeset in 10/12 Times by Mathematical Composition Setters Ltd, Salisbury, Wilts
Printed and bound in Great Britain by Antony Rowe Ltd, Chippenham, Wilts

CONTENTS

FOREWORD

The phenomenal blossoming of the CNN from the inception of its basic architecture in 1987 to the invention of the CNN universal machine and supercomputer[1] in 1992 testifies to its explosive vitality and unprecedented rate of advance. Rarely has an idea triggered so much scientific interest, galvanized so much research activity, and created so many applications in so short a time span.

Born of necessity, the original CNN offers a VLSI compatible neural network of locally interconnected 'neurons'. Because all neurons and their interconnections are identical, each CNN is uniquely specified by at most 19 real numbers (many are equal to zero in practice) in most applications, regardless of the size of the CNN. In other words, the function of a CNN of any size, be it a hundred pixel, or a million pixel, array, can be controlled by changing at most 19 parameters from *one and the same circuit*. This array-size independent feature is the key to the CNN's programmability, both from the software and hardware point of view.

From a software perspective, the 19 numbers can be interpreted as a neural computer program which instructs the neurons to carry out a *specific* task, or algorithm. Implementing the same algorithm on a conventional digital computer would have required writing a computer program with thousands of instructions. With 19! permutations among these 19 parameters, each one corresponding to a distinct CNN, the amount of *compression* in program instructions is truly astonishing when compared to a Von Neumann digital computer. From the CNN perspective, it is instructive to think of each permutation of 19 *parameters* as a *code* for a particular computer program, where each parameter is a real number ranging from $-\infty$ to $+\infty$. It is also instructive to think of these 19! permutations where each parameter can be *any* real number as an infinite family of Green functions or impulse responses for nonlinear systems. But above all, it is instructive to think of each permutation of these 19 parameters as an *analog program* which uses not the number-crunching power of a digital computer to carry out successive sequences of mathematical calculations in binary codes, but rather the laws of nature—Kirchhoff laws and device physics—to implement a *parallel* dynamical process in real time—less than 1 μs in current 2 μm VLSI technology.

From a hardware perspective, only 19 wires need to be brought out from the chip to the pins to allow unrestricted external programming, regardless of the array size. This unique feature of the CNN makes it possible to design a single chip whose function can be programmed externally. Moreover, unlike a *serial* digital computer, where the computing time of an image processing array scales *exponentially* with the array size, the dynamic settling time of a CNN, and hence its image processing time, scales linearly in many cases. In fact, for certain applications, e.g., the noise-removal CNN, the processing time is virtually independent of the

[1] See the Special Issue on Cellular Neural Networks, *IEEE Trans. Circuits and Systems*, **CAS-40**, Part II, pp 163–173, March 1993.

array size. Moreover, because of the uniformity (space-invariance property) of the CNN interconnections, it is possible to build a large array of CNN chips by matching the boundary conditions between chips. This means that the array size of future CNNs is not restricted by the number of neurons that can be crammed into a chip.

The original CNN has been generalized by many researchers in many directions. In one direction, the introduction of *nonlinear* and *time delay* templates has significantly enlarged the scope of CNN applications to include a great variety of time-intensive computing tasks, e.g., global optimization, counting, sorting, median filtering, eigenvalue calculation, etc. In another direction, a one million pixel hardware accelerator has been developed which instantly converts a PC to a CNN workstation capable of emulating a CNN at high speed but at a fraction of the cost of a high-power computer necessary to perform the same calculations. In yet another direction, a discrete-time version of the CNN which can be easily realized with standard switched capacitor technology has been successfully demonstrated in chip form, and which exhibits a number of useful features.

But the most significant generalization of the CNN is the recent invention of the CNN universal machine and supercomputer which incorporates within *each* cell a relatively small chip area for a *local analog memory*, a *local logic memory*, and a communication and control unit for both local and global interfacing. This conceptual breakthrough transforms the previously template programmable CNN to a *stored program* CNN which inherits all of the programmable features of a digital computer except that the CNN universal machine works in real time. Just like the digital computer, a CNN language and a CNN operating system have been developed which allow the user to write a program for processing an input image via a *sequence* of CNNs having different templates, each one tailored for a specific task. Because each CNN template requires less than 1 μs to operate, a very large number of templates can be processed in the sequence prescribed by the stored program so that the final result is still achieved in real time. The analog and logic memory circuitry requires only a relatively small chip area because only a very few memory locations are needed—they are analogous to *registers* and *buffers* in a digital computer. Unlike in a digital memory where each *word* in the register is a string of binary numbers, each *word* in the local logic memory contains only *one* state (e.g., 'YES' or 'NO'). This choice of a 'primitive' logic memory is the key to the invention of the CNN universal machine because the need for including an area-intensive *analog-to-digital converter* within each cell is thereby obviated, and the logic memory can be implemented with very little increase in the chip area.

Because the stored program for the CNN universal machine involves both *analog* (real-time dynamics) and *logic* (AND, OR, NOR, etc.) operations, such a CNN program is aptly called an *Analog-Logic program*, or *Analogic* for simplicity. A rough estimate of a CNN universal machine operating with the CNN language and operating system shows that its computing power is greater than the current generation of general purpose supercomputers. Yet, the CNN universal machine can be realized in a *monolithic* chip. Such a chip is no less than the analog equivalent of the microprocessor whose invention had spawned the personal computer industry. Consequently, there is little doubt that the CNN universal machine will be used as an *analog microprocessor* in a new generation of powerful real-time smart machines. Indeed, if the analogy to the development of the personal computer industry alluded to earlier holds, one can predict the birth of a new industry based on the CNN universal machine and supercomputer where intelligent information and cognition machines which mimic certain known aspects of the human brain will be its mainstay.

The CNN universal machine, with its combined analog and logic circuitry, is the only known neural network whose dynamics resembles the 'analog-like' brain functions implemented in the *right* hemisphere, and the 'logic-like' brain functions implemented by the *left* hemisphere of the human brain. In other words, the CNN universal machine represents a crude attempt to mimic the 'cerebral asymmetry' of the human brain.

The CNN universal machine has recently been proved to be an *analog Turing machine*. In particular, an analog version of the game of Life has been implemented as a CNN machine, thereby proving its *universality* in the sense of Turing. This fundamental result proves that the CNN machine can solve any problem solvable by a digital computer.

With the advent of the CNN universal chip on the horizon, bolstered by the concurrent rapid development of many CNN learning and adaptive algorithms, the time seems to be ripe for an industrial-scale technology transfer where many high-tech innovations from other areas can be brought to bear. These technologies include electromagnetic, optical, and even quantum effects. For example, a novel CNN output interfacing technique has recently been invented whereby the *state* of certain neurons having a prescribed property, e.g., abnormal cancer cells, microscopic defects in machine parts, defects in textile patterns, etc., will emit a high-frequency oscillation whose location can be easily pinpointed by an external antenna, thereby saving precious chip areas for higher cell density. It is also conceivable that the great strides in quantum dot devices will someday make it possible to build super-dense CNN arrays where each neuron will be the size of a quantum dot. It is this vision of the immense potential applications of the CNN universal machine and supercomputer that makes this excellent book both timely and exciting.

Leon O. Chua

INTRODUCTION

The concept of Cellular Neural Networks (CNN) was conceived in 1987 in Professor Leon O. Chua's laboratory in Berkeley. Their seminal publication (L. O. Chua and L. Yang, *IEEE Transactions on Circuits and Systems*, October, 1988) soon convinced many researchers that it is a very powerful new paradigm, and within two years about one hundred authors sent their papers to the first IEEE International Workshop on Cellular Neural Networks and their Applications (CNNA '90). This meeting in Budapest was a real success, and soon the *International Journal of Circuit Theory and Applications* announced that a Special Issue would be devoted to this topic. This Special Issue is now an archival document on CNN containing 13 papers on almost all aspects of CNN. This book is, in fact, an edited version of this issue with two additional papers. One is a tutorial and the other one exposes the biological relations. This book is the first on CNN, and is thus self-contained. However, it is strongly advised that you study in addition the original seminal paper of L. O. Chua and L. Yang.

CNN (also called cellular nonlinear network) is now a powerful new paradigm of 3D locally connected analog processor arrays. Its stored program architecture, the recently invented CNN Universal Machine, marks the first stored program analog array computer.

The articles in this book cover many interesting topics. In addition to the first paper, which is a tutorial with a quite extensive bibliography, the book covers four areas: (i) theory, (ii) VLSI implementations, (iii) CNN design tools, and (iv) applications.

The basic idea of the CNN, exposed in the original paper, is very simple yet powerful. CNN is an analog processor array placed on a regular 2D or 3D geometric grid. Each analog processor (processing continuous valued signals) is interacting with the others within a finite neighborhood only. The radius of this neighborhood (also called the receptive field) is normally much smaller than the linear measure of the whole array. The interaction pattern is called the cloning template. This original idea remained invariant during the last few years when researchers introduced a lot of special cases of CNN for specific purposes. Local connectivity is a must for efficient complex VLSI design. Though it is a restriction of the architecture, yet it was shown that CNN is as general as a Turing machine.

CNN is now used in several application areas. Image processing, artificial vision, solving PDEs, and modeling biological systems are just some of the many possibilities. All problems where the signal array is represented on a geometric grid call for a CNN approach.

The first, *tutorial*, paper recalls the basic CNN definition and shows, through examples, the basic notions and possibilities. These examples include: global optimization in a single transient, color halftoning, and a CNN algorithm containing several cloning templates for detecting almost invisible defects in a textile pattern.

The part on *theory* contains six papers and starts with the description and analysis of discrete-time CNN. In many applications this is the only means to slow down the high-speed

CNN chip to the real life application. In addition it provides additional flexibility and robustness. The next paper shows the various template-type extensions, nonlinear and delay-type, as well as the non-uniform grids and some stability results. The relation between CNN, cellular automaton, and systolic array is analyzed. In the original paper it was also shown how the CNN can be applied to calculate the approximate solution of diffusion-type PDEs (discretized in space). The next paper introduces a CNN for a class of wave-type PDEs and shows an interesting application for signature analysis.

Stability is a key issue in CNN. This is considered next. It is shown how the existing stability results for symmetric and positive cell-linking templates can be extended to many other classes by appropriate and simple template transformations. Another type of qualitative result is presented in the next paper. A path-following method is introduced for finding multiple equilibrium points in cellular neural networks. Finally, this part is completed by a paper showing the CNN in a system setting and giving guidelines for CMOS circuit design.

In the section on *VLSI implementations* three different questions are addressed. The first paper here describes the details of the design of the first fully tested CNN chip. It has an impressive speed, almost one tera (trillion) operations per second calculated on a 1 cm^2 silicon area. The next paper shows the architecture and design of the first programmable CNN silicon chip, including the circuitry for combining logic with the analog array dynamics. The testability issue is exposed briefly in the third paper of this section.

As to the *design tools*, a digital multiprocessor hardware accelerator board is described next. This PC add-on board provides workstation computing power on a PC when CNN algorithms are used. In addition it serves as a development system for CNN template and algorithm design since it has built-in template and image libraries, camera and optical scanner inputs, and several other convenient input/output possibilities, including video.

The last section covers some *applications*. A handwritten character recognizer for Latin and Japanese characters, the first CNN motion detection algorithms, and a template design methodology for shape extractions are considered in detail. Finally, it is shown that the CNN framework is a natural mode for many biological sensory information processing organs and systems.

The index is short but, we hope, instructive. It is the first index of this field. We have tried to summarize the key notions and subjects specifically relevant to CNN.

The book also contains a CNN software simulator diskette. The program is called CNNS, produced and provided by the Research Laboratory of Professor Roska at the Computer and Automation Research Institute of the Hungarian Academy of Sciences in Budapest. This program has already been used in many laboratories all over the world. CNNS runs under MS DOS, and is accompanied by a user's manual. Thus the reader can start using CNN immediately.

This book is a primer and a first reference for the world of CNN. New results are emerging continuously, special sessions and conferences as well as new special issues mark the growing interest. We do hope the readers will be new contributors to this exciting field.

Special thanks are due to many individuals. First of all to the Editor of the *International Journal of Circuit Theory and Applications*, Professor J. O. Scanlan who discovered the impact of this field in its early stage and provided the forum of this distinguished journal. We are indebted to Professor Leon O. Chua, the leader and originator of this area, for his constant help and encouragement. The kind assistance and patience of the people responsible at the Publisher, Ann-Marie Halligan and Laura Denny, is gratefully acknowledged.

Tamas Roska
Joos Vandewalle
Budapest, Leuven, Berkeley, January 1993

THE CNN PARADIGM—A SHORT TUTORIAL

L. O. CHUA†, T. ROSKA‡, T. KOZEK‡, and Á. ZARÁNDY‡

†Department of Electrical Engineering and Computer Sciences and the Electronics Research Laboratory, University of California at Berkeley

‡Computer and Automation Institute of the Hungarian Academy of Sciences, Budapest

1. INTRODUCTION

Since its invention in 1988[1,2] the Cellular Neural/nonlinear Network (CNN) paradigm, though it developed rapidly, retained its two basic concepts: geometrically *local connectedness* of the 3D processor array and *analog dynamics* of the cell processors and interactions. The CNN paradigm has evolved to cover a very broad class of problems and frameworks, and the recent invention of the CNN Universal Machine and Supercomputer[99] provides a unified platform and a prototype architecture for a new type of stored program: the analogic† CNN algorithm.

This tutorial serves two goals: to give a short introduction to CNN, establishing notations for the rest of the book, and to highlight some capabilities of the paradigm through instructive examples. In addition, a precise definition of the CNN architecture is given and the bibliography gives broad coverage of the literature.

The many applications of the paradigm (image processing, pattern analysis including shape, motion, color, and depth, solving various types of partial differential equations, modeling of various biological effects, etc.) that appeared within a few years after the basic invention, and the rigorously proven theoretical capabilities of CNN justify the unusual interest in this new area. The first tested CNN chips promise unprecedented computing power, in the range of tera XPS (10^{12} operations per second on a $2\ \text{cm}^2$ chip).

In Section 2 we introduce the simple CNN framework: the basic circuit implementation, the canonical equations, the role of the cloning template, different types of template elements, and a simple example. The basic notations of this section are used throughout this book. The general definition of the CNN paradigm is given in Section 3. Section 4 contains three instructive examples illustrating the use of a nonlinear propagating template, multilayer CNNs, and an analogic CNN algorithm. The three examples represent, at the same time, useful and difficult problems: calculating global maxima in real-time, color halftoning, and a complex pattern detection task. In the bibliography, though it is not complete, we have tried to collect the major basic references and many application-oriented results. Some other recent results can be found indirectly, in the special issues and conference proceedings listed in the bibliography.

2. THE SIMPLE CNN FRAMEWORK

In the simplest case, the CNN is an array of simple, identical, nonlinear, dynamical circuits placed on a 2D geometric grid.[1,2] Different grid types are shown in Figure 1. (Unless noted otherwise, a rectangular grid is supposed.) Duplicating the grids in 3D, a multi-layer CNN can be constructed. Each node on the grid is occupied by a cell whose structure is shown in Figure 2. The state capacitor C and the resistor R_x constitute the cell kernel. The state voltage v_{xij} across the capacitor at node (i,j) controls the nonlinear

† Analogic is a contraction for 'analog' and 'logic' computation—two key features of our CNN Universal chip which differs from other parallel computers. We emphasize that 'logic' involves only the symbolic variables 'yes' or 'no'. There is *no binary* A/D conversion during the entire computation.

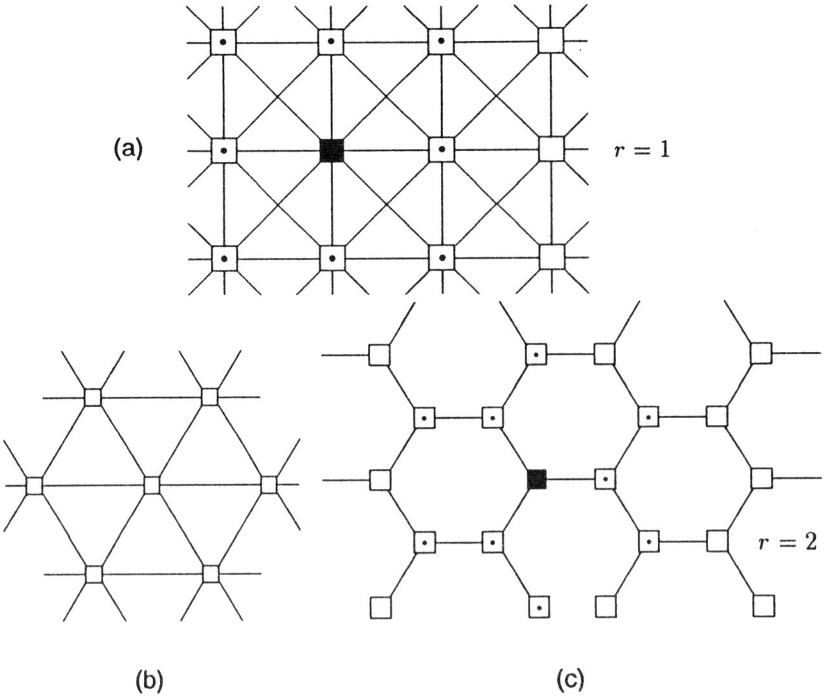

Figure 1. Different grid types. The dotted cells show the $r = 1$ and $r = 2$ neighborhoods of the black cell on different grid types

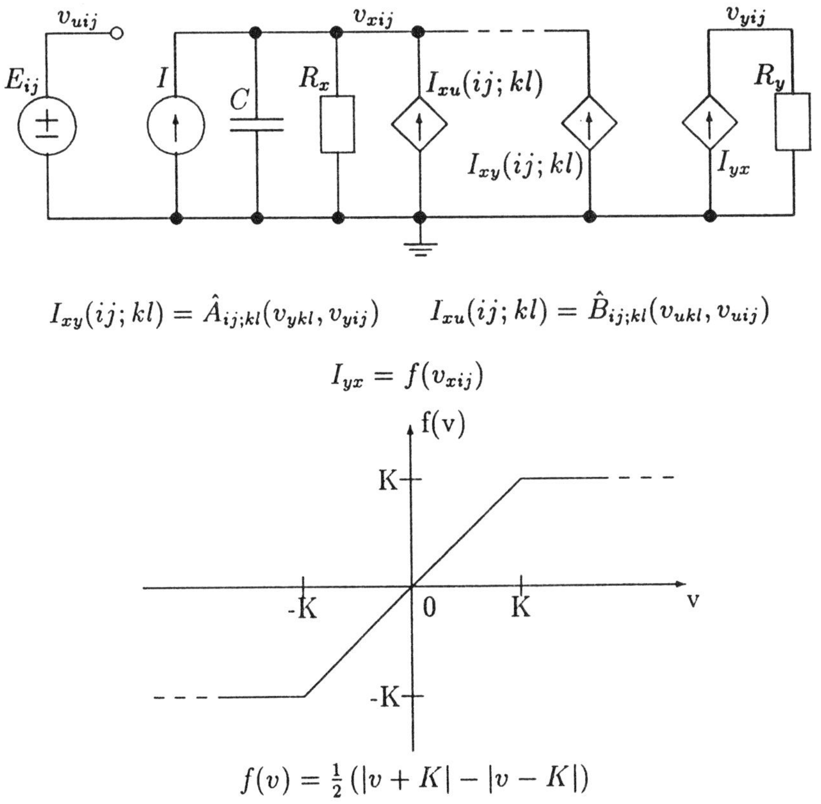

Figure 2. A simple CNN cell with piecewise linear nonlinearity

current source at the output of the cell producing the output voltage v_{yij}. Usually, this source has sigmoid-type piecewise linear transfer characteristics ($f(\cdot)$), also shown in Figure 2. In fact, these elements represent a simple amplifier with input capacitor and resistor. Each cell receives external signals through its input v_{uij} and has an independent current source I_{ij}. As indicated by the transconductances, the state voltage of a given cell is influenced not only by its own input, bias current I_{ij} and, through a feedback, its output, but also by the input and output voltages of the neighboring cells. These interactions are implemented by voltage-controlled current sources as shown in Figure 2.

Assuming normalized parameters (i.e. $C = 1$, $R_x = 1$, and $R_y = 1$) and linear interactions the dynamics of the array of $M \times N$ cells are determined by the state and output equations:

$$\dot{v}_{xij}(t) = -v_{xij}(t) + \sum_{C(kl) \in N_r(ij)} A(ij;kl)v_{ykl} + \sum_{C(kl) \in N_r(ij)} B(ij;kl)v_{ukl} + I_{ij} \tag{1}$$

$$v_{yij} = f(v_{xij}) \tag{2}$$

$$1 \leqslant i \leqslant M \qquad 1 \leqslant j \leqslant N$$

where $v_{uij}(t)$, $v_{xij}(t)$, and $v_{yij}(t)$ are the input, state, and output signals of cell $C(ij)$, respectively. Each cell C_{ij} is interacting with its neighbors (C_{kl}) within a finite neighborhood $N_r(ij)$ defined by

$$C_{kl} \in N_r(ij) \qquad \text{iff} \quad \max\{|k-i|, |l-j|\} \leqslant r \tag{3}$$

Cells corresponding to the $r = 1$ neighborhood in a rectangular grid and those in the $r = 2$ neighborhood in a hexagonal grid are indicated by dotted cells in Figure 1a and c, respectively. At cells near the edges of the array certain cells are missing from the neighborhood. Terms corresponding to the missing cells are omitted in (1).

It is clear from (1) that, given an initial state and an input signal, the behavior of the system is determined exclusively by the interaction coefficients $A(ij;kl)$ and $B(ij;kl)$. If the value of these, in the simplest case linear, interaction terms depends only on the relative position of cells C_{ij} and C_{kl}, the interaction pattern around a given cell is translation-invariant, i.e. a CNN array of arbitrary size is defined by only a few parameters. This set of parameters, arranged in matrices, is called a *cloning template*. In the case of the smallest neighborhood ($r = 1$)

$$A = \begin{bmatrix} a_{i-1,j-1} & a_{i,j-1} & a_{i+1,j-1} \\ a_{i-1,j} & a_{i,j} & a_{i+1,j} \\ a_{i+1,j+1} & a_{i,j+1} & a_{i+1,j+1} \end{bmatrix}, \qquad B = \begin{bmatrix} b_{i-1,j-1} & b_{i,j-1} & b_{i+1,j-1} \\ b_{i-1,j} & b_{i,j} & b_{i+1,j} \\ b_{i+1,j+1} & b_{i,j+1} & b_{i+1,j+1} \end{bmatrix}, \qquad I = c$$

are the feedback template, feedforward template, and the bias term, respectively, constituting a cloning template.

The inherent capabilities of CNN can be extended by introducing nonlinear ($\hat{A}_{ij,kl}$) and delay type ($A^{\tau}_{ij,kl}$) interactions.[3] The state equation is modified then as:

$$\begin{aligned} \dot{v}_{xij}(t) = {} & -v_{xij}(t) + I \\ & + \sum_{C(kl) \in N_r(ij)} \hat{A}_{ij;kl}(v_{ykl}, v_{yij}) + \sum_{C(kl) \in N_r(ij)} \hat{B}_{ij;kl}(v_{ukl}, v_{uij}) \\ & + \sum_{C(kl) \in N_r(ij)} A^{\tau}_{ij;kl} v_{ykl}(t) + \sum_{C(kl) \in N_r(ij)} B^{\tau}_{ij;kl} v_{ukl}(t) \end{aligned} \tag{4}$$

It is also possible to use nonlinear operators in the delay templates which are denoted by $\hat{A}^{\tau}_{ij;kl}(v_{ykl}, v_{yij})$ and $\hat{B}^{\tau}_{ij;kl}(v_{ykl}, v_{yij})$. The use of nonlinear templates is demonstrated by the first example of the next section.

A significant feature of CNN is that it has two independent input capabilities: the generic input and the initial state of the cells. Normally they are bounded by

$$|v_{uij}(t)| \leqslant 1 \quad \text{and} \quad |v_{xij}(0)| \leqslant 1 \tag{5}$$

Similarly, if $|f(\cdot)| \leqslant 1$ then $|v_{yij}(t)| \leqslant 1$.

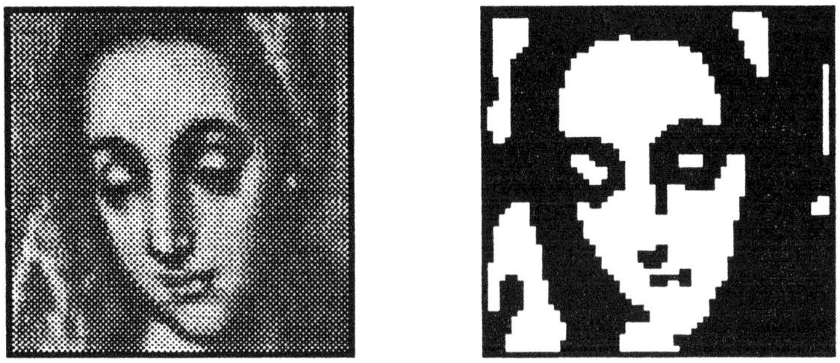

(a) (b)

Figure 3. The input and the output of an averaging operation

When used as an array processing device, the CNN performs a mapping

$$\left.\begin{matrix} v_{xij}(0) \\ v_{uij}(t) \end{matrix}\right\} \overset{F}{\mapsto} v_{yij}(t) \tag{6}$$

where F is a function of the cloning template $T(A, B, I)$. For example, if a grey-scale image is coded such that $+1$ corresponds to black pixels, -1 to white ones, with the grey scale values in between, and this image is loaded into the initial states $v_{xij}(0)$ of the cells, then the mapping $F(T)$ is an image processor with an output image $v_{yij}(\infty)$. The cloning template T defines the image-processing operation. If we define T as

$$A = \begin{bmatrix} 0 & 1 & 0 \\ 1 & 2 & 1 \\ 0 & 1 & 0 \end{bmatrix}, \qquad B = \begin{bmatrix} 0 & 0 & 0 \\ 0 & 0 & 0 \\ 0 & 0 & 0 \end{bmatrix}, \qquad I = 0,$$

then $v_{yij}(t)$ tend to constant values $[-1, 1]$, and our CNN image processor performs averaging[1] over a 3×3 window. An example is shown in Figure 3. (Here the B template is zero, therefore the value of $v_{uij}(t)$ is irrelevant. The generic input is not used here.) In this case the settling time is independent of the number of cells.

3. THE GENERAL DEFINITION OF THE CNN PARADIGM

Following the basic assumptions of Reference 1, and the simple definition given above we can define the generic notion of the CNN[100] as:

Definition The CNN is a

(i) 2-, 3-, or n-dimensional **array of**
(ii) mainly identical **dynamical systems**, called cells, which satisfies two properties:
(iii) most **interactions** between the cells **are local** within a finite radius r, and
(iv) all **state variables are continuously valued functions**.

Remarks

1. The space variable is always discretized: each cell is identified by 2, 3 (or n) integers $(i, j, k \ldots, n)$.
2. The time variable t may be continuous or discrete.
3. The interconnection effect represented by the *cloning template* may be a *nonlinear* function of the state x, output y, and input u of each cell, within the neighborhood N_r of radius r, as well as that

of time t (e.g. *time delay* or time-varying coefficients). The cloning template may be space-variant or space-invariant.

4. The dynamical system is determined uniquely by an evolution law (e.g. ODE, discrete maps, differential-difference equations, functional equations, etc.) such that given $x(t)$ and $u(t)$ for all $t \in (-\infty, t_0)$, or $t \in (-\infty, \ldots -\Delta t, 0, \Delta t, \ldots n_0\Delta t)$, $x(t)$ and $u(t)$ are uniquely determined for all $t \geqslant t_0$, or $n\Delta t$ with $n \geqslant n_0$. This includes the boundary conditions, as well.
5. Occasionally, the dynamical system and/or the interconnections may be perturbed by some noise sources of known statistics.

In general, the cell may have multivariable input, state, and output vectors. Accordingly, the cloning template T contains more interaction terms.

As to the array grid, the cell (or processor) dynamics, the interaction, and the mode of operation, etc., there exist many choices. They are classified e.g. in Reference 97a.

Considering CNN cloning templates as subroutines, and combining them in series or in parallel, we can build up CNN algorithms. In addition, embedding simple logic operations into these algorithms we can define *analogic* CNN algorithms. Although the formal definition is given elsewhere,[99] to illustrate these algorithms, a simple but useful example is shown in Section 4.3. Analogic CNN algorithms can be represented by flowcharts, language, and machine code. Using the CNN Workstation[99] all these are available.

4. APPLICATION EXAMPLES

4.1. A nonlinear propagating template for calculating global maximum

When applied to topological analysis of images, the locally connected CNN offers a flexible and efficient tool for extracting local geometrical features. It is more surprising, however, that global properties can also be detected with propagating-type templates. This first example shows how the value and exact location of the maximum of an arbitrary 3D surface can be determined by a sequence of two templates.

The first template determines the value of the maximum. The basic idea is this: let us consider a CNN grid where the spatially discretized surface is loaded into the initial states ($v_{xij}(0)$) with normalized values in the range $[-1, +1]$. To perform the desired operation the behavior of the cells must obey the following two rules:

1. If a cell has a greater output value than any of its neighbors, the state (and output) value of the cell stays constant.
2. If there is a greater output value in the neighborhood of the cell then its state (and output) value starts rising until they become equal to the greatest value in the neighborhood.

These rules will result in propagating behavior, and after the transient settles, all output values will be equal to the originally highest one, the global maximum. The template which defines such a CNN can be seen in Figure 4. The central +1 value makes the CNN satisfy the first rule, and the nonlinear function is responsible for the second one.

The second step is to detect the exact location of the maximum with the following simple template:

$$A = \begin{bmatrix} 0 & 0 & 0 \\ 0 & 2 & 0 \\ 0 & 0 & 0 \end{bmatrix}, \qquad B = \begin{bmatrix} 0 & 0 & 0 \\ 0 & -1 & 0 \\ 0 & 0 & 0 \end{bmatrix}, \qquad I = 0{\cdot}05$$

This template compares the result of the first one and the original picture loaded into the inputs and the initial states of the CNN array, respectively. Figure 5 shows the stages of the two-step algorithm. It is clear from the above description that, if there are multiple global maxima on the surface, all of them will be detected.

$$A = \begin{bmatrix} 0 & a & 0 \\ a & 1 & a \\ 0 & a & 0 \end{bmatrix}, \quad B = 0, \quad I = 0,$$

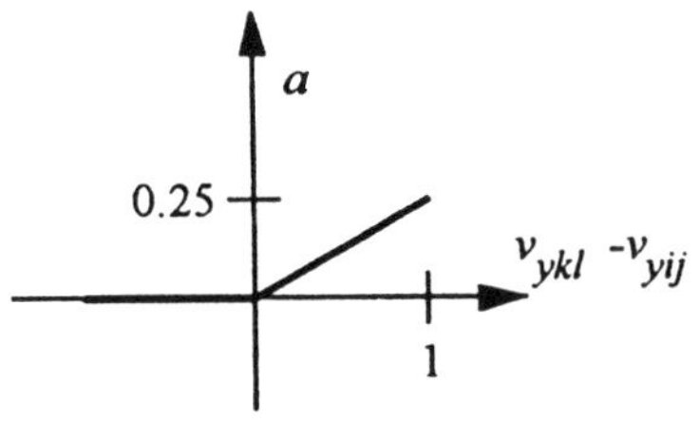

Figure 4. The global maximum value calculating template

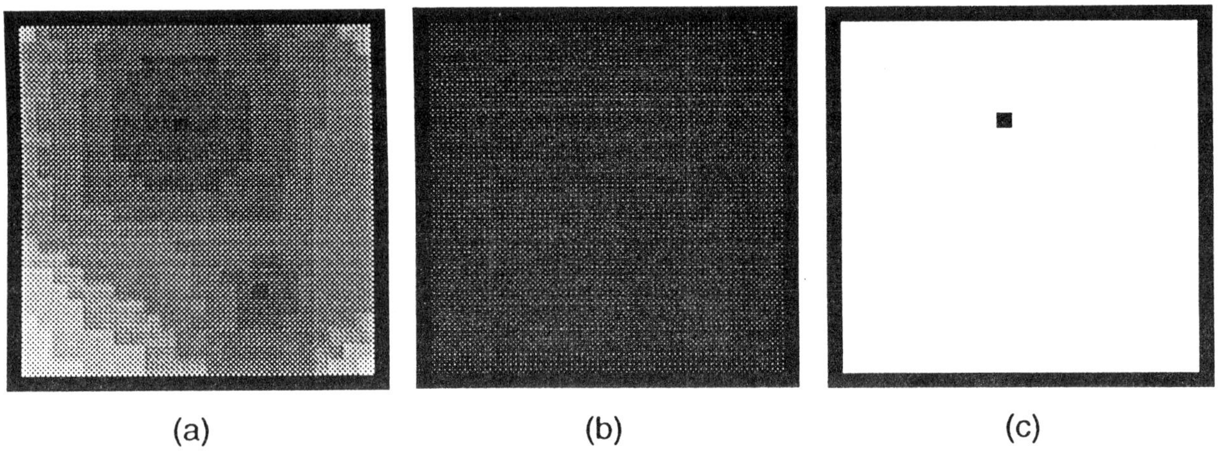

Figure 5. Global maximum calculation: the original surface (a), the value (b), and the location (c) of the global maximum

4.2. A four-layer CNN for color halftoning

Halftoning is the process in which a black-and-white picture is generated from a grey-scale one such that the black-and-white picture, when viewed from the appropriate distance, appears to be identical to the grey-scale original. CNN has already been used for halftoning.[36,67]. The color halftone method is similar, but it starts with the trichromatic decomposition of the original color picture. It then transforms the red, green and blue images into binary images, and the combination of these three bitmaps result in an 8 (2^3) color picture.

Color halftoning is performed by a multi-layer CNN structure that contains four layers. The architecture is shown in Figure 6. The input and the initial state of the first three layers are the three analog monochromatic pictures, and an eight-color image will appear on the output of the fourth layer.

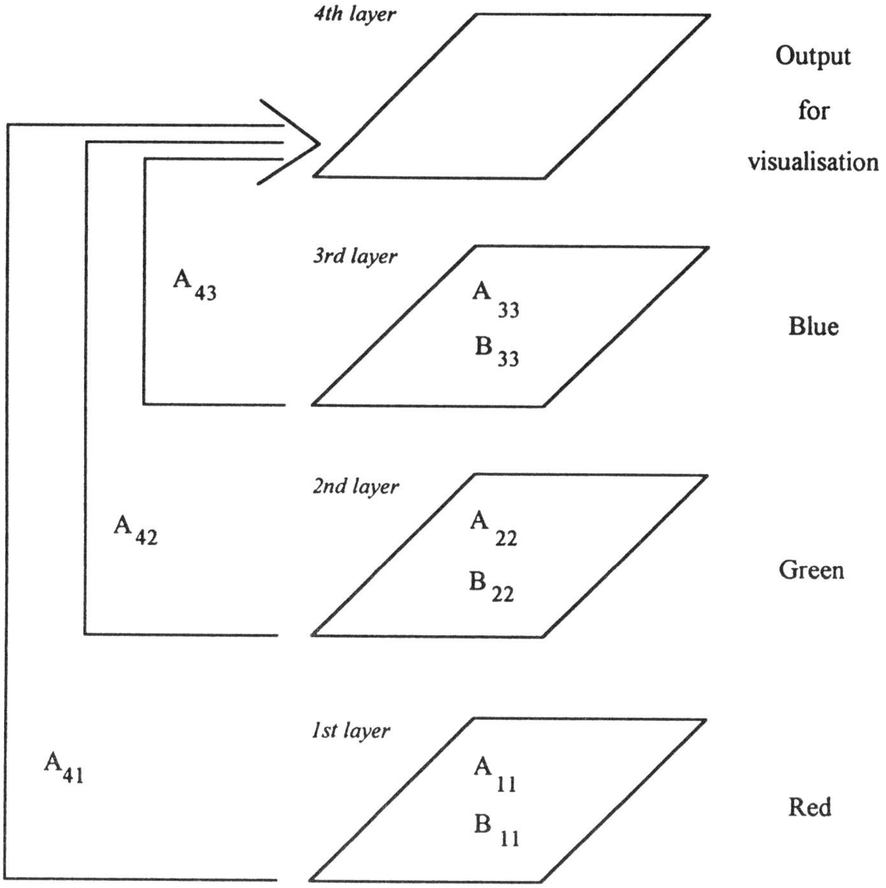

Figure 6. The structure of the multi-layer CNN used for color halftoning

The templates of the multi-layer CNN are the following:

$$A_{11}=A_{22}=A_{33}\begin{bmatrix} -0\cdot03 & -0\cdot08 & -0\cdot13 & -0\cdot08 & -0\cdot03 \\ -0\cdot08 & -0\cdot36 & -0\cdot6 & -0\cdot36 & -0\cdot08 \\ -0\cdot13 & -0\cdot6 & 1\cdot25 & -0\cdot6 & -0\cdot13 \\ -0\cdot08 & -0\cdot36 & -0\cdot6 & -0\cdot36 & -0\cdot08 \\ -0\cdot03 & -0\cdot08 & -0\cdot13 & -0\cdot08 & -0\cdot03 \end{bmatrix}$$

$$B_{11}=B_{22}=B_{33}=\begin{bmatrix} 0 & 0 & 0\cdot33 & 0 & 0 \\ 0 & 0\cdot24 & 0\cdot47 & 0\cdot24 & 0 \\ 0\cdot33 & 0\cdot47 & 1\cdot45 & 0\cdot47 & 0\cdot33 \\ 0 & 0\cdot24 & 0\cdot47 & 0\cdot24 & 0 \\ 0 & 0 & 0\cdot33 & 0 & 0 \end{bmatrix}$$

$$I_1=I_2=I_3=I_4=0,$$

$$A_{41}=\begin{bmatrix} 0 & 0 & 0 \\ 0 & 0\cdot125 & 0 \\ 0 & 0 & 0 \end{bmatrix},\qquad A_{42}=\begin{bmatrix} 0 & 0 & 0 \\ 0 & 0\cdot25 & 0 \\ 0 & 0 & 0 \end{bmatrix},\qquad A_{43}=\begin{bmatrix} 0 & 0 & 0 \\ 0 & 0\cdot5 & 0 \\ 0 & 0 & 0 \end{bmatrix}.$$

The color halftoning (the three bitmaps) is formed on the first three layers. On the fourth layer, we

(a) (b)

Figure 7. Demonstration of color halftoning: the original painting (a) and its color halftone representation (b)

generate an eight-color picture. This image, displayed in appropriate colors, is the visual result of the color halftoning process. Figure 7 (see the color pictures on the front cover) shows an original image and its color halftone representation. Due to the binary output of the first three layers, cells on the fourth layer can output the following values:

$$(\pm 1) \times 0\cdot 125 + (\pm 1) \times 0\cdot 25 + (\pm 1) \times 0\cdot 5 = \{-0\cdot 875;\ -0\cdot 625;\ -0\cdot 375;\ -0\cdot 125; 0\cdot 125; 0\cdot 375; 0\cdot 625; 0\cdot 825\}.$$

The output color coding is as follows:

white (red + green + blue): $-0\cdot 875$
turquoise (green + blue): $-0\cdot 625$
lilac (red + blue): $-0\cdot 375$
blue (blue): $-0\cdot 125$
yellow (red + green): $0\cdot 125$
green (green): $0\cdot 375$
red (red): $0\cdot 625$
black (no colors): $0\cdot 875$

Looking at the color coding and at the A templates connecting the fourth layer, it is clear that, e.g., a blue color appears in a pixel only if the corresponding pixel on the third layer (the blue layer) is +1.

Though the result contains only eight colors, from a large enough distance it looks like the original color picture with many hues and shades.

4.3. An analogic CNN algorithm for textile fault detection

Many complex image processing tasks can be solved by CNN, but usually it is not possible to find a single template for a given problem. We can decompose the task, however, and define intermediate stages, using several templates to solve the subtasks. In this way analogic CNN algorithms can be built up from analog or logical templates/operations performed sequentially or in parallel.

There are two main types of textile defects: knots and fiber breaks. Here an analogic CNN algorithm is presented which is able to detect both fault types. The algorithm works on a simple, loose woven textile

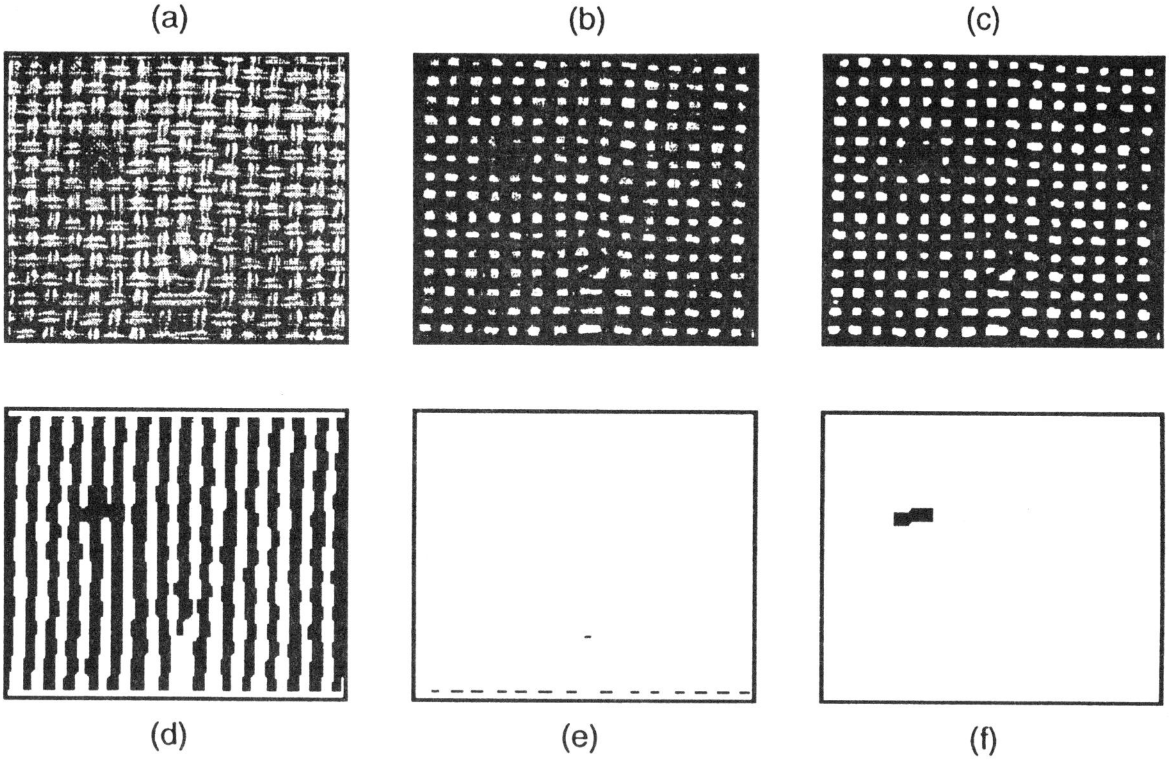

Figure 8. The procedure for textile fault detection: the original picture (a), its inverted version (b), the black-and-white fiber grid (c), the vertical fibers (d), the fiber ends (e), and the knot (f)

(Figure 8a). Both types of defects can be located by extracting the parallel fibers in one orientation and then finding their ends. The end of broken fibers will appear inside the picture, while those of the good ones are on the borders.

The grey-scale picture of a piece of textile with a broken fiber and a knot is shown in Figure 8a. The templates work on black fibers with a white background; therefore the inverted image shown in Figure 8b is used as the input to the algorithm. To extract the parallel fibers in a given orientation, first we have to transform the grey-scale picture to a black-and-white one. The averaging template introduced in Section 1 is suitable for this task. The result of the averaging (the black-and-white fiber grid) is shown in Figure 8c. To extract, for example, the vertical fibers, we only have to delete the horizontal ones. The vertical 'Cut-seven' template was designed for just this task:

$$A=\begin{bmatrix}0&0&1&0&0\\0&0&0\cdot5&0&0\\0&0&2&0&0\\0&0&0.5&0&0\\0&0&1&0&0\end{bmatrix},\quad B=\begin{bmatrix}0&0&1&0&0\\0&0&1&0&0\\0&0&1&0&0\\0&0&1&0&0\\0&0&1&0&0\end{bmatrix},\quad I=-5\cdot5$$

It is called 'Cut-seven', because it deletes each vertical connected component containing less than eight pixels, and it removes two pixels from both ends (the top and the bottom) of the longer components. The vertical fibers extracted by this template are shown in Figure 8d. On this image it is easy to locate the knot and the broken fiber. Finding the ends of the extracted lines with the following template shows the location of the fiber break:

$$A=\begin{bmatrix}0&0&0\\0&2&0\\0&0&0\end{bmatrix},\quad B=\begin{bmatrix}0&0&0\\0&2&0\\-1&-1&-1\end{bmatrix},\quad I=-5\cdot5$$

In Figure 8e the local southern places of the vertical fibers can be seen. The fiber end lying inside the picture signals the broken fiber.

From the same image with a different template the knots can be extracted, as well. Deleting the vertical fibers with a rotated 'Cut-seven' template results in an image where only the knot will remain, because it is the thickest horizontal connected component (Figure 8f). The flow diagram of the whole process is shown in Figure 9.

Complex analogic CNN algorithms, similar to this example, can be run on the CNN Workstation,[79] a hardware–software toolkit installed in a standard PC configuration. Figure 10 shows a CNN Workstation output with four displays showing the outputs of four different templates acting on the same image.

The algorithm of Figure 9 in a special language designed for analogic CNN algorithms is as follows:

```
& input_pic=''text grey.img''          ; picture definition
& aver_tem=''average.tem''             ; template definitions
& ver_cut7_tem=''cut7v.tem''
& hor_cut7_tem= ''cut7h.tem''
& southern_tem=''south.tem''
B&W_pic=sub(aver_tem,input_pic,*)     ; analogic CNN operation
display_output                         ; displays the result
ver_fibers_pic=sub(ver_cut7_tem,B&W_pic,*)
display_output
fiber_ends_pic=sub(southern_tem,ver_fibers_pic,*)
display_output
knots=sub(hor_cut7_tem,ver_fibers_pic,*)
display_output
```

where '&' introduces the picture and the template declarations, and a CNN operation (subroutine) is invoked with the following parameters:

```
output=sub(template,input initial state)
```

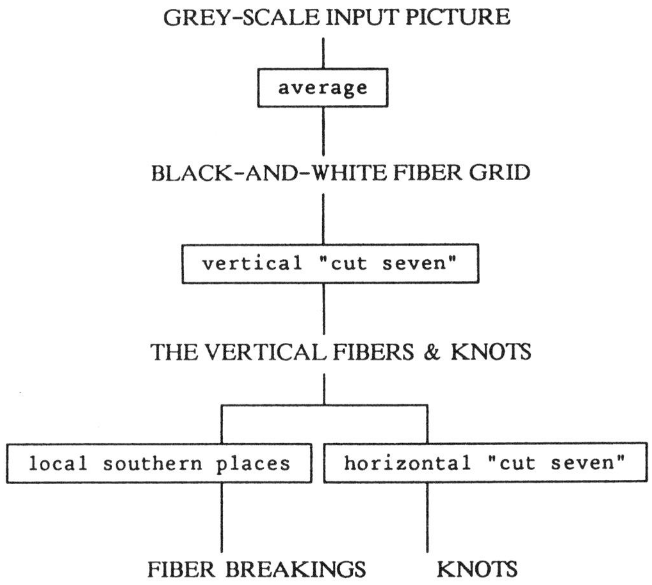

Figure 9. Flow diagram of the fault detection algorithm

Figure 10. The CNN workstation with four monitors, two video cameras, and a color image scanner

Comments on the right hand side explain the operations. More details of this language can be found in Reference 79.

ACKNOWLEDGEMENTS

This work has been supported by the Office of Naval Research, under grant N00014-89-J-1402, by the National Science Foundation, under grant MIP-89-12639, and by a joint research grant No. INT 90-01336 of the National Science Foundation and the Hungarian Academy of Sciences, as well as by the research grant No. 2578/91 of the National Research Fund of Hungary (OTKA).

BIBLIOGRAPHY

1. L. O. Chua and L. Yang, 'Cellular neural networks: theory,' *IEEE Trans. Circuits and Systems*, **CAS-35**, 1257–1272 (1988).
2. L. O. Chua and L. Yang, 'Cellular neural networks: applications,' *IEEE Trans. Circuits and Systems*, **CAS-35**, 1273–1290 (1988).
3. T. Roska and L. O. Chua, 'Cellular neural networks with nonlinear and delay-type template elements,' in *IEEE International Workshop on Cellular Neural Networks and Their Applications, Proceedings*, pp. 12–25, 1990. Extended version in *Int. J. Circuit Theory and Applications*, **20**, 469–481 (1992).
4. L. O. Chua, T. Roska, P. L. Venetianer and A. Zarandy, 'Some novel capabilities of CNN: Game of life and examples of mulitpath algorithms,' Report DNS-3-1992, Dual and Neural Computing Systems Laboratory, Hungarian Academy of Sciences, 1992.
5. J. A. Nossek, G. Seiler, T. Roska and L. O. Chua, 'Cellular neural networks: theory and circuit design,' Report TUM-LNS-TR-90-7, Technische Universität München, Dec. 1990. Also in *Int. J. Circuit Theory and Applications*, **20**, 533–553 (1992).
6. H. Harrer and J. A. Nossek, 'Discrete time cellular neural networks: architecture, applications and realization,' Report TUM-LNS-TR-90-12, Technische Universität München, Nov. 1990. Also in *Int. J. Circuit Theory and Applications*, **20**, 453–467 (1992).
7. J. Henseler and P. J. Braspennig, 'Membrain: a cellular neural network model based on a vibrating membrane,' *Int. J. Circuit Theory and Applications*, **20**, 483–496 (1992).

8. A. Radványi, K. Halonen and T. Roska, 'The CNNL simulator and some time-varying CNN templates,' Report DNS-9-1991, Dual and Neural Computing Systems Laboratory, Hungarian Academy of Sciences, 1992.
9. L. O. Chua and B. E. Shi, 'Exploiting cellular automata in the design of cellular neural networks for binary image processing,' Memorandum UCB/ERL M89/130, University of California at Berkeley Electronics Research Laboratory, Nov. 1989.
10. L. O. Chua and T. Roska, 'Stability of a class of nonreciprocal cellular neural networks,' *IEEE Trans. Circuits and Systems*, **CAS-37**, 1520–1527 (1990).
11. V. Cimagalli, 'A neural network architecture for detecting moving objects II.,' in *IEEE International Workshop on Cellular Neural Networks and Their Applications, Proceedings*, pp. 124–126, 1990.
12. N. Frühauf and E. Lüder, 'Realizations of CNNs by optical parallel processing with spatial light valves,' in *IEEE International Workshop on Cellular Neural Networks and Their Applications, Proceedings*, pp. 281–291, 1990.
13. K. Halonen, V. Porra, T. Roska and L. Chua, 'VLSI implementation of a reconfigurable cellular neural network containing local logic (CNNL),' in *IEEE International Workshop on Cellular Neural Networks and Their Applications, Proceedings*, pp. 206–215, 1990. Extended version in *Int. J. Circuit Theory and Applications*, **20**, 573–582 (1992).
14. T. Matsumoto, L. O. Chua and R. Furukawa, 'CNN cloning template: hole-filler,' *IEEE Trans. Circuits and Systems*, **CAS-37**, 635–638 (1990).
15. T. Matsumoto, L. O. Chua and H. Suzuki, 'CNN cloning template: shadow detector,' *IEEE Trans. Circuits and Systems*, **CAS-37**, 1070–1073 (1990).
16. T. Matsumoto, L. O. Chua and H. Suzuki, 'CNN cloning template: connected component detector,' *IEEE Trans. Circuits and Systems*, **CAS-37**, 633–635 (1990).
17. T. Matsumoto, T. Yokohama, H. Suzuki and R. Furukawa, 'Several image processing examples by CNN,' in *IEEE International Workshop on Cellular Neural Networks and Their Applications, Proceedings*, pp. 100–111, 1990.
18. T. Matsumoto, L. O. Chua and T. Yokohama, 'Image thinning with a cellular neural network,' *IEEE Trans. Circuits and Systems*, **CAS-37**, 633–635 (1990).
19. J. A. Nossek and G. Seiler, 'An equivalence between multi-layer perceptrons with step function type nonlinearity and a class of cellular neural networks,' Report TUM-LNS-TR-90-7, Technische Universität München, Dec. 1990.
20. T. Roska, T. Boros, P. Thiran, and L. O. Chua, 'Detecting simple motion using cellular neural networks,' in *IEEE International Workshop on Cellular Neural Networks and Their Applications, Proceedings*, pp. 127–138, 1990.
21. T. Roska, G. Bártfai, P. Szolgay, T. Szirányi, A. Radványi, T. Kozek, Z. Ugray and A. Zarándy, 'A digital multiprocessor hardware accelerator board for cellular neural networks: CNN-HAC,' in *IEEE International Workshop on Cellular Neural Networks and Their Applications, Proceedings*, pp. 160–168, 1990. Extended version in *Int. J. Circuit Theory and Applications*, **20**, 589–599 (1992).
22. T. Roska and A. Radványi, *CNND Simulator, Cellular Neural Network Embedded in a Simple Dual Computing Structure, User's Guide Version 3.0* Computer and Automation Institute, Hungarian Academy of Sciences, Budapest, 1990. Report No. 37/1990.
23. G. Seiler, 'Small object counting with cellular neural networks,' in *IEEE International Workshop on Cellular Neural Networks and Their Applications, Proceedings*, pp. 114–123, 1990.
24. K. Ślot, 'Determination of cellular neural network parameters for feature detection of two dimensional images,' in *IEEE International Workshop Cellular Neural Networks and Their Applications, Proceedings*, pp. 82–91, 1990.
25. L. Vandenberghe and J. Vandewalle, 'Finding multiple equilibrium points of cellular neural networks without enumeration,' in *IEEE International Workshop on Cellular Neural Networks and Their Applications, Proceedings*, pp. 45–55, 1990. Also in *Int. J. Circuit Theory and Applications*, **20**, 519–531 (1992).
26. J. E. Varrientos, J. Ramirez-Angulo and E. Sanchez-Sinencio, 'Cellular neural networks implementation: a current-mode approach,' in *CNNA*, pp. 216–225, 1990.
27. L. Yang, L. O. Chua and K. R. Krieg, 'VLSI implementation of cellular neural networks,' in *Proceedings of the IEEE International Symposium on Circuits and Systems*, pp. 2425–2427, IEEE, 1990.
28. G. G. Yang, 'Optical associative memory with invariance,' in *IEEE International Workshop on Cellular Neural Networks and Their Applications, Proceedings*, pp. 291–302, 1990.
29. F. Zou, S. Schwarz and J. A. Nossek, 'Cellular neural network design using a learning algorithm,' in *IEEE International Workshop on Cellular Neural Networks and Their Applications, Proceedings*, pp. 73–81, 1990.
30. F. Zou and J. A. Nossek, 'Stability of cellular neural networks with opposite sign templates,' Report TUM-LNS-TR-15, Technische Universität München, Dec. 1990.
31. M. Balsi, 'Remarks on the stability and functionality of CNN's with one-dimensional templates,' Report DNS-5-1991, Dual and Neural Computing Systems Laboratory, Hungarian Academy of Sciences, 1991. Under publication in *Int. J. Circuit Theory and Applications.*
32. T. Boros, K. Lotz, A. Radványi and T. Roska, 'Some useful, new, nonlinear and delay-type templates,' Report DNS-1-1991, Dual and Neural Computing Systems Res. Lab., Comp. Aut. Inst., Hung. Acad. Sci. (MTA SzTAKI), Budapest, 1991.
33. L. O. Chua and B. E. Shi, 'Multiple layer cellular neural networks: a tutorial,' in *Algorithms and Parallel VLSI Architectures* (E. F. Deprettere and A. van der Veen, eds.), vol. A: Tutorials, pp. 137–168, New York: Elsevier Science Publishers B. V., 1991.
34. L. O. Chua and P. Thiran, 'An analytic method for designing simple cellular neural networks,' *IEEE Trans. Circuits and Systems*, **CAS-38**, 1332–1341 (1991).
35. L. O. Chua, L. Yang and K. R. Krieg, 'Signal processing using cellular neural networks,' *Journal of VLSI Signal Processing*, **3**, 25–52 (1991).
36. P. P. Civalleri, M. Gilli and L. Pandolfi, 'On stability of cellular neural networks with delay,' *PROC. IEEE CNNA-92*, pp. 94–99, 1992.
37. K. R. Crounse, T. Roska and L. O. Chua, 'Image halftoning with cellular neural networks, *IEEE Trans. Circuits and Systems*, **CAS-40a** (1993).

38. J. M. Cruz and L. O. Chua, 'A CNN chip for connected component detection,' *IEEE Trans. Circuits and Systems*, **CAS-38**, 812–817 (1991).
39. N. Frühauf, E. Lüder, M. Gaiada and G. Bader, 'An optical implementation of space invariant cellular neural networks,' in *European Conference on Circuit Theory and Design*, pp. 42–51, 1991.
40. S. Fukuda, T. Boros and T. Roska, 'A new efficient analysis of thermographic images by using cellular neural networks,' Technical Report DNS-11-1991, Dual and Neural Computing Systems Laboratory, Hungarian Academy of Sciences, 1990. Under publication.
41. C. Guzelis and L. O. Chua, 'Stability analysis of generalized cellular neural networks,' Tech. Rep. UCB/ERL M91/23, University of California at Berkeley Electronics Research Laboratory, Mar. 1991.
42. H. Harrer and J. A. Nossek, 'An analog CMOS compatible convolution circuit for analog neural networks,' Report TUM-LNS-TR-91-11, Technische Universität München, 1991.
43. H. Harrer, J. A. Nossek and R. Stelzl, 'An analog implementation of discrete-time cellular neural networks, Report TUM-LNS-TR-91-14, Technische Universität München, June 1991. Extended version in *IEEE Trans. Neural Networks*, **3**, 466–477 (1992).
44. V. I. Krinsky, V. N. Biktashev and I. R. Efimov, 'Autowave principles for parallel image processing,' *Physica D*, **49**, 247–253 (1991).
45. A. Radványi and T. Roska, 'The CNN workstation—CNND version 4.1,' Technical Report DNS-12-1991, Dual and Neural Computing Systems Laboratory, Hungarian Academy of Sciences, 1991.
46. T. Roska, A. Radványi, T. Kozek and T. Boros, 'Dual CNN software library,' Report DNS-7-1991, Dual and Neural Computing Systems Laboratory, Hungarian Academy of Sciences, 1991.
47. T. Roska, 'Cellular neural networks: a state-of-the-art review,' in *Proc. European Conference on Circuit Theory and Design (ECCTD-91)*, pp. 1–9, 1991.
48. T. Roska and P. Szolgay, 'A comparison of various cellular neural network (CNN) realizations—a review,' in *Proc. 2nd Int. Conf. on Microelectronics for Neural Networks*, pp. 423–431, 1991.
49. T. Roska, 'Dual computing structures containing analog cellular neural networks and digital decision units,' in *Proc. IFIP Workshop on Silicon Architectures for Neural Nets, (Nice, 1990)*, pp. 233–244, North Holland, Amsterdam, 1991.
50. S. Schwarz and W. Mathis, 'A design algorithm for cellular neural networks,' in *Proc. 2nd Int. Conf. on Microelectronics for Neural Networks*, pp. 53–59, 1991.
51. G. Seiler and M. Hasler, 'Convergence of reciprocal cellular neural networks,' Report TUM-LNS-TR-91-12, Technische Universität München, June 1991.
52. B. E. Shi and L. O. Chua, 'A generalized cellular neural network of a novel edge detection algorithm,' Memorandum UCB/ERL M91/25, University of California at Berkeley Electronics Research Laboratory, Apr. 1991.
53. P. Szolgay and T. Kozek, 'Optical detection of layout errors of printed circuit boards using learned CNN templates,' Report DNS-8-1991, Dual and Neural Computing Systems Res. Lab., Comp. Aut. Inst., Hung. Acad. Sci. (MTA SzTAKI), Budapest, 1991.
54. T. Roska, C. W. Wu, M. Balsi and L. O. Chua, 'Stability and dynamics of delay-type and nonlinear cellular neural networks,' Memorandum UCB/ERL M91/110, University of California at Berkeley Electronics Research Laboratory, Dec. 1991.
55. L. O. Chua and C. W. Wu, 'On the universe of stable cellular neural networks,' ERL Memorandum UCB/ERL M91/31, University of California, Berkeley, 1991. Extended version in *Int. J. of Circuit Theory and Applications*, **20**, 497–517 (1992).
56. F. Zou and J. A. Nossek, 'A chaotic attractor with cellular neural networks,' *IEEE Trans. Circuits and Systems*, **CAS-38**, 811–812 (1991).
57. J. M. Cruz and L. O. Chua, 'Design of high speed high density CNNs in CMOS technology,' *Int. J. of Circuit Theory and Applications*, **20**, 555–572, (1992).
58. G. Erőss, A. Radványi, T. Boros, Á. Kiss, P. Lantos, J. Bitó, T. Roska and J. Vass, 'Optical tracking system for automatic guided vehicle (AGV) using cellular neural networks,' Report DNS-15-1992, Dual and Neural Computing Systems Laboratory, Hungarian Academy of Sciences, 1992. Also in *Proc. IEEE CNNA-92*, pp. 216–221, 1992.
59. W. Heiligenberg and T. Roska, 'On biological sensory information processing principles relevant to dually computing CNNs,' Report DNS-4-1992, Dual and Neural Computing Systems Laboratory, Hungarian Academy of Sciences, 1992.
60. T. Kozek, T. Roska and L. O. Chua, 'Genetic algorithm for CNN template learning,' *IEEE Trans. Circuits and Systems*, **CAS-40** (1993).
61. V. Pèrez-Muñuzuri, V. Pèrez-Villar and L. O. Chua, 'Autowaves for image processing on a two-dimensional CNN array of Chua's circuits: flat and wrinkled labyrinths,' *IEEE Trans. Circuits and Systems*, **CAS-40** (1993).
62. T. Roska, J. Hámori, E. Lábos, K. Lotz, L. Orzó, J. Takács, P. Venetianer, Z. Vidnyánszky and A. Zarándy, 'The use of CNN models in the subcortical visual pathway,' Report DNS-16-1992, Dual and Neural Computing Systems Laboratory, Hungarian Academy of Sciences, 1992.
63. T. Roska and L. O. Chua, 'The dual CNN analog software,' Technical Report DNS-2-1992, Dual and Neural Computing Systems Laboratory, Hungarian Academy of Sciences, 1992.
64. T. Roska, 'Programmable cellular neural networks—a state-of-the-art,' in *State-of-the-art in Computer Systems and Software Engineering* (P. Dewilde and J. Vandewalle, eds.), pp. 151–168, Kluwer Academic Publishers, 1992.
65. T. Roska, C. W. Wu, M. Balsi and L. O. Chua, 'Stability and dynamics of delay-type general and cellular neural networks,' *IEEE Trans. Circuits and Systems*, **CAS-39**, 487–490 (1992).
66. T. Roska, T. Boros, A. Radványi, P. Thiran and L. O. Chua, 'Detecting moving and standing objects using cellular neural networks,' *Int. J. Circuit Theory and Applications*, **20**, 613–628 (1992).
67. M. Tanaka, K. Crounse and T. Roska, 'Template synthesis of cellular neural networks for information coding and decoding,' *Proc. IEEE CNNA-92*, pp. 29–35, 1992.
68. B. E. Shi, T. Roska and L. O. Chua, 'Design of linear cellular neural networks for motion sensitive filtering,' submitted for publication, 1992.

69. P. Szolgay, I. Kispál and T. Kozek, 'An experimental system for optical detection of layout errors using analog software on a dual CNN,' *Proc. IEEE CNNA-92*, pp 203–209, 1992.
70. P. Szolgay, G. Vörös and G. Erőss, 'On the application of cellular neural network (CNN) paradigm in mechanical vibrating systems,' *IEEE Trans. Circuits and Systems*, **CAS-40** (1993).
71. L. Vandenberghe, J. Vandewalle, S. Daenen and M. Verduyn, 'Measuring the size of an object by a sequence of CNNs,' Memorandum, Catholic University, Leuven, 1992.
72. *Proceedings of the IEEE International Workshop on Cellular Neural Networks and their Applications (CNNA-90)*, 1990. Budapest, IEEE Cat. No. 90TH0312-9.
73. *Proceedings of the 2nd IEEE International Workshop on Cellular Neural Networks and their Applications (CNNA-92)*, 1992. Munich.
74. *Proceedings of the Special Session on Cellular Neural Networks, European Conference on Circuit Theory and Design (ECCTD-91)*, Sept. 1991.
75. *Int. J. Circuit Theory and Applications*, Special issue on Cellular Neural Networks, **20**, 1992.
76. A. G. Radványi, 'A dual CNN model of cyclopean perception and its application potentials in artificial stereopsis,' *Proc. IEEE CNNA-92*, pp. 222–227, 1992.
77. T. Roska, 'Analog events and a dual computing structure using analog and digital circuits and operators,' in *Discrete Event Systems: Models and Applications* (P. Varaiya and A. B. Kurzhanski, eds.), pp. 225–238, Springer Verlag, 1988.
78. C. W. Wu, T. Roska and L. O. Chua, 'Detecting features by CNNs in oscillating or chaotic mode,' Tech. Rep., 1992. In preparation.
79. 'CNN Workstation (version 5.1),' users' documentation, SZTAKI, 1992. Budapest.
80. T. Szirányi and J. Csicsvári, 'High speed character recognition using a dual cellular neural network architecture (CNND),' *IEEE Trans. Circuits and Systems*, **CAS-40** (1993).
81. Á. Csurgay, 'Fundamental limits in large-scale circuit modelling,' in *Proc. European Conference on Circuit Theory and Design (ECCTD-83)*, pp. 454–457, VDE-Verlag, Berlin, 1983.
82a. C. Mead, 'Analog VLSI and neural systems,' Addison Wesley, Reading, MA, 1989.
82b. *Analog VLSI Implementation of Neural Systems* (C. Mead and M. Ismail, eds.), Boston: Kluwer Academic Publ., 1989.
82c. M. Mahowald and C. Mead, 'The silicon retina,' *Scientific American*, **264**, no. 5, 76–82 (1991).
83. V. I. Krinsky, V. N. Biktashev and I. R. Efimov, 'Autowave principles for parallel image processing,' *Physica D*, **49**, 247–253 (1991).
84. C. B. Price, P. Wambacq and A. Osterlinck, 'Image enhancement and analysis with reaction–diffusion paradigm,' *IEE Proceedings*, **137**, 136–145 (1990).
85a. W. Heiligenberg, 'The neural basis of behavior: a neuroethological view,' *Ann. Rev. Neurosci.*, **14**, 247–267 (1991).
85b. P. Baldi and W. Heiligenberg, 'How sensory maps could enhance resolution through ordered arrangements of broadly tuned receivers,' *Biol. Cybern.*, **59**, 313–318 (1988).
86a. M. Konishi, 'Deciphering the brain's codes,' *Neural Computation*, **3**, 1–18 (1991).
86b. M. Konishi *et al.*, 'Neurophysiological and anatomical substrates of sound localization in the owl,' in *Auditory Function* (G. M. Edelman, W. E. Gall and W. M. Cowan, eds.), pp. 721–745, New York: J. Wiley, 1988.
87a. J. Allmann, F. Miezin and E. McGuiness, 'Stimulus specific responses beyond the classical receptive field: neurophysiological mechanisms for local-global comparisons in visual neurons,' *Ann. Rev. Neurosci.*, **8**, 407–430 (1985).
87b. G. Montgomery, 'The mind's eye,' *Discover*, May (1991).
88. J. Hámori, P. P. T. Pasik and J. Szentágothai, 'Triadic synaptic arrangements and their possible significance in the lateral geniculate nucleus of the monkey,' *Brain Research*, **80**, 379–393 (1974).
89. E. Lábos, 'Theoretical considerations of local neuron circuits and their triadic synaptic arrangements (TSA) in subcortical nuclei,' *J. Neuroscience Research*, **3**, 1–10 (1977).
90. D. C. V. Essen, C. H. Anderson and D. J. Felleman, 'Information processing in the primate visual system: an integrated systems perspective,' *Science*, **255**, 419–423 (1992).
91. J. L. Bradshaw and N. C. Nettleton, 'Human cerebral asymmetry (Chapter Nine),' (Englewood Cliffs), Prentice Hall, 1983.
92. E. R. Kandel, J. H. Schwarz and T. M. Jessel, *Principles of Neural Science, Third Edition*, Amsterdam, Elsevier, 1991.
93. 'Special issue on mind and brain,' *Scientific American*, September (1992).
94a. F. S. Werblin, 'Synaptic connections, receptive fields, and pattern of activity in the tiger salamander retina,' *Investigative Opthalmology and Visual Science*, **32**, 459–483 (1991).
94b. J. L. Teeters and F. S. Werblin, 'Real-time simulation of the retina allowing visualization of each processing stage,' *SPIE*, **1472** (1991). Image Understanding and the Man-machine Interface III.
95a. L. O. Chua and T. Roska, 'The CNN universal machine—part 1: the architecture,' in *Proc. IEEE CNNA-92*, pp. 1–10, 1992.
95b. T. Roska and L. O. Chua, 'The CNN universal machine—part 2: programmability and applications,' in *Proc. IEEE CNNA-92*, pp. 181–190, 1992.
96. T. Roska, A. Zarándy and L. O. Chua, 'Color image processing by CNN.' In preparation.
97. *Workshop on Unifying models of Biological Sensory Information Processing via Cellular Neural Networks (CNNs), Aug. 3 1992*. Nonlinear Electronics Laboratory, U. C. Berkeley.
98. J. Buhmann and K. Schulten, 'Influence of noise on the function of a "physiological" neural network,' *Biological Cybernetics*, **56**, 313–327 (1987).
99. T. Roska and L. O. Chua, 'The CNN universal machine: an analogic array computer,' *IEEE Trans. Circuits and Systems*, **40**, Mar. 1993. Accepted for publication.
100. L. O. Chua and T. Roska, 'The CNN paradigm,' *IEEE Trans. Circuits and Systems*, **CAS-40** (1993).

DISCRETE-TIME CELLULAR NEURAL NETWORKS

HUBERT HARRER AND JOSEF A. NOSSEK

Institute for Network Theory and Circuit Design, Technical University Munich, Arcisstr. 21, D-8000 Munich 2, Germany

SUMMARY

A network structure called a discrete-time cellular neural network is introduced. It is derived from cellular neural networks and feedback threshold networks. The architecture is discussed and its advantages are presented. Convergence is proved for a large class of templates and applications are given for the following image-processing tasks: linear thresholding, connected component detection, hole filling, concentric contouring, increasing and decreasing objects step by step, searching for objects with minimal distance, and oscillation.

1. INTRODUCTION

Cellular neural networks (CNNs) as introduced in Reference 1 have recently received growing attention. They are an efficient architecture for performing image processing and pattern recognition. The simple structure of a single cell and the local connectivity make CNNs well suited for analogue VLSI implementation. The network is described by a system of non-linear differential equations[2]

$$\frac{\mathrm{d}x^c}{\mathrm{d}t} = -x^c(t) + \sum_{d \in N_r(c)} a^c_d y^d(t) + \sum_{d \in N_r(c)} b^c_d u^d + i^c \tag{1}$$

$$y^c(t) = \tfrac{1}{2}(|x^c(t)+1| - |x^c(t)-1|) \tag{2}$$

The variable x^c denotes the state of the cell c, y^c its output and u^c its input. The state of a single cell is controlled by the inputs and outputs of adjacent cells within an r-neighbourhood $N_r(c)$. The outputs are fed back multiplied by the coefficients a^c_d, and the inputs are multiplied by the control parameters b^c_d. The value of i^c is constant and is used for adjusting a threshold. A set of feedback, control coefficients and the thresholds is called a template. The template coefficients are translationally invariant. The r-neighbourhood $N_r(c)$ is the set of all cells within a distance r including cell c. There are problems concerning the boundary of the grid, because not all neighbouring cells exist. To resolve these, dummy cells are inserted which possess constant values for u^c and y^c. Below, $u^c = -1$ and $y^c = -1$ are used for dummy cells, which means that the input and initial array are surrounded by a ring of white pixels. An r-neighbourhood can be defined similarly on a hexagonal grid.

The output y^c is obtained by a piecewise linear function of the cell state x^c in (2). Depending on the initial state $\mathbf{x}(t_0)$, the system converges to a final state $\mathbf{x}(t_\infty)$ with binary outputs if the feedback coefficients are symmetric and the self-feedback is large enough ($a^c_c > 1$).

When applying the Einstein summation convention, a summation is performed automatically for identical indices in the lower and upper positions of variables in a product term. A summation is not done if the indices are parenthesized. This abbreviation will be used throughout this text. The index d is always applied within the r-neighbourhood of a cell c, where the index c is used for all cells on the grid. With that, (1) is written as

$$\frac{\mathrm{d}x^c}{\mathrm{d}t} = -x^c(t) + a^c_d y^d(t) + b^c_d u^d + i^c \tag{3}$$

Complex dynamics can be applied to image processing. If global tasks are performed by a wave-like propagation of the signals, the system can solve such *global* tasks with only *local* interconnections. Many useful applications of this type have been found recently, e.g. the connected component detector,[3] the hole filler,[4] the shadow detector[5] and linear thresholding.[6] A big problem of CNNs is the design of the template coefficients. Normally they are determined only by a trial-and-error method. The learning algorithm described in Reference 7 fails if the system dynamics is too complex. Hence a system is required which solves global tasks despite local connectivity, where the coefficients can be derived by a logic formulation of the problem. This system should also possess reliable convergence properties[1,2,8,9] and a simple cell structure to make it well suited for VLSI implementation.

2. ARCHITECTURE

Discrete-time cellular neural networks (DTCNNs) are a special kind of feedback threshold network[10–12] where the local interconnections and the translationally invariant weights are transferred from CNNs. They are completely described by a recursive algorithm. The dynamic behaviour is based on the feedback of clocked and binary outputs. A single cell is influenced by the inputs and outputs of adjacent cells within an r-neighbourhood as defined for CNNs. The architecture is closely related to cellular automata[13,14] but differs from them in having continuous-valued inputs and weights.

Definition 1. Discrete-time cellular neural networks

Discrete-time cellular neural networks (DTCNNs) are defined by the following algorithm using the index d within the r-neighbourhood of cell c:

$$x^c(k) = a^c_d y^d(k) + b^c_d u^d + i^c \tag{4}$$

$$y^c(k) = f(x^c(k-1)) = \begin{cases} 1 & \text{if } x^c(k-1) > 0 \\ -1 & \text{if } x^c(k-1) < 0 \end{cases} \tag{5}$$

In contrast to CNNs, the system is clocked and only binary values $(-1, +1)$ are weighted by the coefficients $a^c_d \in \mathbb{R}$ of the feedback operator. Before an iteration is started, the initial values $y^c(0) \in \{-1, 1\}$ have to be defined. They are of crucial importance for the dynamic behaviour of the system and can be considered as a second cell input. The value $+1$ denotes a black pixel, -1 a white pixel. The inputs $u^c \in D^c$ are restricted to a continuous definition set, which is necessary for the convergence proof in Section 3. The inputs are multiplied by the control parameters $b^c_d \in \mathbb{R}$ for all neighbouring cells d within $N_r(c)$. A constant value $i^c \in \mathbb{R}$ is added to adjust a threshold of cell c. A non-indexed value $i \in \mathbb{R}$ means that there are equal values for all cells. The continuous template coefficients and inputs are assumed to be time-invariant, but an extension to time-dependent parameters[15] is also possible. The binary output state y^c is determined by the sign of x^c in the previous iteration step. It is undefined for $x^c = 0$. In practice there is always some noise, which implies that $|x^c| > 0$.

If the template coefficients are chosen such that

$$\Delta = \min_{c,k} | a^c_d y^d(k) + b^c_d u^d + i^c | \tag{6}$$

is large enough, then the algorithm is relatively insensitive to parameter tolerances of a^c_d, b^c_d or i^c smaller than Δ, because individually they cannot cause any changes in the output states.

Abbreviating the notation, the time-independent part of (4) is defined as the cell bias $k^c(\mathbf{u})$, which again is a function of the constant inputs and the threshold values:

$$k^c = k^c(\mathbf{u}) = b^c_d u^d + i^c \tag{7}$$

The advantage of DTCNNs is the description of the next output state through a set of linear inequalities. The template coefficients are computed by solving this system of linear inequalities, where in each time step the subsequent output state is postulated. This can be done iteratively using the design algorithm described in Reference 16. The simulation on a digital computer is very simple, since the system is time-discrete and a sophisticated integration algorithm is not necessary.

For the algorithm (4), (5) the network structure of a single cell is given in Figure 1. The variables are substituted by voltages and the iteration steps by discrete-time instances.

The next iteration step of the system is obtained after one period of the clock signals φ_1 and φ_2 (Figure 2). Neglecting parasitic transients, the time instance kT can be chosen arbitrarily during the 'high' phase of φ_1.

The multiplications are done by linear voltage-controlled current sources. The total current sum is transformed into a voltage $v_x^c(kT)$ by the resistor R_x with

$$v_x^c(kT) = R_x(A_d^c v_y^d(kT) + B_d^c v_u^d + I^c) \tag{8}$$

The voltage-controlled voltage sources $F(v_x^c(kT))$ and $F(v_z^c(kT))$ have the non-linear characteristic

$$F(v) = \begin{cases} V_{\text{sat}} & \text{if } v > 0 \\ -V_{\text{sat}} & \text{if } v < 0 \end{cases} \tag{9}$$

Here the next output state is determined and the capacitor C_1 is charged during the 'high' phase of φ_1 to V_{sat} or $-V_{\text{sat}}$. During the 'high' phase of φ_2 (φ_1 'low') the capacitors C_1 and C_2 are connected in parallel. The voltage $v_z^c(kT)$ has settled after the transient to

$$v_z^c(kT) = \frac{C_1 F(v_x^c((k-1)T)) + C_2 v_z^c((k-1)T)}{C_1 + C_2} \tag{10}$$

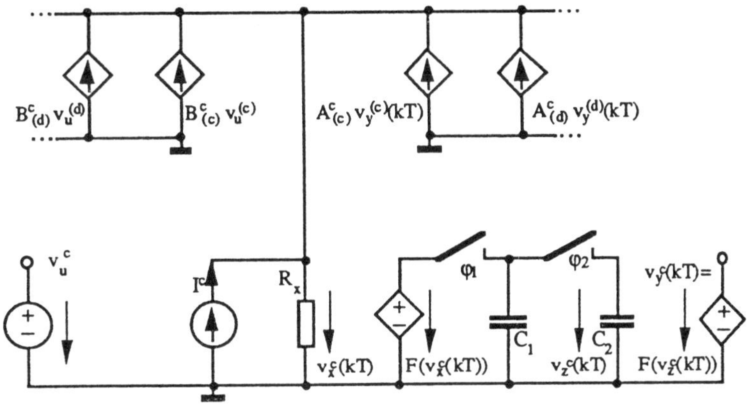

Figure 1. Network structure of a single cell

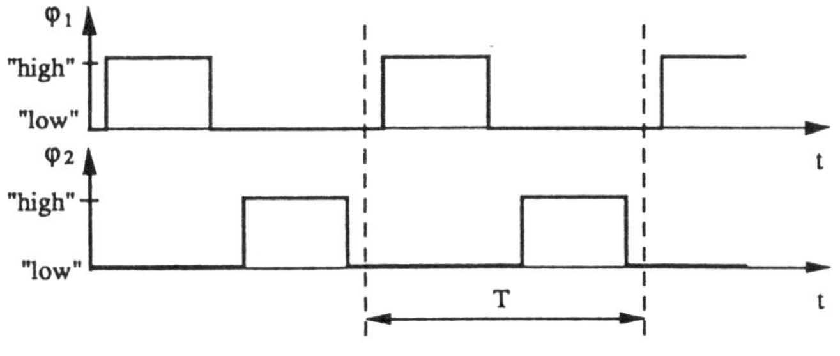

Figure 2. Clock diagram of discrete-time cellular neural networks

Table I. Normalization of the variables and network elements

$u^c = v_u^c / V_{sat}$	$a_d^c = A_d^c R_x$
$x^c(k) = v_x^c(kT)/ V_{sat}$	$b_d^c = B_d^c R$
$z^c(k) = v_z^c(kT)/ V_{sat}$	$i^c = I^c R_x / V_{sat}$
$y^c(k) = v_y^c(kT)/ V_{sat}$	

Because $|F(v_x^c)| = V_{sat}$ and $|v_z^c| \leqslant V_{sat}\ \forall k$, it follows for $C_1 > C_2$ that

$$v_y^c(kT) = F(v_x^c((k-1)T)) = \begin{cases} V_{sat} & \text{if } v_x^c((k-1)T) > 0 \\ -V_{sat} & \text{if } v_x^c((k-1)T) < 0 \end{cases} \tag{11}$$

Inserting (8) into (11) leads to

$$v_y^c(kT) = F(R_x(A_d^c v_y^d((k-1)T) + B_d^c v_u^d + I^c)) \tag{12}$$

Applying the normalization in Table I, the algorithm (4), (5) is obtained.

Compared to CNNs, an analogue realization of DTCNNs has some important advantages.

(i) Because of the binary outputs, the interconnection of several chips is very simple. For CNNs, continuous output signals have to be propagated during the transient.
(ii) The parameter insensitivity of (6) makes the network robust against fabrication tolerances if the templates are designed appropriately.
(iii) The propagation speed can be controlled within a large range only by changing clock rate. This also simplifies the testability of a chip.

Efficient analogue circuit structures are given in References 17 and 18 for this architecture.

Remarks A DTCNN containing a large enough neighbourhood can be considered as a single-layer perceptron with a comparator non-linearity and feedback connections or a discrete Hopfield net which is updated in parallel. However, large neighbourhoods are not well suited for simple VLSI implementation. For the latter, strong restrictions are given for the weight coefficients (local connectivity and translational invariance). If translational invariance is abolished, the network structure can be used for an associative memory trained by the Hebb rule.[19]

3. CONVERGENCE

In the following sub-sections theoretical aspects concerning the convergence of DTCNNs are given. Until now, general conditions to obtain a convergent output pattern have not been found. However, for a large class of templates, convergent behaviour can be shown. It is also proved that in contrast to CNNs, oscillation is possible for symmetric feedback coefficients and large enough self-feedback. In this case the cycle length is limited to a period of two.

3.1. Eigendominant templates

Convergence can be proved for dominant self-feedback and cell bias of a DTCNN. Such templates are used for a large class of applications, where a transition from white to black or from black to white cannot be undone. Examples of such tasks are the hole filler, increasing and decreasing of objects or the search for objects with minimal distance in Section 4. First, the following definitions are necessary.

Definition 2. Convergence

A DTCNN is said to be convergent if

$$\exists k_0: \quad (\forall k > k_0: \mathbf{y}(k) = \mathbf{y}(k_0)) \tag{13}$$

The obtained output state $\mathbf{y}(k_0)$ is called a fixed point.

Remark It is obvious that such a fixed point is reached for any two subsequent iterations with identical output states for *all* cells and $\Delta > 0$. The latter condition makes sure that no cell state can be zero, which is always assumed throughout this paper.

Definition 3. Minimum absolute cell bias

A minimum absolute cell bias $k^c_{\min}$ is defined as the smallest absolute value of the cell bias $k^c(\mathbf{u})$ for all possible input patterns:

$$k^c_{\min} = \min_{\mathbf{u} \in D} |k^c(\mathbf{u})| \tag{14}$$

Remark If the inputs u^c are not restricted to a definition set D^c, it is usually possible to choose a combination for $\mathbf{u}$ that implies $\forall_c$: $k^c = 0$ for any $b^c_d \neq 0$.

Definition 4. Eigendominance

A DTCNN is said to be eigendominant for a template if

$$\forall_c: \quad (a^{\{c\}}_{\{c\}} + k^c_{\min}) > \sum_{\substack{d \in N_r(c) \\ d \neq c}} |a^c_d| \tag{15}$$

Remarks Eigendominant templates with $k^c_{\min} = 0$ imply the diagonal dominance

$$a^{\{c\}}_{\{c\}} > \sum_{\substack{d \in N_r(c) \\ d \neq c}} |a^c_d| \tag{16}$$

If $\forall_c$: $k^c = 0$, the next output state is determined by the self-feedback coefficient and no cell can change its output state. The initial pattern is also the convergent output pattern and therefore such networks make no sense for image processing.

Theorem 1. Convergence for eigendominant templates

An eigendominant DTCNN converges in n or fewer iterations, where n denotes the number of cells on the grid.

Proof To prove the theorem, let us first consider any cell with $k^c > 0$ ($k^c > k^c_{\min}$). For the output state we have two cases.

Case 1: $y^c(k) = +1$

The sign of $x^c(k)$ is positive according to Definition 4, because

$$a^{\{c\}}_{\{c\}} y^c(k) + k^c > - \sum_{\substack{d \in N_r(c) \\ d \neq c}} a^c_d y^d(k) \tag{15}$$

for any $y^d \in \{-1, 1\}$. Hence the output state of cell c cannot be influenced by adjacent cells and the output state is $+1$ for *all* subsequent iterations.

Case 2: $y^c(k) = -1$

$$\text{(a)}\quad x^c(k) < 0 \Rightarrow y^c(k+1) = -1$$

The white output state in the subsequent iteration step is caused by the neighbouring cells.

$$\text{(b)}\quad x^c(k) > 0 \Rightarrow y^c(k+1) = +1$$

The output state changes from white to black. For the following iterations Case 1 applies and the transition cannot be undone.

Similar considerations apply sign-symmetrically to a cell with $k^c < 0$ ($k^c \leqslant -k^c_{\min}$). In this case a transition is only allowed once from black to white and cannot be undone afterwards. Because each cell can change its output state only *once* during the transient, an eigendominant DTCNN is forced to converge in n or fewer iterations. It is not necessarv to satisfy Case 1 ($k^c > 0$, $y^c(k) = +1$) or the sign-symmetric case ($k^c < 0$, $y^c(k) = -1$) for a fixed point, because convergence is also obtained if an output state is determined by Case 2(a) and *no* cell changes its output value in the subsequent iteration. □

Examples of eigendominant templates are given in Section 4. It is possible to embed the condition of eigendominance in the learning algorithm described in Reference 16 to make sure that only convergent templates are utilized. However, the class of tasks for which a solution exists at all is then crucially restricted. The proof can also be applied to general feedback threshold networks as perceptrons with feedback connections or discrete Hopfield nets.

3.2. Symmetric feedback templates

It is proved in Reference 10 that feedback threshold networks with symmetric weights are either convergent or oscillate with a cycle length of two. This can also be concluded for DTCNNs containing symmetric feedback coefficients.

Theorem 2. Symmetric feedback templates

A DTCNN with symmetric feedback coefficients

$$a^c_d = a^d_c \tag{17}$$

is either convergent or oscillates with a period of two.

Proof The idea of the proof is based on the Lyapunov theory for discrete systems. The function

$$E(k) = -y_c(k)a^c_d y^d(k-1) - (y_c(k) + y_c(k-1))k^c \tag{18}$$

with

$$y_c = y^c \tag{19}$$

is used as an energy function and can only decrease during the transient. It is restricted to $|E(k)| \leqslant E_{\max}$, where

$$E_{\max} = n\,|a^c_d|\,1^d + 2n(u_{\max}\,|b^c_d|\,1^d + i_{\max}) \tag{20}$$

$$u_{\max} = \max_c |u^c|, \qquad i_{\max} = \max_c |i^c| \tag{21}$$

Then we obtain for two subsequent iterations

$$\begin{aligned}\Delta E &= E(k+1) - E(k)\\ &= -y_c(k+1)a^c_d y^d(k) - (y_c(k+1) + y_c(k))k^c + y_c(k)a^c_d y^d(k-1) + (y_c(k) + y_c(k-1))k^c \end{aligned}\tag{22}$$

Using the symmetry condition (17), it is easy to see that

$$y_c(k)a_d^c y^d(k-1) = y_c(k-1)a_d^c y^d(k)$$

and thus it follows immediately that

$$\Delta E = -(y_c(k+1) - y_c(k-1))(a_d^c y^d(k) + k^c) = -(y_c(k+1) - y_c(k-1))x^c(k)$$

Because the sign of $x^c(k)$ determines the value of the following output state, we obtain with (5)

$$x^c(k) = |x^{(c)}(k)|\, y_{(c)}(k+1)$$

$$\Delta E = -|x^c(k)|(y_c(k+1)^2 - y_c(k-1)y_c(k+1)) \leqslant 0$$

Hence E *decreases* if at least one cell changes its output state from iteration $k-1$ to $k+1$. Thus $\Delta E = 0$ is only possible if

$$\forall c: \quad y^c(k-1) = y^c(k+1) \tag{23}$$

Then two cases have to be considered:

(a) $\mathbf{y}(k) = \mathbf{y}(k-1)$

This is a fixed point in accordance with Theorem 2.

(b) $\mathbf{y}(k) \neq \mathbf{y}(k-1)$

Here we have with (23) an oscillating system of period two. Larger cycle lengths cannot appear, because the energy function is restricted by (20) and cannot steadily decrease. □

The number of iterations before a fixed point or a limit cycle is obtained is estimated in Reference 10.

4. APPLICATIONS

Most applications of CNNs[3-6,20] could be performed by DTCNNs too. A novel task of image processing is the transformation of objects to their concentric contours and the search for objects with minimal distance to a fixed point. For CNNs no template has been found for these tasks up to now. As an example for an oscillating system, a template with symmetric feedback coefficients is introduced. The applications are defined on a square grid but can easily be mapped onto a hexagonal grid.

The following definitions are used throughout this section.

Definition 5. Four-connected

Two pixels (i, j) and (m, n) of the same colour are called four-connected if

$$|i-m| + |j-n| \leqslant 1 \tag{24}$$

Definition 6. Eight-connected

Two pixels (i, j) and (m, n) of the same colour are called eight-connected if

$$|i-m| \leqslant 1 \wedge |j-n| \leqslant 1 \tag{25}$$

Definition 7. Four-connected path

A four-connected path P is defined as a set of r pixels where two following pixels p_i and p_{i+1} for $i = 1, \ldots, r-1$ are four-connected.

Definition 8. Four-connected object

A four-connected object O is defined as the maximum set of identical coloured pixels where for any two pixels $(i, j) \in O$ and $(m, n) \in O$ a four-connected path $P \subseteq O$ does exist.

Remarks The 1-neighbourhood was defined for eight-connected pixels by Definition 6. A four-connected neighbourhood consists of cells which are only influenced by four-connected pixels. Four-connected pixels are always eight-connected. For one-dimensional problems or on a hexagonal grid, this distinction is superfluous. The definition of an eight-connected object is similar to Definition 8.

4.1. Linear thresholding

Problem description For linear thresholding, only local features of the image to be processed are of importance. It can be considered as a convolution. The feedback coefficients are set to zero and the image is applied to the cell inputs. The control coefficients are used for extracting particular features of the input image within the r neighbourhood of each cell. This corresponds to the cell bias k^c performing a discrete convolution with a kernel **b**:

$$x^c = k^c = b^c_d u^d + i^c \tag{26}$$

The value of i^c adjusts a threshold for accepting the decision of the detected feature.[2,6,20]

Convergence proof A DTCNN used as a linear threshold network is convergent after a single iteration for any parameters b^c_d, u^c and i^c, because the network has *no feedback*. The subsequent output state depends only on the inputs.

As an example for local image processing, the edge detection of black eight-connected objects is demonstrated. This task can be solved by the template

$$a = \begin{array}{|c|c|c|}\hline 0 & 0 & 0 \\ \hline 0 & 0 & 0 \\ \hline 0 & 0 & 0 \\ \hline \end{array}\,, \qquad b = \begin{array}{|c|c|c|}\hline 0 & -1 & 0 \\ \hline -1 & 4\cdot5 & -1 \\ \hline 0 & -1 & 0 \\ \hline \end{array}\,, \qquad i = -1\cdot5 \tag{27}$$

The condition for obtaining a black output state of cell i, j is

$$4\cdot5u^{i,j} - u^{i-1,j} - u^{i,j-1} - u^{i,j+1} - u^{i+1,j} - 1\cdot5 > 0 \tag{28}$$

i.e. a cell only stays black if there is at least one white neighbouring cell by Definition 5. This is true for all pixels in the edge set. It is proved in the Appendix that the edges of noisy input patterns are recognized correctly if they are restricted to a definition set $D = \{u^c \mid 0\cdot89 < |u^c| < 1\cdot11\}$.

4.2. Connected component detector (CCD)

Problem description For pattern recognition and data compression, black objects within a single line should be compressed to the size of one pixel and moved to the right boundary, where black and white pixels are alternating depending on the number of objects. The black pixels in the final output pattern correspond to the number of connected components in the initial pattern.[4]

For this, the DTCNN has to be programmed as a cellular automaton. The output state of the middle cell is *only* allowed to change if one of the following situations occurs:

(29)

(30)

Table II. Computation of $x^c(k)$ for all combinations $y^{i,j-1}$, $y^{i,j}$ and $y^{i,j+1}$

$y^{i,j-1}(k)$	$y^{i,j}(k)$	$y^{i,j+1}(k)$	$x^{i,j}(k)$	$y^{i,j}(k+1)$
−1	−1	−1	−1	−1
−1	−1	+1	−3	−1
−1	+1	−1	+1	+1
−1	+1	+1	−1	−1
+1	−1	−1	+1	+1
+1	−1	+1	−1	−1
+1	+1	−1	+3	+1
+1	+1	+1	+1	+1

Initializing the state $\mathbf{y}(0)$ with the pattern to be analysed, the task can be solved by the template

$$a = \boxed{+1 \mid +1 \mid -1}, \qquad b = \boxed{0 \mid 0 \mid 0}, \qquad i = 0 \tag{31}$$

This is proved by computing the sign of $x^c(k)$ for all possible combinations of $y^{i,j-1}$, $y^{i,j} = y^c$ and $y^{i,j+1}$(Table II).

The CCD is an example of global image processing with only local cell connectivity.

Convergence proof A DTCNN with the template coefficients of (31) is convergent for all initial conditions, because a changing output state (from black to white or from white to black) can *only* be propagated to the right side. After n or fewer iterations a fixed point is achieved because of the bounded size of the field. Considering the change of cell b from black to white in Figure 3, cell a has to remain white. The transition (30) is only allowed if the right neighbour of cell a is white. Cell c must stay black, because (29) is *only* possible if the left neighbour is white. Therefore white pixels can only move to the right. Similar considerations can be made for a transition from white to black.

Figure 3. Transition from black to white

Remark A rotation of the parameters in the feedback operator by a multiple of 45° allows connected component detection in different directions.

4.3. Hole filler

Problem description White pixels completely enclosed by a black eight-connected object O should be black in the final output state.[5]

This task can be solved by a propagation starting at the boundaries of the grid. The initial state is black (+1). The pattern is used as input of the cells. Transitions from black to white in the output are only allowed if the cell input is white and there is a four-connected white neighbour cell by Definition 5. This is satisfied for

$$y^{i,j}(k+1) = -1 \Rightarrow u^{i,j} = -1 \wedge y^{i-1,j}(k) + y^{i,j-1}(k) + y^{i,j+1}(k) + y^{i+1,j}(k) < +4 \tag{32}$$

$$y^{i,j}(k+1) = +1 \Rightarrow u^{i,j} = +1 \vee y^{i-1,j}(k) + y^{i,j-1}(k) + y^{i,j+1}(k) + y^{i+1,j}(k) = +4 \tag{33}$$

and can be done using the template

$$a=\begin{array}{|c|c|c|}\hline 0 & 1 & 0\\\hline 1 & 2 & 1\\\hline 0 & 1 & 0\\\hline\end{array}\,, \qquad b=\begin{array}{|c|c|c|}\hline 0 & 4\cdot5 & 0\\\hline\end{array}\,, \qquad i=-1 \tag{34}$$

Convergence proof A DTCNN using the template (34) is convergent by Theorem 1 if the inputs are restricted to a definition set

$$D=\{u^c \mid |u^c|>\tfrac{2}{3}\} \tag{35}$$

The template is eigendominant by Definition 4 because $k^c_{\min}>2$ and the condition

$$\min_c(2+k^c_{\min})>4$$

is satisfied for all combinations of the input pattern.

4.4. Concentric contours

Problem description A concentric contour of a black eight-connected object O is defined as a set of alternating black and white rings starting from the boundary to the interior. Notice that complex object structures may have more than one interior point. Figure 4 shows a black square and its concentric contour.

This can be used for a special kind of size-independent object recognition. Objects of different size but with identical interior contour rings could be detected by linear thresholding.

A DTCNN solves this task by extracting the outer contour ring of the object and inverting its interior pixels. The contour ring remains stable during the following iterations. A new concentric ring is built up every iteration step. The arising rings can be interpreted as a propagation to the interior of the objects. A cell output $y^{i,j}$ is only allowed to change its state if there are four neighbours $y^{i-1,j}$, $y^{i,j-1}$, $y^{i,j+1}$ and $y^{i+1,j}$ by Definition 5 which have the same colour. Applying the pattern to the initial state $\mathbf{y}(0)$ *and* the cell inputs $\mathbf{u}$, this task is performed by the template

$$a=\begin{array}{|c|c|c|}\hline 0 & -1 & 0\\\hline -1 & 3\cdot5 & -1\\\hline 0 & -1 & 0\\\hline\end{array}\,, \qquad b=\begin{array}{|c|c|c|}\hline 0 & 0 & 0\\\hline 0 & 4 & 0\\\hline 0 & 0 & 0\\\hline\end{array}\,, \qquad i=-4 \tag{36}$$

Choosing $b^{i,j}_{i,j}=4$ and $i=-4$ assures that $x^{i,j}$ can never be positive if $u^{i,j}=-1$. On top of this the following conditions are required:

$$u^{i,j}=+1 \wedge y^{i,j}(k)=y^{i-1,j}(k)=y^{i,j-1}(k)=y^{i,j+1}(k)=y^{i+1,j}(k)=+1 \Rightarrow x^{i,j}<0 \tag{37}$$

$$u^{i,j}=+1 \wedge y^{i,j}(k)=y^{i-1,j}(k)=y^{i,j-1}(k)=y^{i,j+1}(k)=y^{i+1,j}(k)=-1 \Rightarrow x^{i,j}>0 \tag{38}$$

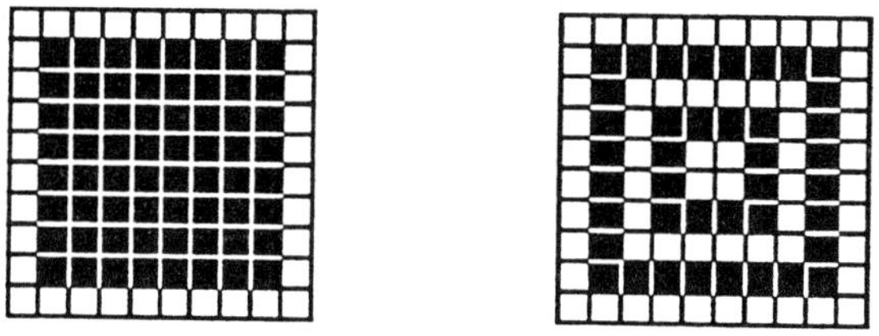

Figure 4. Square and its concentric contour

Convergence proof A DTCNN with the template (36) is convergent for all input patterns. In every iteration step the number of cells possessing four equal neighbours decreases, because the extracted contour ring always becomes stable. In fewer than $n/2$ iterations there is no cell with four equal neighbours.

Remark For 'decomposing' an object of any size in its concentric rings, only a template of neighbourhood $r = 1$ is necessary. For this application a template for CNNs has not been found up to now.

4.5. Increasing and decreasing objects step by step

Problem description A black object O should be increased or decreased every iteration step by a one-pixel layer.

The DTCNN has to be programmed as a cellular automaton. The pattern is used as initial state. Increasing the object, *only* a transition from white to black of the output state is allowed if the cell has at least one black neighbour. Doing this, the object is increased every iteration by a new ring of black pixels.

The problem is solved by the template

$$a = \begin{array}{|c|c|c|}\hline 0{\cdot}5 & 0{\cdot}5 & 0{\cdot}5 \\ \hline 0{\cdot}5 & 0{\cdot}5 & 0{\cdot}5 \\ \hline 0{\cdot}5 & 0{\cdot}5 & 0{\cdot}5 \\ \hline \end{array}, \qquad b = \begin{array}{|c|c|c|}\hline 0 & 0 & 0 \\ \hline 0 & 0 & 0 \\ \hline 0 & 0 & 0 \\ \hline \end{array}, \qquad i = +4 \tag{39}$$

because of

$$\exists d \in N_r(c): \quad y^d(k) = +1 \Rightarrow y^c(k+1) = +1 \tag{40}$$

Convergence proof A DTCNN using the template (39) is convergent by Theorem 1. The template is eigendominant by Definition 4 because $k^c_{\min} = 4$ and the condition

$$\min_c(a^{\{c\}}_{\{c\}} + k^c_{\min}) = 4{\cdot}5 > 4$$

is satisfied for all combinations of the initial pattern.

Decreasing a black object O, *only* transitions from black to white are allowed if the cell has at least one white neighbour. Therefore the object is decreased every iteration by a single ring of black pixels.

This is done by changing the threshold i of (39) to -4.

Increasing and decreasing is another example of global image processing using wave propagation. Here the propagation is done orthogonal to the edges of the object. Decreasing an object, the propagation is directed towards the interior point(s); for an increase the opposite direction is used.

Remark The template (39) increases an object corresponding to an eight-connectivity. It can also be done with respect to a four-connectivity using the template

$$a = \begin{array}{|c|c|c|}\hline 0 & 0{\cdot}5 & 0 \\ \hline 0{\cdot}5 & 0{\cdot}5 & 0{\cdot}5 \\ \hline 0 & 0{\cdot}5 & 0 \\ \hline \end{array}, \qquad b = \begin{array}{|c|c|c|}\hline 0 & 0 & 0 \\ \hline 0 & 0 & 0 \\ \hline 0 & 0 & 0 \\ \hline \end{array}, \qquad i = +2 \tag{41}$$

4.6 Search for objects with minimal distance

Problem description The increase of an object in Section 4.5 should be modified to control the propagation by the cell inputs. A transition from white to black is only allowed if the cell has at least one black neighbour and the input of the cell is white. Therefore black pixels in the input pattern stop the propagation. The input pattern can be considered as a maze where a wave spreads through the lanes beginning from a starting point. This can be applied for the search of objects with minimal distance to a given pixel if the propagating wave is compared to the position of distributed targets on the grid, which are only allowed on white pixels in the input pattern. For this, an additional logic layer is used to perform an AND function between the propagating wave and the time-dependent positions of the targets. If an object is detected, the propagation has to be stopped. The detected object has the minimal distance to the starting point.

Figure 5 shows the time-dependent development of the output pattern for the first four iterations. Black pixels correspond to black pixels in the input pattern and grey pixels denote black pixels in the output pattern.

This can be solved by the template

$$a = \begin{array}{|c|c|c|}\hline 0\cdot 5 & 0\cdot 5 & 0\cdot 5 \\ \hline 0\cdot 5 & 0\cdot 5 & 0\cdot 5 \\ \hline 0\cdot 5 & 0\cdot 5 & 0\cdot 5 \\ \hline \end{array}, \qquad b = \begin{array}{|c|c|c|}\hline 0 & -4 & 0 \\ \hline \end{array}, \qquad i = 0 \tag{42}$$

The starting point, which is only allowed to be a white pixel in the input pattern, is the only pixel that is black-coloured in the initial state. For the next output state the following requirements are given:

$$u^c = +1 \wedge y^c(k) = -1 \Rightarrow y^c(k+1) = -1 \tag{43}$$

$$u^c = -1 \wedge \exists d \in N_r(c)\colon\ y^d(k) = +1 \Rightarrow y^c(k+1) = +1 \tag{44}$$

Convergence proof A DTCNN using the template (42) is convergent by Theorem 1 if the inputs are restricted to a definition set

$$D = \{u^c \mid |u^c| > \tfrac{7}{8}\} \tag{45}$$

The template is eigendominant by Definition 4 because $k^c_{\min} > 3\cdot 5$ and the condition

$$\min_c(0\cdot 5 + k^c_{\min}) > 4$$

is satisfied for all combinations of the input pattern.

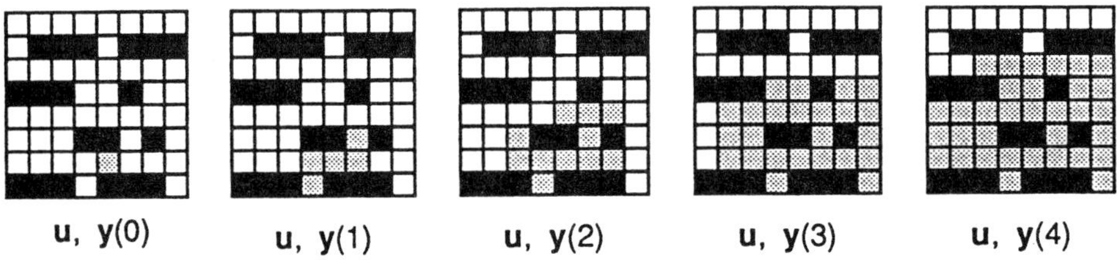

Figure 5. Time-dependent development of the output pattern (grey pixels) for the illustrated input pattern (black pixels)

4.7. Example of a limit cycle using a symmetric template

Convergence cannot be guaranteed for DTCNNs with symmetric feedback templates and a self-feedback coefficient larger than unity as proved for Theorem 2. Using the template

$$a = \begin{array}{|c|c|c|}\hline 0 & -1 & 0 \\ \hline -1 & 1\cdot 5 & -1 \\ \hline 0 & -1 & 0 \\ \hline \end{array}, \qquad b = \begin{array}{|c|c|c|}\hline 0 & 0 & 0 \\ \hline \end{array}, \qquad i = 0 \tag{46}$$

an oscillating output pattern of period two is obtained for the initial pattern in Figure 6.

Figure 6. Oscillating output pattern

5. CONCLUSIONS

In contrast to CNNs, discrete-time cellular neural networks have clocked variables and a comparator characteristic for the non-linear function. The distinctions from cellular automata are the continuous-valued weight coefficients and inputs in the linear part of the cell which allow linear thresholding and processing of noisy or grey-level input signals.

Global image processing with only local cell connectivity is obtained by clocked and binary output states which are fed back within an r-neighbourhood. In contrast, CNNs do this by way of an analogue dynamic transient using continuous feedback.

For CNNs the propagation speed depends crucially on the template coefficients and the input or output signals, because these influence the derivative of the state variable x^c in (1). Large values of the derivative can change the state variables within a short time and cause fast propagation. Hence the CNN structure is able to do image-processing tasks *where the propagation speed is controlled by both the input and initial pattern*. Such tasks cannot be done by DTCNNs because of the clocked and binary outputs.

However, for some applications a constant signal propagation is required. The search for objects with minimal distance described in Section 4.6 cannot be done by a CNN, because there propagation through wide lanes of the maze is faster than through small lanes. This is caused by larger derivatives of the state variables depending on the number of black neighbours to a cell.

The DTCNN architecture guarantees a constant propagation speed. Here *only the sign* of the variable x^c is of importance and *not its absolute value*.

Another example is the concentric contouring described in Section 4.4. The DTCNN is able to decide in a *binary* fashion if a cell has only neighbours of the same colour. For the CNN the output of a cell has to change *continuously*. This change in the output values is *not equal* for all 'interior' pixels, because corner cells of a new concentric ring possess more neighbours with constant output values and therefore change their values at different speeds. This leads to 'disturbed' rings for complex object structures because of the different propagation speeds.

Further applications-oriented research has to be done, especially to decide which classes of tasks can *only* be solved by CNNs and which only by DTCNNs.

Simulation on digital computers is far less expensive, because a numerical integration algorithm is not needed.

A learning algorithm[16] based on the relaxation method performs the computation of the template coefficients if the logical problem formulation is given. This method is well suited for DTCNNs, because the following output state of a cell is described by an inequality. Another advantage is the insensitivity of the parameters if the coefficients are chosen appropriately.

Analogue implementation of the DTCNN has the properties of being insensitive against fabrication tolerances, simple interconnection of several chips, control of the propagation speed by the clock rate and good testability.

A comparison with cellular automata would be of importance, because most applications are taken outside this area. For special problems they are certainly more efficient and simpler to implement. However, the advantage of CNNs and DTCNNs is their flexibility. The programmable structure can solve a large class of problems simply by changing the parameters.

ACKNOWLEDGEMENT

The authors are very grateful to Professor Dr. Tamás Roska for stimulating discussions and helpful comments.

APPENDIX: DETERMINATION OF THE DEFINITION SET FOR THE EDGE DETECTOR

The choice of the definition set for the inputs has to guarantee that the value of the state variable x^c cannot be zero. This is identical to the condition of $k^c_{\min} > 0$, because we have no feedback for edge detection. Hence we must restrict the absolute values of the inputs to a lower and an upper bound by

$$D = \{u^c \mid u_{\min} < \mid u^c \mid < u_{\min}\} \tag{47}$$

where $u_{\min} = 1 - \delta$ and $u_{\max} = 1 + \delta$ has to be satisfied for symmetric limits. This implies

$$u_{\min} + u_{\max} = 2 \tag{48}$$

For the 'worst case' (only *one* white neighbour, e.g. $u^{i-1,j}$) the condition

$$4\cdot 5 \mid u^{i,j} \mid + \mid u^{i-1,j} \mid - |u^{i,j-1}| - \mid u^{i,j+1} \mid - |u^{i+1,j}| - 1\cdot 5 > 0 \tag{49}$$

has to be true. This is achieved by transforming (49) to an equality where positive inputs are replaced by the lower bound $u_{\min}$ and negative inputs by the upper bound $u_{\max}$.

$$4\cdot 5u_{\min} + u_{\min} - u_{\max} - u_{\max} - u_{\max} - 1\cdot 5 = 0, \quad \text{i.e. } 5\cdot 5u_{\min} - 3u_{\max} - 1\cdot 5 = 0 \tag{50}$$

Inserting (48) into (50) leads to

$$u_{\min} = 7\cdot 5/8\cdot 5 \approx 0\cdot 89, \qquad u_{\max} = 9\cdot 5/8\cdot 5 \approx 1\cdot 11 \tag{51}$$

REFERENCES

1. L. O. Chua and L. Yang, 'Cellular neural networks: theory', *IEEE Trans. Circuits and Systems*, **CAS-35**, 1257–1272 (1988).
2. J. A. Nossek, G. Seiler, T. Roska and L. O. Chua, 'Cellular neural networks: theory and circuit design', *Int j. circ. theor. appl.*, **20**, 523–543, (1992).
3. T. Matsumoto, L. O. Chua and H. Suzuki, 'CNN cloning template: shadow detector', *IEEE Trans. Circuits and Systems*, **CAS-37**, 1070–1073 (1990).
4. T. Matsumoto, L. O. Chua and H. Suzuki, 'CNN cloning template: connected component detector', *IEEE Trans. Circuits and Systems*, **CAS-37**, 633–635 (1990).
5. T. Matsumoto, L. O. Chua and R. Furukawa, 'CNN cloning template: hole-filler', *IEEE Trans. Circuits and Systems*, **CAS-37**, 635–638 (1990).
6. L. O. Chua and B. E. Shi 'Exploiting cellular automata in the design of cellular neural networks for binary image processing', *ERL Memo UCB/ERL M89/130*, University of California, Berkeley, November 1989.
7. F. Zou, S. Schwarz and J. A. Nossek, 'Cellular neural network design using a learning algorithm', *Proc. First IEEE Int. Workshop on Cellular Neural Networks and Their Applications, CNNA-90*, Budapest, 1990, IEEE, New York, 1991, pp. 73–81.
8. L. O. Chua and T. Roska 'Stability of a class of nonreciprocal cellular neural networks', *IEEE Trans. Circuits and Systems*, **CAS-37**, 1520–1527 (1990).

9. T. Roska, 'On the qualitative and quantitative relationships between the analog and digital realizations of "neural" computing circuits', *Report 2/89*, Hungarian Academy of Sciences, Budapest, 1989.
10. E. Goles, F. Fogelman and D. Pellegrin, 'Decreasing energy functions as a tool studying threshold networks', *Discrete Appl. Math.*, **12**, 261–277 (1985).
11. E. Goles, 'Fixed point behaviour of threshold functions on a finite set', *SIAM J. Alg. Appl. Math.*, **3**, 529–531 (1982).
12. J. Bruck, 'On the convergence properties of the Hopfield model', *Proc. IEEE,* **78**, 1579–1585 (1990).
13. K. Preston Jr. and M. J. B. Duff, *Modern Cellular Automata: Theory and Applications*, Plenum, New York, 1984.
14. T. Toffoli and N. Margolus, *Cellular Automata Machines*, MIT Press, Cambridge, MA, 1987.
15. T. Roska and L. O. Chua, 'Cellular neural networks with nonlinear and delay-type template elements', *Proc. First IEEE Int. Workshop on Cellular Neural Networks and Their Applications, CNNA-90*, Budapest, 1990, IEEE, New York, 1991, pp. 12–25.
16. H. Harrer, J. A. Nossek and F. Zou, 'A learning algorithm for time-discrete cellular neural networks', *Proc. IJCNN-91*, Singapore, 1991, pp. 717–722.
17. H. Harrer, J. A. Nossek, G. Seiler and R. Stelzl, 'An analog CMOS compatible convolution circuit for analog neural networks', *Proc. Micro Neuro-91*, Munich, 1991, pp. 231–241.
18. H. Harrer, J. A. Nossek and R. Stelzl, 'An analog implementation of discrete-time cellular neural networks', *Report TUM-LNS-TR-91-14*, Technical University of Munich, June 1991.
19. S. Tan, J. Hao and J. Vandewalle, 'Cellular neural networks as a model of associative memories', *Proc. First IEEE Int. Workshop on Cellular Neural Networks and Their Applications, CNNA-90*, Budapest, 1990, IEEE, New York, 1991, pp. 26–35.
20. L. O. Chua and L. Yang, 'Cellular neural networks: applications', *IEEE Trans. Circuits and Systems*, **CAS-35**, 1273–1290 (1988).
21. N. Frühauf and E. Lüder, 'Realization of CNNs by optical parallel processing with spatial light valves', *Proc. First IEEE Int. Workshop on Cellular Neural Networks and Their Applications, CNNA-90*, Budapest, 1990, IEEE, New York, 1991, pp. 281–290.

CELLULAR NEURAL NETWORKS WITH NON-LINEAR AND DELAY-TYPE TEMPLATE ELEMENTS AND NON-UNIFORM GRIDS†

TAMÁS ROSKA‡ AND LEON O. CHUA

Department of Electrical Engineering and Computer Sciences and the Electronics Research Laboratory, University of California, Berkeley, CA, 94720, U.S.A.

SUMMARY

The cellular neural network (CNN) paradigm is a powerful framework for analogue non-linear processing arrays placed on a regular grid. In this paper we extend the current repertoire of CNN cloning template elements (atoms) by introducing additional non-linear and delay-type characteristics. In addition, architectures with non-uniform processors and neighbourhoods (grid sizes) are introduced. With this generalization, several well-known and powerful analogue array-computing structures can be interpreted as special cases of the CNN. Moreover, we show that the CNN with these generalized cloning templates has a general programmable circuit structure (a prototype machine) with analogue macros and algorithms. The relations with the cellular automaton (CA) and the systolic array (SA) are analysed. Finally, some robust stability results and the state space structure of the dynamics are presented.

1. INTRODUCTION

The cellular neural network (CNN) paradigm[1] provides a powerful analogue non-linear computing structure for a variety of array computations. Array computations can be defined as the parallel execution of complex operations on a large number of processors placed on a geometrically regular grid. If the operations are logical (involving only a few bits), the classical cellular automaton (CA) of John von Neumann and its recent variants are the ideal tools. If the operations are numerical (typically 8, 16 or 32 bits), the systolic arrays are the best current approach. In both cases the computing devices are digital processors. If, however, the signal values are continuous and/or analogue real-time operations are performed, then the CNN is the optimal solution with respect to both area and speed. Its structural simplicity, robust stability and function variability are all ideal for VLSI implementation.[2–4]

In this paper we show that by retaining all the important features of the original CNN structure and by introducing very simple non-linear and delay-type templates, the CNN becomes a powerful framework for *general* analogue array dynamics. Architectures with a few different types of processors as well as with a few different CNN neighbourhood sizes are also defined.

From a theoretical point of view a CNN has well-defined qualitative properties and contains as special cases a variety of powerful practical image-processing solutions (both artificial and biologically motivated applications, e.g. the 'silicon retina' and other resistive grids).[5–9] From a practical point of view the CNN provides a natural paradigm for designing programmable analogue VLSI or optoelectronic real-time computing arrays.

In Section 2 we present the *general framework* and show the simple programmability properties inherent in the CNN structure. In Section 3 we present some special cases which are practically important and promising. In Section 4 we show that our CNN is a general *programmable analogue* array computer.

---

† Part of this research has been reported in the Proceedings of the 1990 International Workshop on Cellular Neural Networks and Their Applications held in Budapest.

‡ Visiting scholar from the Computer and Automation Institute of the Hungarian Academy of Sciences, Uri-u. 49, H-1014 Budapest, Hungary.

With their simplest template elements (atoms), e.g. the gates, we can define the CNN templates as macros and introduce the notion of CNN algorithms based on these macros.

In Section 5 we present some robust *stability results* for a class of non-symmetric non-linear CNN templates as well as some other qualitative properties. In Section 6 we summarize our conclusions and pose some open questions and new directions.

2. THE GENERAL CNN FRAMEWORK WITH NON-LINEAR AND DELAY-TYPE TEMPLATES

Consider a two-dimensional grid such as those shown in Figure 1. The small squares represent the processors and their connections show the possible communication lines (feedback and feedforward connections) with their neighbours. Here the solid and dashed lines correspond to the nearest-neighbour interconnections. This two-dimensional grid of processors is called a *layer*. Multiple layers of processors are also allowed and can be defined similarly, as the multilayer cellular neural networks in Reference 1.

The analogue processor element, the cell, is shown in Figure 2. It differs from that of the original cell (Figure 3 on p. 1258 of Reference 1) basically in the controlled sources I_{xy} and I_{xu} only. Namely, instead

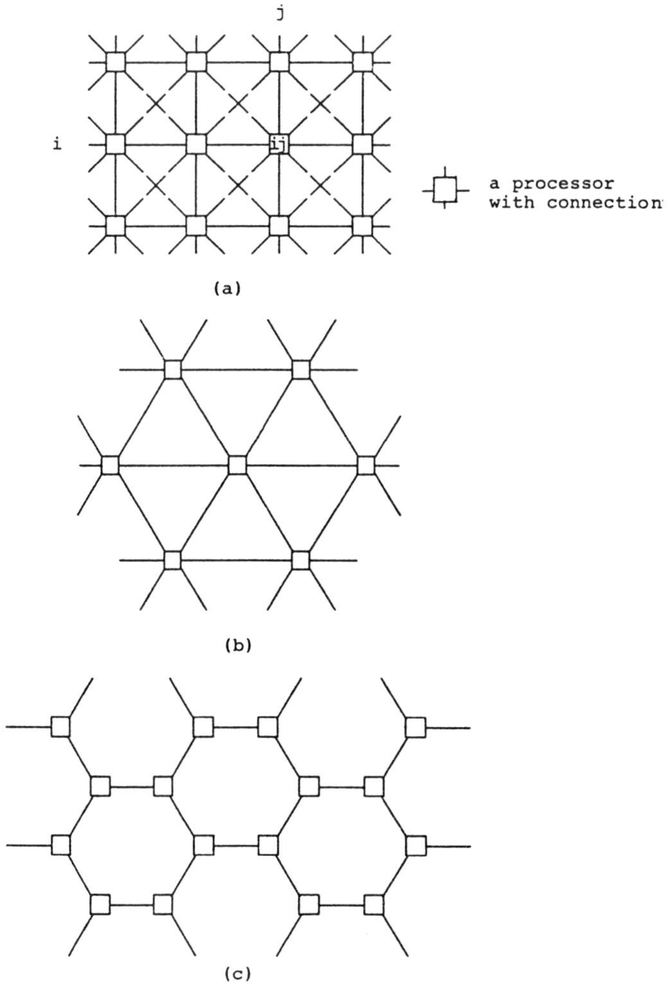

Figure 1. Typical regular grids: (a) rectangular (four or eight neighbours); (b) triangular (six neighbours); (c) hexagonal (three neighbours)

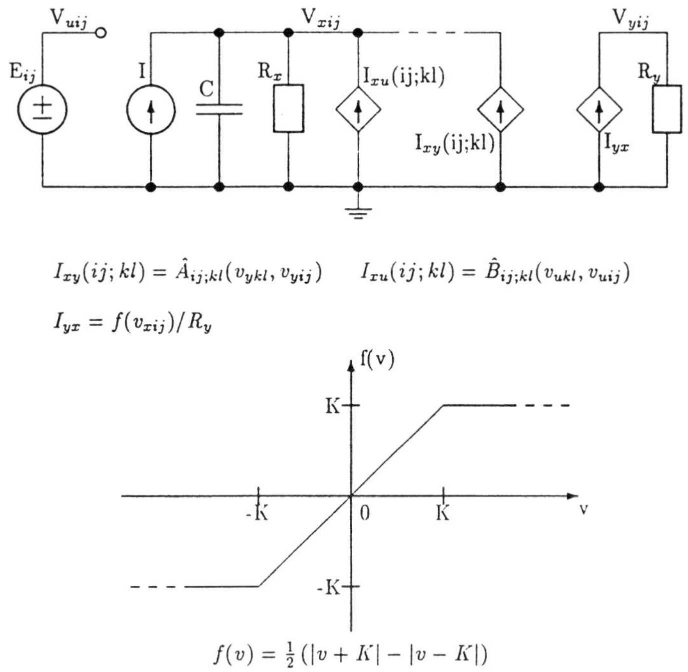

Figure 2. (a) The analogue processor cell circuit and (b) the non-linear output function

of the two linear controlled sources defined by

$$A(ij; kl)v_{ykl} \quad \text{and} \quad B(ij; kl)v_{ukl}$$

associated with a typical cell C_{ij} and a typical interacting cell C_{kl}, we allow non-linear and delayed controlled sources defined by

$$\hat{A}_{ij;\,kl}(v_{ykl}, v_{yij}) + A^{\tau}_{ij;\,kl}v_{ykl}(t-\tau) \quad \text{and} \quad \hat{B}_{ij;\,kl}(v_{ukl}, v_{uij}) + B^{\tau}_{ij;\,kl}v_{ukl}(t-\tau)$$

where u, x and y denote the input, state and output variables respectively. Possibly $\tau = \tau_{kl}$.

That is, instead of having a linear VCCS (voltage-controlled current source) in the A and B cloning templates, we now have a *non-linear* and/or *delay-type* VCCS. The structure of the non-linearity in the templates is also important: it is a function of *at most two variables*, namely the output voltage of cell C_{ij} and that of a neighbour C_{kl}.

We also allow the output function to have a wider range of output voltages so that the saturation voltage is $\pm K$ instead of ± 1. However, these slight changes do extend the class of CNN dynamics significantly.

Again, with $M \times N$ cells in an array, we define the r-neighbourhood of cell C_{ij} as

$$N_r(ij) = \{C_{kl}\colon \max(|k-i|, |l-j|) = r(\text{integer})\} \tag{1}$$

Let us now formulate the *canonical equations* describing the CNN analogue non-linear processor array with non-linear and delay-type templates. Henceforth we will use the same term CNN in this *generalized* concept and refer to a *linear* template CNN when only *linear* controlled sources are being used in the cloning templates. Without loss of generality we assume that the processor cells are drawn on a rectangular grid with $M \times N$ cell units, $1 < i < M$, $1 < j < N$. We can call this CNN simply a *delay CNN* or a *non-linear CNN*. Similarly, when we introduce the possibility of having more than one type of processor and/or more than one neighbourhood size in a CNN, we can call the CNN a *non-uniform processor CNN* or a *multiple-neighbourhood-size CNN* respectively.

Referring to Figures 1 and 2, we can define a CNN mathematically as follows.

State equation

$$C\dot{v}_{xij} = -(1/R_x)v_{xij}(t) + I + \sum_{C_{kl} \in N_r(ij)} \hat{A}_{ij;\,kl}(v_{ykl}(t), v_{yij}(t)) + \sum_{C_{kl} \in N_r(ij)} \hat{B}_{ij;\,kl}(v_{ukl}(t), v_{uij}(t))$$

$$+ \sum_{C_{kl} \in N_r(ij)} A^{\tau}_{ij;\,kl} v_{ykl}(t-\tau) + \sum_{C_{kl} \in N_r(ij)} B^{\tau}_{ij;\,kl} v_{ukl}(t-\tau) \qquad (2a)$$

where $\hat{A}$, $\hat{B}$ and A^{τ}, B^{τ} are associated with the non-linear and the delay-type cloning templates respectively. In particular, $\hat{A}_{ij;\,kl}$ and $\hat{B}_{ij;\,kl}$ are continuous functions of at most two variables and $A^{\tau}_{ij;\,kl}$ and $B^{\tau}{}_{ij;\,kl}$ are real constants. If we write $A(ij;kl)$ and $B(ij;kl)$, this means a simple linear template.

For example, with $d_1 = c_1[\exp(v_{ykl}) - 1]$ and $d_2 = c_2(v_{ykl} - v_{yij})$ a cloning template can be defined symbolically as

$$\hat{A} = \begin{bmatrix} 0 & d_1 & 0 \\ d_1 & 2 & d_1 \\ 0 & d_1 & 0 \end{bmatrix}, \qquad \hat{B} = \begin{bmatrix} 0 & 0 & 0 \\ d_2 & 1 & d_2 \\ 0 & 0 & 0 \end{bmatrix}, \qquad I = 0$$

Here we have tacitly assumed that the templates are space-invariant, though they could be space-variant too. Henceforth we use space-invariant templates.

If I_{ij} is space-variant, we have a space-variant template.

Output equation

The memoryless output function

$$v_{yij}(t) = f(v_{xij}(t)) \qquad (2b)$$

is depicted in Figure 2(b). It is a piecewise linear characteristic with unity slope in the range $(-K, K)$, $K > 0$. It can be approximated to within any precision by a *smooth* (C^1) strictly monotone-increasing sigmoid function. This approximation is sometimes desirable in analytical proofs where the C^1 condition is more convenient to work with. It is also a more realistic assumption since the physically realized characteristic is actually C^1.

As a further generalization we can allow the output function to have its own dynamics too. For example, equation (2b) can be replaced by a first-order state equation

$$\dot{v}_{yij} = -a v_{yij} + f(v_{xij}(t)) \qquad (2b')$$

or by a higher-order state equation.

Moreover, remaining monotone-increasing, f might be more complicated.

Input equation

$$v_{uij} = E_{ij} \qquad (2c)$$

Constraint equations

$$|v_{xij}(0)| \leqslant 1 \qquad (2d)$$

$$|v_{uij}(t)| \leqslant 1 \qquad (2e)$$

Parameter assumptions

$$K \geqslant 0, \qquad C \geqslant 0, \qquad R_x > 0, \qquad \tau \geqslant 0 \qquad (2g)$$

In the case when $\hat{A}_{ij;\,kl}$ and $\hat{B}_{ij;\,kl}$ are linear functions (of a single variable) and $\tau = 0$, or equivalently when

$$\hat{A}_{ij;\,kl} = A(ij;\,kl)v_{ykl}, \qquad \hat{B}_{ij;\,kl} = B(ij;\,kl)v_{ukl}$$

and $A^{\tau}_{ij;\,kl}$ and $B^{\tau}_{ij;\,kl}$ are equal to zero, we have the original CNN[1] with linear cloning templates.

Motivated partly by neurobiological structures, it is worthwhile to introduce non-uniform grid CNN architectures having more than one type of cell processor and/or more than one size of neighbourhood. We will call such a CNN a *non-uniform processor CNN* (NUP-CNN) or a *multiple-neighbourhood-size CNN* (MNS-CNN) respectively.

A NUP-CNN with two processor types is shown in Figure 3(a).[24] The two types of processors are denoted by empty and black boxes. A special case of the dual-type CNN (including logic processors as well) is the CNNL architecture.[20] This is also a NUP-CNN (Figure 3(b)). NUP-CNN architectures contain more than one processor type placed on a regular grid where the structures of all the interprocessor connections are space-invariant.

An MNS-CNN with two sizes of neighbourhoods is shown in Figure 3(c). All the processors are the same. However, grid A (on layer A) is a fine grid with neighbourhood size $r = 1$, while grid B on the next layer (dashed line) is a course grid for the processors with neighbourhood size $r = 3$ (connected to layer A). On layer B connections to one processor only are shown. MNS-CNN architectures contain layers with different grids and neighbourhood sizes. They reflect some characteristic structures found in living visual systems[27]. A special case of the MNS-CNN with two sizes of neighbourhoods contains only one processor on layer B. In this case the connections to this processor are global lines from each processor.

With regard to the information-processing or computing function of the CNN array, it has *two input ports* and one output port per cell; namely, besides the generic input $E_{ij}(t) = v_{uij}(t)$ the initial value of the state, i.e. $v_{xij}(0)$, can also be used as an input. The output $v_{yij}(t)$ is defined as either the DC steady state $v_{yij}(\infty)$ or as a snapshot at a given finite time instant $t = T$, i.e. $v_{uij}(T)$. In the former case the successive inputs can be applied in a sampled data mode, allowing a sufficient settling time to elapse between the samples. It is clear that the most direct application area of the CNN is in image processing (where the light intensity of the pixel, I_{ij}, is the input). However, other applications such as solving special types of partial differential equations are also potentially important. Even the two input possibilities can be combined, e.g. for processing successive images of a motion picture. Using the A^{τ} and B^{τ} delay-type templates, various forms of successive (moving) images can be generated within the CNN.

In picture-processing applications the grey-scale image pixel values are coded between $+1$ and -1, which correspond to black and white extremes.

Let us derive next *an estimate for the dynamic range.*

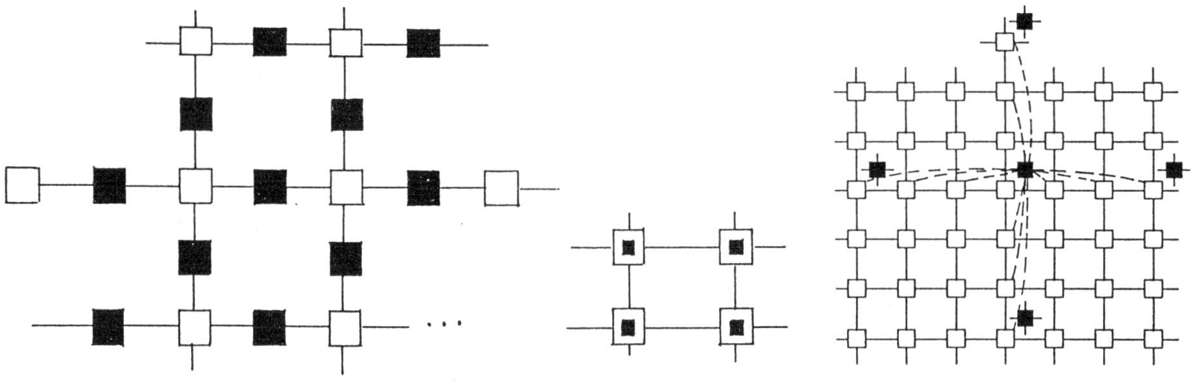

(a) (b) (c)

Figure 3. Non-uniform processors and grid sizes: (a) non-uniform processor CNN; (b) a special case; (c) multiple-neighbourhood-size CNN

Proposition 1

For a non-linear CNN characterized by bounded non-linear cloning templates (but without delay) and a memoryless, piecewise linear output equation (equation (2b)) in the canonical equations (2), all states v_{xij} are bounded for all times $t > 0$ and the bound $v_{\max}$ can be computed by the formula

$$v_{\max} = 1 + R_x \mid I \mid + R_x \max\left(K \sum_{C_{kl} \in N_r(ij)} \max \mid \hat{A}_{ij;\, kl} \mid + \sum_{C_{kl} \in N_r(ij)} \max \mid \hat{B}_{ij;\, kl} \mid \right), \quad 1 \leqslant i \leqslant M, \quad 1 \leqslant j \leqslant N \tag{3}$$

The proof follows in the same way as in Section III of Reference 1, except that the bounds $\mid A(ij; kl) \mid$ and $\mid B(ij; kl) \mid$ are replaced by $\max \mid \hat{A}_{ij;\, kl} \mid$ and $\max \mid \hat{B}_{ij;\, kl} \mid$ respectively and the bound for $v_{yij}(t)$ is equal to K instead of one.

3. SPECIAL CASES

As mentioned in Section 1, a CNN with non-linear and delay-type cloning templates as defined by the canonical equations (2) can be considered as a unifying paradigm for analogue non-linear processor arrays where the processors are placed on a regular three-dimensional geometrical grid. One layer of this array is placed on a two-dimensional regular grid (see Figure 1). We will now describe only one important special case.

A special case: non-linear resistive grid (with capacitors)

Several recent analogue processor arrays which mimic very simple models of the vertebrate retina[5] or some functional model of the visual process[7] make use of the circuit structure depicted in Figure 4. Observe that this structure is a simple special form of the CNN defined in Section 2.

Namely, choose

$$I = 0, \qquad A^{\tau} = B^{\tau} = 0$$

Choose $I_{xu} = I_b = B_{ij;\, ij}(E_{ij}) = g(E_{ij})$ so that

$$\hat{B} = \begin{bmatrix} 0 & 0 & 0 \\ 0 & g & 0 \\ 0 & 0 & 0 \end{bmatrix}$$

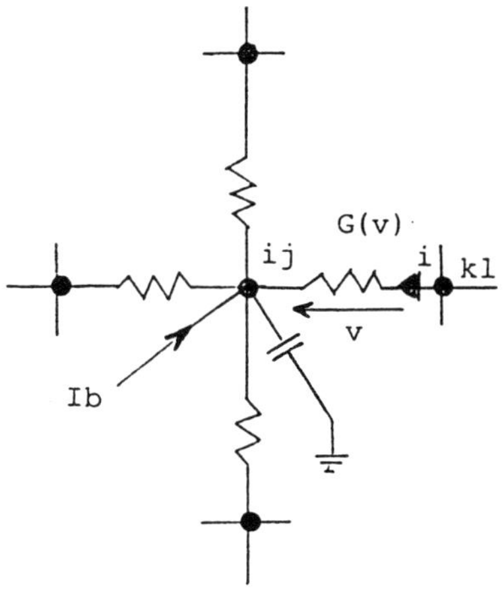

Figure 4. A resistive grid-type circuit structure. All resistors are non-linear and characterized by $i = G(v)$

in Figure 1(a) or

$$\hat{B} = \begin{array}{c} \overline{0\quad 0} \\ [0 \quad g \quad 0] \\ \underline{0\quad 0} \end{array}$$

in Figure 1(b). Choose $v_{yij} = f(v_{xij})$ with $K \gg 1$; hence $v_{yij} = v_{xij}$, $R_x = 0$, $\hat{A}_{ij;\,kl} = G(v)$ or 0 and $G = G(v_{ykl} - v_{yij})$, so that

$$\hat{A} = \begin{bmatrix} 0 & G & 0 \\ G & 0 & G \\ 0 & G & 0 \end{bmatrix}$$

in Figure 1(a) or

$$A = \begin{array}{c} \overline{G\quad G} \\ [G \quad 0 \quad G] \\ \underline{G\quad G} \end{array}$$

in Figure 1(b).

The non-linear resistor v–i characteristics can be defined according to the various applications.[5–9,25] The type of '*fading*' *resistor characteristic* in Figure 5 is used in several vision-related applications having different names.

We wish to emphasize that the CNN defined here can *operate in two modes*, either in the *transient mode* or in the DC *steady state mode*. In the transient mode the output is the snapshot at a given time instant $t = T$ (e.g. in the scale-spacing method associated with the anisotropic diffusion equation[8]). In the DC steady state mode the output is the DC steady state of the circuit. Equivalently, we can choose the speed of the input to be much smaller than the settling time of the circuit (e.g. the speed is chosen equal to the reciprocal time constant in Reference 6). Furthermore, these circuits can be operated in discrete time as well.[22]

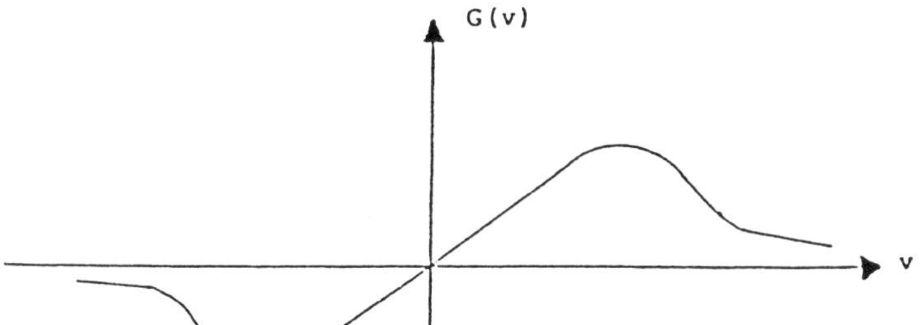

Figure 5. The v–i characteristic of a 'fading' resistor

Remark 1

The above cases are special cases indeed. The choice of many different resistive grids with other non-linear characteristics such as square, square root, absolute value, etc. can lead to many interesting results.

As an *example* consider the input picture of Figure 6. This picture is applied as the initial state ($v_{xij}(0)$)

Figure 6. An input picture

to a connected component detection CNN[19] with the following linear templates on a square grid:

$$\hat{A} = A = \begin{bmatrix} 0 & 0 & 0 \\ 1 & 2 & -1 \\ 0 & 0 & 0 \end{bmatrix}, \qquad \hat{B} = 0, \qquad I = 0$$

The output $v_{yij}(\infty)$ shown in Figure 7(a) gives the number of connected components when Figure 6 is projected along the horizontal axis. Now suppose we choose next

$$\hat{A} = A = \begin{bmatrix} 0 & 0 & 0 \\ 1 & 2 & -1 \\ 0 & 0 & 0 \end{bmatrix}, \qquad \hat{B} = 0, \qquad I = 0, \qquad A^{\tau} = \begin{bmatrix} 0 & 1 & 0 \\ 0 & 1 & 0 \\ 0 & 1 & 0 \end{bmatrix}, \qquad B^{\tau} = 0$$

and $\tau = 3$. The output shown in Figure 7(b) gives a combination of a connected component detector and a vertical line detector with a delay. This new CNN therefore gives a qualitatively new image-processing effect.

An example for a *non-linear CNN* is shown in Figure 8.[25] The cloning template is

$$\hat{B} = \begin{bmatrix} 0 & g & 0 \\ g & 0 & g \\ 0 & g & 0 \end{bmatrix}, \qquad \hat{A}_{ij,ij} = 2, \qquad \hat{A}_{ij,kl} = 0 \quad (ij \neq kl), \qquad I = 0$$

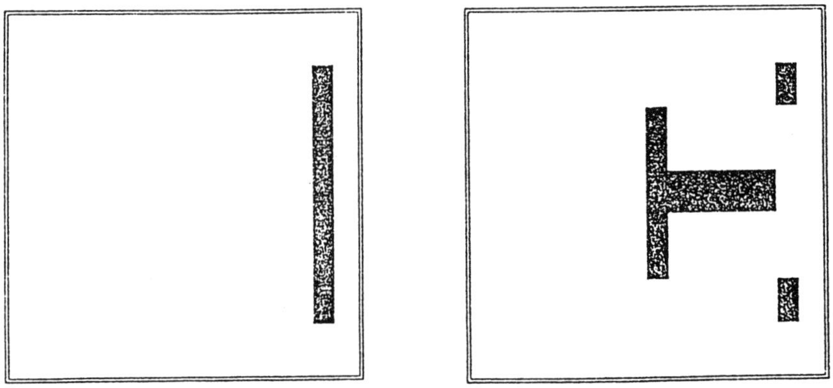

(a) (b)

Figure 7. The output pictures using (a) a linear cloning template and (b) a delay-type cloning template

(a) (b)

Figure 8. (a) The input picture (part of a painting by El Greco) and (b) the output image using a non-linear template

where $g(v_{ukl} - v_{yij})$ is a constant value of 0·5 except between −0·18 and 0·18 where the value is −1. The input image is a grey-scale version of part of an El Greco painting and the output shows the contours.

All these calculations were performed using the CNND simulator programme[15] running on an IBM PS/2 Model 60 computer.

4. CNN AS ANALOGUE PROGRAMMABLE COMPUTING ARRAY

We can compare the *input* and *specification* of a cellular automaton (CA), a systolic array (SA), and a cellular neural network (CNN) as follows:

	CA	SA	CNN
Input	Logic values (1–4 bits)	Numerical values (8–32 bits)	Analogue values
Specification	truth table	Numerical algorithm	Cloning templates

Some relations between a CA and a CNN with linear cloning templates have been studied in Reference 13.

The computing efficiency in terms of finite area/time/dissipation varies from problem to problem. All three classes of arrays have their own advantages and disadvantages. Observe that the analogue CNN solves the problem in a one-step transient process while in general the other two architectures make several iterations.

A crucial problem, howver, is the *programmability* question. Both the CA and the SA can be programmed easily. The programmability of the CNN may not be obvious. We will show that owing to its simple *current-additive structure* (even with non-linear and delay-type templates), the *CNN is in fact a programmable analogue VLSI array computer.*

The basic building blocks (atoms) of the CNN are the coefficients in the cloning templates, namely

$$\hat{A}_{ij;\,kl},\quad \hat{B}_{ij;\,kl},\quad A^{\tau}_{ij;\,kl},\quad B^{\tau}_{ij;\,kl},\quad I$$

These coefficients are realized by voltage-controlled current sources. Note that I can be considered as a VCCS with a constant controlling voltage.

These *analogue template atoms* are really programmable, since each additive constant term can be easily selected by switching on or off the relevant component from a bank of parallel connected linear and non-linear controlled sources. The practical solutions and the method of programmability differ depending on the actual design.[4,10,20,23]

The cloning template is built up from the above template atoms, thereby forming an elementary processing function or an *analogue macro* instruction.

The sequential or parallel applications of these analogue macro instructions or cloning templates constitute the *analogue algorithm* or *analogue software* of the CNN. In fact, this notion of the software is in good agreement with both the basic physical meaning[16] and the recursive function analogy.[17] Hence the following correspondence gives not only a clear structure of the hierarchy of both types of computation but also its engineering realization:

Analogue CNN	Digital CA or SA
	Algorithm
Sequence of templates (series of parallel)/analogue software	Recursive function/algorithm/software
	Network
Cloning template	Machine instruction
Analogue 'macro'	Macro
	Physical circuit
Cloning template atom	'Gate'

Finally, we remark that in some recent practical applications various cloning template sequences are used successfully for different tasks e.g. 1 and even the analogue and digital functions can be used jointly in a dual computing structure.[11,17]

5. SOME QUALITATIVE PROPERTIES

The canonical equations (2) describe the dynamics of the CNN. In order to apply specific qualitative results from dynamical systems theory, it is convenient to reformulate these equations in a closed vector–matrix form. Without loss of generality, suppose $R_x = 1$, $C = 1$ and $M = N$.

In the case of linear cloning templates we have the simple form

$$\dot{\mathbf{v}}_x = -U\mathbf{v}_x + A^{\mathrm{L}}\mathbf{v}_y + B^{\mathrm{L}}\mathbf{v}_u + \mathbf{I} \tag{4}$$

where $\mathbf{v}_x(t)$, $\mathbf{v}_y(t)$, $\mathbf{v}_u(t) \in \mathbb{R}^{N^2} \times \mathbb{R}^1$ (i.e. N^2-vectors), $\mathbf{I} \in \mathbb{R}^{N^2}$ with all elements equal to I, U is an $N^2 \times N^2$ unit matrix, $\mathbf{v}_y = \mathbf{f}(\mathbf{v}_x)$ is a diagonal mapping with $v_{yi} = f(v_{xi})$ and A^{L}, $B^{\mathrm{L}} \in \mathbb{R}^{N^2 \times N^2}$ (i.e. $(N^2 \times N^2)$ matrices).

We have N^2 cells, i.e. N^2 elements in the vectors. The question is how to order them. Figure 9 shows three possibilities for the case of a square grid with $N = 5$. Observe that many nearest-neighbour cells will not be placed next to each other in this reordering. For example, for the three reorderings in Figures

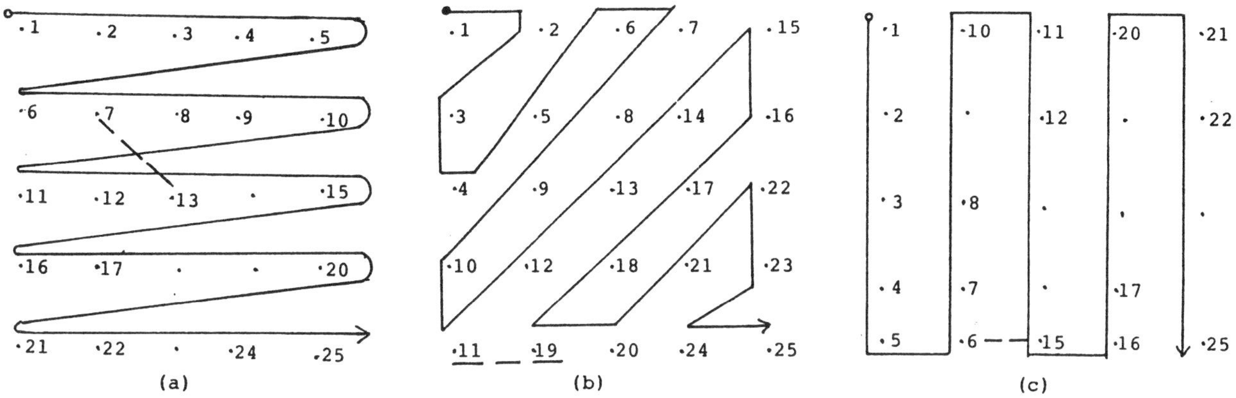

Figure 9. The distinct ordering schemes

9(a)–9(c) the nearest neighbours can be found in the vectors as far as the following respective index distances from each other:

$$N+1, \qquad 2N-2, \qquad 2N-1 \tag{5}$$

For example, we have $13-7=6$ in Figure 9(a), $19-11=8$ in Figure 9(b) and $15-6=9$ in Figure 9(c).

For example, consider the linear cloning template

$$A=\begin{bmatrix} 0 & p_N & 0 \\ p_W & 2 & p_E \\ 0 & p_S & 0 \end{bmatrix}, \qquad B=0, \qquad I=0 \tag{6}$$

and choose the diagonal ordering of Figure 9(b). The matrix A^L associated with this ordering is shown below, where the entries in some selected rows and columns are given:

$$A^L = \begin{array}{c|ccccccccccccccc|} & 1 & 2 & 3 & 4 & 5 & 6 & \cdots & 9 & 10 & 11 & \cdots & 19 & \cdots & 25 \\ \hline 1 & 2 & p_E & p_S & & & & & & & & & & & \\ 2 & p_W & 2 & & & p_S & p_E & & & & & & & & \\ 3 & p_N & & 2 & p_S & p_E & & & & & & & & & \\ 4 & & & p_N & 2 & & & & p_E & p_S & & & & & \\ 5 & & & & & & & & & & & & & & \\ 6 & & & & & & & & & & & & & & \\ \vdots & & & & & & & & & & & & & & \\ 9 & & & & p_W & p_N & & & 2 & & & & & & \\ 10 & & & & p_N & & & & & 2 & p_S & & & & \\ 11 & & & & & & & & & p_N & 2 & & p_E & & \\ \vdots & & & & & & & & & & & & & & \\ 19 & & & & & & & & & & p_W & & 2 & & \\ \vdots & & & & & & & & & & & & & & \\ 25 & & & & & & & & & & & & & & 2 \end{array} \tag{7}$$

The selected rows are complete. Observe that if A is centrally symmetric, then A^L is symmetric too.

For the *general case* of a CNN with non-linear and delay-type templates and using the same variables and vectors as above, the canonical matrix–vector description of the dynamics is as follows (again assuming a square grid with $r = 1$, $R_x = 1$, $C = 1$):

$$\dot{\mathbf{v}}_x = -U\mathbf{v}_x + \hat{A}(\mathbf{v}_y)\mathbf{e} + \hat{B}(\mathbf{v}_u)\mathbf{e} + A^{\tau}\mathbf{v}_y(t-\tau) + B^{\tau}\mathbf{v}_u(t-\tau) + \mathbf{I} \tag{8}$$

where $\mathbf{e} = [1, 1, 1, \ldots, 1]^{\mathrm{T}} \in \mathbb{R}^{N^2}$, $\hat{A}$ and $\hat{B}$ are $(N^2 \times N^2)$-matrices with elements $\hat{A}_{ij;\,kl}(v_{ykl}, v_{yij})$ and $\hat{B}_{ij;\,kl}(v_{ukl}, v_{uij})$ respectively and A^{τ} and B^{τ} are $(N^2 \times N^2)$-matrices with elements $A^{\tau}_{ij;\,kl}$ and $B^{\tau}_{ij;\,kl}$ respectively.

The ordering of the cell variables into the vectors $\mathbf{v}_x$, $\mathbf{v}_y$ and $\mathbf{v}_u$ is chosen in the same way as above for the linear templates.

On the basis of the structure of the above canonical equations, several important stability results can be proved using similar to those in References 1, 2 and 12. Here, as a representative result, we will prove only our next proposition.

Proposition 2

If a CNN has a non-linear cell-linking template[2] which has template atoms with strictly monotone-increasing C^1 characteristics in the non-local variable (i.e. for $p(v_{ykl}, v_{yij})$, $\partial p/\partial v_{ykl} > 0$), then the CNN is completely stable except possibly from a set of initial conditions of measure zero. A circuit is said to be completely stable iff every trajectory tends to an equilibrium state. Consequently, such a circuit cannot oscillate or become chaotic (it is called convergent in Reference 12).

Proof In view of the cell-linking property, the dynamics described by (8) is irreducible. Moreover, the hypothesis of Proposition 2 guarantees that all the off-diagonal terms of the associated Jacobian matrix are positive. Therefore[18] the dynamics is completely stable (convergent) except for a starting set with measure zero.

Remark 2

In the case of a delay CNN a recent strong result[21] can be applied to show that the positive cell-linking templates with finite uniform delay are completely stable. This result is developed in detail in a forthcoming paper.[26]

6. CONCLUSIONS

The cellular neural network with non-linear and delay-type cloning templates as well as the NUP-CNN and MNS-CNN form a unifying paradigm for analogue (non-linear) processing arrays. It not only provides a rigorous theoretical framework, but at the same time many practical solutions to various image-processing problems can be solved by these arrays in real time. In particular, the first two phases of many image-processing problems (e.g. noise removal and feature extraction of grey-scale pictures) belong to these classes of problems. In addition to many new applications, the *non-linear cloning templates* allow us to model some biological properties of the retina, but they can also be used for modelling motion dynamics. The *delay-type templates* provide us with even more flexibility and new applications, including the detection of some motion features. Multiple-neighbourhood-size and non-uniform processor CNNs are natural variants of the CNN paradigm.

(i) It is a programmable, analogue, regular, non-linear, dynamic processing array model with local connectivity the processors are placed on a 3D geometrical grid.
(ii) By a sequence of different templates analogue software can be generated.
(iii) Algorithms, circuits and physical implementation are considered jointly.
(iv) Two independent images are combined dynamically.

(v) Delay templates provide unprecedented effects, including dynamically programmable detection and estimation of motion.
(vi) Combining analogue dynamics with local logic in the dual computing structure provides new capabilities.
(vii) It is a natural and promising complex model of the vertebrate retina.

ACKNOWLEDGEMENTS

This work has been supported by the Office of Naval Research grant N00014-89-5-1402, the National Science Foundation grant MIP-8614000 and by a joint grant of the National Science Foundation and the Hungarian Academy of Sciences No. NSF INT-90-01336.

REFERENCES

1. (a) L. O. Chua and L. Yang, 'Cellular neural networks: Theory', *IEEE Transactions on Circuits and Systems*, **35**, 1257–1272, 1988. (b) L. O. Chua and L. Yang, 'Cellular neural networks: Applications', ibid., 1273–1290.
2. L. O. Chua and T. Roska, 'Stability of a class of nonreciprocal cellular neural networks', ibid., **37**, pp. 1520–1527, December, 1990.
3. J. M. Cruz and L. O. Chua, 'A CNN chip for connected component detection', ibid., **38**, pp. 812–817, 1991.
4. L. Yang, L. O. Chua and K. R. Krieg, 'VLSI Implementation of Cellular Neural Networks', *Proc. IEEE ISCAS-90*, pp. 2425–27, May 1990.
5. C. Mead and M. Ismail (eds), Analog VLSI implementation of neural systems, Kluwer Academic Publ., Boston, 1989.
6. T. Roska, T. Boros, P. Thiran and L. O. Chua, 'Detecting simple motion using Cellular Neural Networks', *Proc. CNNA-90*, pp. 127–138, 1990.
7. C. Mead, Analog VLSI and neural systems, Addison-Wesley, Reading, MA, 1989.
8. P. Perona and J. Malik, 'Scale-space and edge detection using anisotropic diffusion', *IEEE Trans. Pattern Rec. and Machine Intelligence*, **12**, pp. 629–639, 1990.
9. D. Anastassiou, 'Error diffusion coding for A/D conversion', *IEEE Transactions on Circuits and Systems*, **36**, pp. 1175–1186, 1989.
10. N. Frühauf and E. Lüder, Realization of CNNs by optical parallel processing with spatial light valves', Proc. IEEE CNNA-90, pp. 281–290, 1990.
11. T. Roska, 'Analog events and a dual computing structure using analog and digital circuits and operators', pp. 225–238 in Discrete Event Systems: Models and Applications (eds P. Varaiya and A. B. Kurzhanski), Spinger-Verlag, Berlin, 1988.
12. M. W. Hirsch, 'Convergent activation dynamics in continuous time networks', *Neural Networks*, **2**, pp. 331–349, 1989.
13. L. O. Chua and B. Shi, 'Exploiting cellular automata in the design of cellular neural networks for binary image processing', ERL memo UCB/ERL M89/130, University of California, Berkeley, Nov. 15, 1989.
14. G. Bártfay and T. Roska, 'A new cellular neural network template and its application to skeletonization', Report, No. 24/1989, Computer and Automation Institute, Hungarian Academy of Sciences (MTA SzTAKI), Budapest, 1989.
15. CNND simulator, Cellular Neural Network embedded in a simple Dual computing structure, User's guide Version 3.0, Report, No. 37/1990 (ed. T. Roska and A. Radványi), Computer and Automation Institute, Hungarian Academy of Sciences (MTA SzTAKI), Budapest, 1990.
16. Á. Csurgay, Fundamental limits in large-scale circuit modelling, *Proc. ECCTD-83*, pp. 454–457, VDE-Verlag, Berlin, 1983.
17. T. Roska, 'A dual way of computing-learning from cerebral asymmetry', Proc. Symposium on Computer and Cognitive Sciences, Tampere University of Technology, pp. 101–108, Tampere, 1989.
18. M. W. Hirsch, 'Systems of differential equations that are competitive or cooperative II.: Convergence almost everywhere', SIAM *J. Math. Anal.*, **16**, pp. 423–439, 1985.
19. T. Matsumoto, T. Yokohama, H. Suzuki, R. Furukawa, A. Oshimoto, T. Shimmi, Y. Matsushita, T. Seo, and L. O. Chua, 'Several image processing examples by CNN', Proc. CNNA-90, pp. 100–111, 1990.
20. K. Halonen, V. Porra, T. Roska and L. O. Chua, 'VLSI Implementation of a reconfigurable CNN containing local logic', Proc. CNNA-90, pp. 206–215, 1990.
21. H. L. Smith, 'Monotone semiflows generated by functional diff.egs., *J. Diff. Eg.* **66**, pp. 120–442, 1987.
22. H. Harrer and J. A. Nossek, 'Time discrete cellular neural networks: architecture, applications and realization', Report No. TUM-LNS-TR-90-12, Technical University of Munich, November 1990.
23. J. E. Varrientos, J. Ramírez-Angulo and E. Sánchez-Sinencio, 'Cellular neural networks implementation: a current-mode approach, Proc. IEEE CNNA-90, pp. 216–225, 1990.
24. L. O. Chua and B. Shi, private communication, 1991.
25. T. Boros, K. Lotz, A. Radványi and T. Roska, 'Some useful new nonlinear and delay-type templates', Report DNS-1-1990, Computer and Automation Institute of the Hungarian Academy of Sciences, Budapest 1990.
26. T. Roska, C. W. Wu, M. Balsi and L. O. Chua, 'Stability and dynamics of delay-type and nonlinear cellular neural networks', Memo. No. UCB/ERL M91/110, University of California at Berkeley, December 1991.
27. T. Roska, K. Lotz, J. Hámori, E. Lábos and J. Takács, 'The CNN model in the visual pathway—Part I: The CNN-Retina and some direction- and length- selective mechanisms', Report DNS-8-1991, Computer and Automation Institute of the Hungarian Academy of Sciences, Budapest, 1990.

MEMBRAIN: A CELLULAR NEURAL NETWORK MODEL BASED ON A VIBRATING MEMBRANE†

J. HENSELER AND P. J. BRASPENNING

Department of Computer Science, University of Limburg, PO Box 616, NL-6200 MD Maastricht, Netherlands

SUMMARY

This paper introduces the Membrain model describing a neural network architecture which is similar to the architecture underlying the class of cellular neural networks (CNNs). The main difference pertains to the characteristic processing equation, which is based on a wave equation instead of a heat equation. Within the CNN framework, a cellular Membrain model may be obtained by replacing the neuron output function by a first-order state equation. Furthermore, the network-cloning templates are chosen such that the CNN behaves like a system of coupled harmonical oscillators. Since the energy of such a system is bounded, the piecewise linear neuron characteristic function may be chosen such that it always operates in the linear regime. Our starting point is the analytical and general solution for forced vibrations with damping. This solution applies to a Membrain neural network whose functional architecture is based on the specialized solution for a network of coupled harmonic oscillators. In particular, we present a Membrain CNN (MCNN) having a toroidal connection structure such that the natural modes of vibration of the net are translation-invariant. Moreover, depending on the point group of the network, some rotation invariance can also be obtained. Identifying the input of such a network with the *initial* state of the oscillators gives rise to an output which is in essence a transversally travelling wave made up of components which are *coupled* harmonic neuronal oscillators; that is, the wave is a *superposition* of natural modes of vibration of the network. The temporal wave pattern may be transformed into a one-dimensional temporal signal which is invariant under translation of the initial deflection pattern of the MCNN. The amplitudes of the components in the temporal signal correspond to the power spectrum of the natural vibration modes in the MCNN. Interpreting the initial deflection pattern as a grey-level image, the temporal signal can be viewed as a modulation of a translation-invariant 'fingerprint' of the image. The signal may be sampled such that the modulated 'fingerprint' can be classified using some of the traditional neural network models. In particular we show that (1) a self-organizing feature map clusters correlated images and (2) a back-propagation neural network extracts position-invariant features.

1. INTRODUCTION

When a smooth water surface is stirred, wave fronts emerge which travel across the surface in all directions. The waves are scattered owing to reflection, for instance against a bank, and interfere with other waves. Such an emerging wave pattern can be viewed as a temporal representation of the initial stirring of the surface, whilst depending further on the actual boundary conditions.

In Figure 1 a simulation of such a system is shown in the form of an arbitrary pattern of travelling waves at four different time instances. For each time instance a grey-level image is displayed as well as a projection of the grid. In the image, black represents a surface deflection $\geqslant 1$ whereas white represents a deflection $\leqslant -1$. Figures 1(a) and 1(e) show the chosen *initial* deflection pattern of network activity consisting of two peaks. Each peak acts as the source of a wave front which is propagated radially outwards as shown in Figure 1(b). In Figure 1(c) the transversally travelling waves interfere for the first time. After a while, waves are reflected at the boundary of the network and a complex pattern of continuously interfering waves evolves as shown in Figure 1(d). Figures 1(e)–1(h) show the two-dimensional projections of the surface at the same time instances. The simulation also illustrates the

† Part of this research has been reported in the Proceedings of the 1990 International Workshop on Cellular Neural Networks and Their Applications held in Budapest.

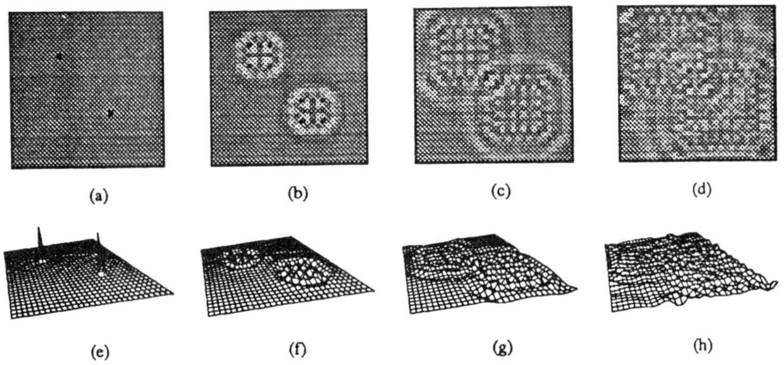

Figure 1. Sequence of travelling wave patterns evolving from an initial deflection at $t = 0$ shown in (a)

energy conservation which is typical for a wave equation;[1] that is, the tall peaks shown in Figure 1(e) are spread over the surface, resulting generally in much lower amplitudes.

Initially, the system may be formalized by a two-dimensional wave equation describing the appearance of transversally travelling waves in a continuous vibrating medium, e.g. a membrane. However, the continuous wave equation may be *discretized* into small spatial intervals, yielding a system of coupled oscillators. The oscillators vibrate orthogonally to the surface, i.e. in a vertical direction if the surface is placed horizontally. Owing to the (vertical) oscillations, waves emerge which travel *transversally* to the oscillating direction, i.e. in horizontal directions. In this paper we present a neural network model which is based on such a system of coupled oscillators. Neural networks based on this model resemble a vibrating membrane, since their dynamics is based on wave propagation in a two-dimensional *discrete* lattice. Therefore we have called this model the *Membrain* model.

The system equations underlying the Membrain model are formally described in Section 2. Since the system is linear, its analytical solution is presented as well. We show that a cellular Membrain neural network strongly resembles a cellular neural network[2] (CNN) if and only if the CNN neuron output function $f(v_{xij})$ is replaced by a first-order differential equation.[3] Thus the Membrain CNN (MCNN) is a second-order CNN operating in *transient mode* such that the internal neuron state v_{xij} remains within the linear part of the sigmoid output function. Rather than the generic input, the initial value of the network state, i.e. $v_{xij}(0)$, is used as input.[3] Interpreting the initial deflection pattern of an MCNN (see Figure 1(e)) as a grey-level image (see Figure 1(a)), the wave patterns are a temporal representation of the image. A practical application of the MCNN model, namely preprocessing of grey-scale images, is presented in Section 3. In particular, we show that an MCNN with a toroidal connection structure yields position-invariant image representations which may be classified using e.g. a self-organizing feature map[4] or a back-propagation[5,6] neural network. Finally, in Section 4 conclusions regarding the MCNN are presented and in Section 5 some interesting possibilities for future research are mentioned.

2. THE MEMBRAIN CELLULAR NEURAL NETWORK MODEL

Chua and Yang[2] have identified the relationship between CNNs and the heat equation from physics. The heat equation is

$$\frac{1}{k}\frac{\partial u(x,y,t)}{\partial t} = \frac{\partial^2 u(x,y,t)}{\partial x^2} + \frac{\partial^2 u(x,y,t)}{\partial y^2} \tag{1}$$

Now, instead of using a heat equation, we will consider a wave equation which is comparable to equation (1). Such a substitution has been suggested in an earlier paper of the authors.[7] The difference between *heat diffusion* and *propagating waves* is that the left-hand side of the wave equation contains a second-

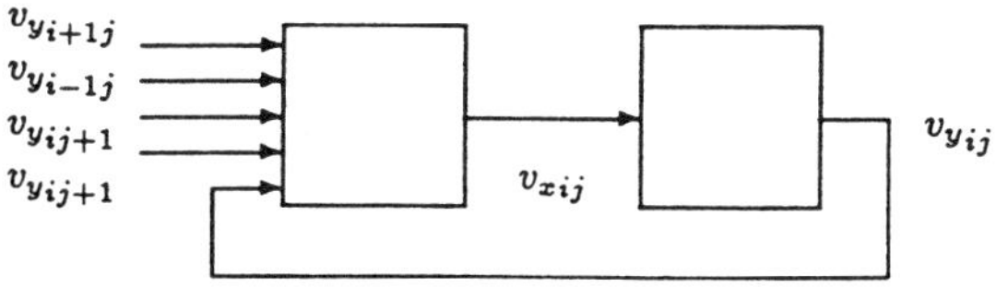

Figure 2. Block diagram of the second-order processing model of an oscillator neuron constructed from two first-order processing stages

order time derivative rather than a first-order time derivative:

$$\frac{1}{a^2}\frac{\partial^2 u(x,y,t)}{\partial t^2}=\frac{\partial^2 u(x,y,t)}{\partial x^2}+\frac{\partial^2 u(x,y,t)}{\partial y^2} \tag{2}$$

The velocity of the waves on the membrane surface is $a=\sqrt{(\tau/\rho)}$, where τ stands for tension (force) per unit length and ρ stands for a mass per unit area. In the same way as for the heat equation, the wave equation underlying the continuous vibrating membrane may be discretized. The result is a system of coupled *oscillators*

$$\frac{1}{a^2}\frac{\mathrm{d}^2 u_{ij}(t)}{\mathrm{d}t^2}=u_{i-1j}(t)+u_{i+1j}(t)+u_{ij-1}(t)+u_{ij+1}(t)-4u_{ij}(t) \tag{3}$$

This particular processing equation represents the processing model of a Membrain CNN (MCNN) with a *specific* template. However, in the general Membrain model defined below (see equation (9)) the right-hand side of equation (3) may contain inputs from arbitrary neurons as long as the characteristic oscillatory properties of the network are maintained. It can be shown that the MCNN specified in equation (3) is a CNN if and only if the CNN output formula is determined by a first-order differential equation[3] together with the assumption that the neuron state always remains in the linear part of the sigmoid output function. This is illustrated in Figure 2, which depicts a second-order neuron-processing equation (see equation (3)) as two first-order processing stages in series. Interpreting $u_{ij}(t)$ as neuron output $v_{y_{ij}}(t)$ and its first-order derivative $\dot{u}_{ij}(t)$ as internal neuron state $v_{x_{ij}}(t)$, equation (3) may be described using the generalized CNN neuron model[3] as

$$\frac{\mathrm{d}v_{x_{ij}}(t)}{\mathrm{d}t}=v_{y_{i-1j}}(t)+v_{y_{i+1j}}(t)+v_{y_{ij-1}}(t)+v_{y_{ij+1}}(t)-4v_{y_{ij}}(t),\qquad \frac{\mathrm{d}v_{y_{ij}}(t)}{\mathrm{d}t}=v_{x_{ij}}(t) \tag{4}$$

In contrast to a CNN, a Membrain neural network is described by a system of coupled *linear* differential equations. This means that an estimate of the dynamic range[2,3] may be replaced by an accurate analytical solution using a standard mathematical procedure.[8] For this purpose we introduce a vector notation representing the neuron states in a two-dimensional grid. Let us assume the grid consists of $w\times h$ neurons; then the double index (i,j) of a neuron is replaced by a single index n with $1\leqslant n\leqslant N=wh$ such that $n=(j-1)w+i$. Just to illustrate the *sparse* connection structure, an example is presented below.

Example

For a network with 3×3 neurons the representation of the network activity looks as follows:

$$\boxed{\begin{matrix} u_1(t) & u_2(t) & u_3(t)\\ u_4(t) & u_5(t) & u_6(t)\\ u_7(t) & u_8(t) & u_9(t)\end{matrix}},\qquad \mathbf{u}(t)=\begin{pmatrix}u_1(t)\\ \vdots\\ u_9(t)\end{pmatrix} \tag{5}$$

In this example we assume that the network has *toroidal* connection structure, i.e. the neurons at the borders are connected to the neurons at the opposite border. Owing to the vector representation, the total system is described by the following set of second-order linear differential equations:

$$
\begin{aligned}
\frac{d^2u_1(t)}{dt^2} &= -4u_1(t) + u_2(t) + u_3(t) + u_4(t) + u_7(t) \\
\frac{d^2u_2(t)}{dt^2} &= u_1(t) - 4u_2(t) + u_3(t) + u_5(t) + u_8(t) \\
\frac{d^2u_3(t)}{dt^2} &= u_1(t) + u_2(t) - 4u_3(t) + u_6(t) + u_9(t) \\
\frac{d^2u_4(t)}{dt^2} &= u_1(t) - 4u_4(t) + u_5(t) + u_6(t) + u_7(t) \\
\frac{d^2u_5(t)}{dt^2} &= u_2(t) + u_4(t) - 4u_5(t) + u_6(t) + u_8(t) \\
\frac{d^2u_6(t)}{dt^2} &= u_3(t) + u_4(t) + u_5(t) - 4u_6(t) + u_9(t) \\
\frac{d^2u_7(t)}{dt^2} &= u_1(t) + u_4(t) - 4u_7(t) + u_8(t) + u_9(t) \\
\frac{d^2u_8(t)}{dt^2} &= u_2(t) + u_5(t) + u_7(t) - 4u_8(t) + u_9(t) \\
\frac{d^2u_9(t)}{dt^2} &= u_3(t) + u_6(t) + u_7(t) + u_8(t) - 4u_9(t)
\end{aligned}
\tag{6}
$$

This system of equations can be described as a matrix equation by introducing a *connectivity matrix* **A**. The matrix for the toroidal connection structure (using the template defined in equation (3)) for a 3×3 network is given by

$$
\mathbf{A} = \begin{bmatrix}
-4 & 1 & 1 & 1 & 0 & 0 & 1 & 0 & 0 \\
1 & -4 & 1 & 0 & 1 & 0 & 0 & 1 & 0 \\
1 & 1 & -4 & 0 & 0 & 1 & 0 & 0 & 1 \\
1 & 0 & 0 & -4 & 1 & 1 & 1 & 0 & 0 \\
0 & 1 & 0 & 1 & -4 & 1 & 0 & 1 & 0 \\
0 & 0 & 1 & 1 & 1 & -4 & 0 & 0 & 1 \\
1 & 0 & 0 & 1 & 0 & 0 & -4 & 1 & 1 \\
0 & 1 & 0 & 0 & 1 & 0 & 1 & -4 & 1 \\
0 & 0 & 1 & 0 & 0 & 1 & 1 & 1 & -4
\end{bmatrix}
\tag{7}
$$

We draw attention to the sparseness and symmetry (both along the main and other diagonal) of this matrix. □

In general, using the connectivity matrix **A** and the vector notation $\mathbf{u}(t)$ for the network state, the system of coupled linear differential equations (see equation (3)) may be written as

$$
\frac{d^2\mathbf{u}(t)}{dt^2} = \mathbf{A}\mathbf{u}(t) \tag{8}
$$

This differential equation is *homogeneous*, i.e. the system described by the equation receives no external input. However, we may generalize this processing equation by introducing an external force $\mathbf{f}(t)$ and a damping coefficient α. In addition to the interacting forces between neurons, the acceleration of $\mathbf{u}(t)$ is determined by $\mathbf{f}(t)$, which as an external driving force 'pushes' against the membrane. Further, the coefficient α can be viewed as indicating the amount of absorption of travelling waves by the system. Thus the general Membrain model is defined as

$$\frac{d^2\mathbf{u}(t)}{dt^2} = \mathbf{A}\mathbf{u}(t) - \alpha\frac{d\mathbf{u}(t)}{dt} + \mathbf{f}(t) \tag{9}$$

Matrix **A** is the connectivity matrix describing a cellular connection structure with arbitrary dimensions. We note, however, that **A** is always a *square* matrix. Moreover, we require that the connections be symmetric, i.e. the connection between neurons i and j is the same as the opposite connection between j and i. Hence **A** is always a square, symmetric matrix. The distinct eigenvalues of such a matrix are real and the linear spaces labelled by them are orthogonal to each other. Orthogonality implies that the related coupled system of differential equations may be transformed into a system of uncoupled equations by a projection of $\mathbf{u}(t)$ into the distinct orthogonal spaces. This projection is accomplished by an operator **E**, which is a matrix whose elements in the ith row correspond to the elements of $\mathbf{e}_i$. Vectors $\mathbf{e}_1, \ldots, \mathbf{e}_N$ are N orthonormal eigenvectors of matrix **A** ($\mathbf{e}_i^T\mathbf{e}_j = \delta_{ij}$). Let the distinct eigenvalues of **A** be $\lambda_1, \ldots, \lambda_n$ and let $\mathscr{L}_1, \ldots, \mathscr{L}_n$ be all linear subspaces of **A** such that $\mathscr{L}_i$ corresponds to $\{\mathbf{x} \mid \mathbf{A}\mathbf{x} = \lambda_i\mathbf{x}\}$. Hence the dimension of $\mathscr{L}_i$ is determined by the multiplicity of λ_i. Let $\boldsymbol{\Gamma}$ be a diagonal matrix with diagonal elements $\gamma_1, \ldots, \gamma_N$ such that $\gamma_i \in \{\lambda_1, \ldots, \lambda_n\}$ and $\mathbf{e}_i \in \mathscr{L}_j$ if and only if $\gamma_i = \lambda_j$. Then **A** may be written as

$$\mathbf{A} = \mathbf{E}^T\boldsymbol{\Gamma}\mathbf{E}, \qquad \mathbf{E}^T = [\mathbf{e}_1, \mathbf{e}_2, \ldots, \mathbf{e}_N], \qquad \boldsymbol{\Gamma} = \text{diag}\{\gamma_1, \gamma_2, \ldots, \gamma_N\} \tag{10}$$

Equation (9) may be transformed into a system of N *uncoupled* second-order differential equations by multiplying both sides by **E** and substituting $\mathbf{E}\mathbf{u}(t) = \mathbf{x}(t)$ and $\mathbf{E}\mathbf{f}(t) = \mathbf{y}(t)$:

$$\frac{d^2\mathbf{x}(t)}{dt^2} = \boldsymbol{\Gamma}\mathbf{x}(t) - \alpha\frac{d\mathbf{x}(t)}{dt} + \mathbf{y}(t) \tag{11}$$

Because $\boldsymbol{\Gamma}$ is a diagonal matrix, the components of $\mathbf{x}(t)$ are not coupled and their solutions may be derived independently. In the Appendix the second-order linear differential equation described by (11) is solved for an arbitrary component $x(t)$ of $\mathbf{x}(t)$ using a Laplace transform. Subsequently, by the inverse transformation the general solution $\mathbf{u}(t)$ of equation (9) may be obtained from $\mathbf{x}(t)$ by multiplication by $\mathbf{E}^T$:

$$\mathbf{u}(t) = \mathbf{E}^T\mathbf{x}(t) = \sum_{i=1}^{N} x_i(t)\mathbf{e}_i \tag{12}$$

We note that the result in the Appendix shows three different regimes for the solution of equation (9). We will only consider the regimes where $\forall i$: $\gamma_i \leqslant 0$, because if $\gamma_i > 0$ is also non-oscillating, exponential solutions exist. As an illustration we will derive the solution of a freely vibrating MCNN, i.e. there is no damping and there is no external force. This solution can be obtained from the general solution by letting $\alpha = 0$ and $\mathbf{f}(t) = 0$. In other words we return here to the homogeneous system described by equation (8).

Let $\dot{\mathbf{u}}(t)$ be the first-order time derivative of $\mathbf{u}(t)$. Using the inverse transformation and the result for $x_i(t)$ (see Appendix) for $\alpha = 0$ and frequency $\omega_i = \sqrt{(-\gamma_i)}$, the (piecewise-linearized) wave $\mathbf{u}(t)$ can be written explicitly as

$$\mathbf{u}(t) = \sum_{i=1}^{N}\left(\frac{1}{\omega_i}\sin(\omega_i t)\dot{x}_i(0) + \cos(\omega_i t)x_i(0)\right)\mathbf{e}_i \tag{13}$$

We recall that $\mathbf{x}(t)$ was obtained from $\mathbf{u}(t)$ by projection with operator **E**. Hence the initial values $x_i(0)$ and $\dot{x}_i(0)$ can be determined

$$x_i(0) = \mathbf{e}_i^T\mathbf{u}(0), \qquad \dot{x}_i(0) = \mathbf{e}_i^T\dot{\mathbf{u}}(0) \tag{14}$$

We conclude this section with a visualization of the mathematical analysis presented above. Figure 3 depicts a pattern corresponding to one of the *eigenvectors* of the connectivity matrix describing a 10×10 network. Such a pattern is called a *natural mode of vibration* which vibrates with its *natural frequency*. When a single natural vibration is present at the surface of a membrane, there are no transversally

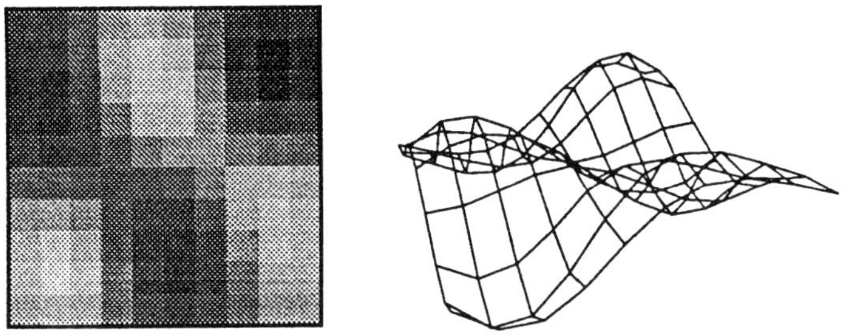

Figure 3. A natural vibration mode with frequency 1·0078 in a 10×10 network

travelling waves on the surface. Such a natural mode of vibration corresponds to a *standing wave*. A *travelling wave*, however, is a *multimode oscillation*, i.e. a superposition of standing waves having different frequencies. Indeed, when a superposition of several natural vibrations with distinct frequencies is present, the network of oscillators gives the impression of waves being propagated along the surface (see Figure 1). Superposition of distinct natural vibration modes is an interpretation of the linear transformation effected by operator **E**.

3. APPLICATION OF THE MCNN

A well-known problem from pattern recognition concerns recognition of patterns regardless of their position. A pixel representation is not fit for this purpose because a translation of the image also effects the pixel representation. In this section, however, an invariant representation that does not change under translation is described. It is based on the toroidal connection structure described in the example presented in the previous section (see equation (7)). Simulation results are presented showing that the toroidal MCNN can be combined with a traditional neural network for position-invariant pattern classification.

3.1. Position invariance

Consider again matrix **A** presented in equations (7) and (8). We recall that **A** describes the toroidal connection structure of a two-dimensional network. Since the surface of a torous is periodic and the cell structure is uniform, it is evident that the whole connection structure as expressed by matrix **A** is invariant under any translation over the grid points of the network. Formally, we can introduce a translation operator **T** which is a permutation matrix, i.e. each row and column of **T** contains exactly one 1 while all other elements are 0. The inverse translation $\mathbf{T}^{-1}$ is $\mathbf{T}^{\mathrm{T}}$, i.e. the transpose of the permutation matrix. The invariance of **A** under translation may be expressed as

$$\mathbf{T}^{-1}\mathbf{A}\mathbf{T} = \mathbf{A} \tag{15}$$

Using this property, we can show that the linear subspace $\mathscr{L}_\lambda$ that corresponds to $\mathbf{Ax} = \lambda\mathbf{x}$ is closed under the translation operation. Let **e** be any eigenvector of **A**:

$$\mathbf{Ae} = \lambda\mathbf{e} \tag{16}$$

Translation of **e** yields **Te**. Using equation (15), we can show that

$$\mathbf{ATe} = \mathbf{TAe} = \mathbf{T}\lambda\mathbf{e} = \lambda\mathbf{Te} \tag{17}$$

Consequently, although **e** is translated, it remains associated with the same eigenvalue λ. Moreover, depending on the amount of point group symmetry of the network, this is also true for rotations. For example, a square network is also invariant under rotation over 90°, 180° and 270°.

3.2. *Toroidal MCNN*

In the case of an MCNN we can achieve position-invariant classification by classifying the temporal representation of the initial network input. In order to do this, we require that the connection structure of the network be toroidal. If the structure is not toroidal but e.g. flat, then the network surface will not be periodic. Having boundaries, any translation of $\mathbf{u}(0)$ results in an altogether different wave pattern owing to different reflections against the network boundaries. In a toroidal connection structure, however, the position of the interference pattern will be different but the pattern itself will be unchanged.

This property results from the fact that the linear subspace $\mathscr{L}_\lambda$ that corresponds to $\mathbf{Ax} = \lambda\mathbf{x}$ is closed under the translation operation which was shown in the previous subsection (see equation (17)). In order to explore this property, we will examine a freely vibrating, toroidal MCNN, i.e. no damping and no external force (processing model in equation (8) and analytical solution in equation (13)). If we assume additionally that the initial oscillator velocity $\dot{\mathbf{u}}(0) = 0$ and the initial deflection is $\mathbf{u}(0)$, then $\mathbf{u}(t)$ becomes

$$\mathbf{u}(t) = \sum_{i=1}^{N} \mathbf{e}_i^{\mathrm{T}}\mathbf{u}(0)\mathbf{e}_i \cos(\omega_i t) \tag{18}$$

The number of terms in equation (18) may be reduced if eigenvalues with multiplicity greater than unity exist. Let $\mathscr{L}_i = \{\mathbf{x} \mid \mathbf{Ax} = \lambda_i\mathbf{x}\}$ (see Section 2). We denote the projection of the initial deflection $\mathbf{u}(0)$ on $\mathscr{L}_i$ by $\mathbf{p}_i$. Let $\nu_i = \sqrt{(-\lambda_i)}$; then equation (18) can be written as a *natural mode expansion* of the system:

$$\mathbf{u}(t) = \sum_{i=1}^{n} \mathbf{p}_i \cos(\nu_i t), \qquad \mathbf{p}_i = \textit{projection of } \mathbf{u}(0) \textit{ on } \mathscr{L}_i, \qquad \nu_i = \sqrt{(-\lambda_i)} \tag{19}$$

Since $\mathscr{L}_i$ is closed under translation, the projection of $\mathbf{Tu}(0)$, i.e. the translation of $\mathbf{u}(0)$, on $\mathscr{L}_i$ equals $\mathbf{Tp}_i$! Consequently, the behaviour of the system with initial deflection $\mathbf{Tu}(0)$ is $\mathbf{Tu}(t)$, i.e. the interference pattern of waves has been translated without destroying its shape.

3.3. *Translation-invariant representation*

We note that the projections $\mathbf{p}_1, \ldots, \mathbf{p}_n$ (see equation (19)) form a complete description of $\mathbf{u}(0)$. Because they result from a projection on linear spaces which are closed under translation, they are useful for invariant pattern classification. The square Euclidean length of the projections, i.e. $\|\mathbf{p}_1\|^2, \ldots, \|\mathbf{p}_n\|^2$, is a *power spectrum* of the natural modes of vibration and may be interpreted as a ‘fingerprint’ of $\mathbf{u}(0)$. By ‘fingerprint’ we mean that the description is not complete, i.e. it will not be sufficient to reconstruct a pattern but it is sufficient to *discriminate* between different patterns. Moreover, because of translation invariance of $\mathscr{L}_1, \ldots, \mathscr{L}_n$, the fingerprint also does not change when $\mathbf{u}(0)$ is translated. This follows from the observation that $\|\mathbf{Tp}_i\| = \|\mathbf{p}_i\|$, i.e. the length of $\mathbf{p}_i$ is constant under translation. Hence the *power spectrum* is a translation-invariant representation of $\mathbf{u}(0)$.

We will now show that the power spectrum of $\mathbf{u}(0)$ in terms of $\mathbf{p}_1, \ldots, \mathbf{p}_n$ is temporally encoded in a one-dimensional signal $\hat{u}(t) = \|\mathbf{u}(t)\|^2$:

$$\begin{aligned}\hat{u}(t) = \|\mathbf{u}(t)\|^2 &= \sum_{k=1}^{N} u_k^2(t) = \sum_{k=1}^{N} \sum_{i=1}^{n} p_{ik} \cos(\nu_i t) \sum_{j=1}^{n} p_{jk} \cos(\nu_j t) \\ &= \sum_{i,j=1}^{n} \cos(\nu_i t)\cos(\nu_j t) \underbrace{\sum_{k=1}^{N} p_{ik}p_{jk}}_{=0 \text{ if } i \neq j} = \sum_{i=1}^{n} \|\mathbf{p}_i\|^2 \cos^2(\nu_i t), \qquad \nu_i = \sqrt{(-\lambda_i)}\end{aligned} \tag{20}$$

From the discussion above it follows immediately that $\hat{u}(t)$ is also invariant under translation of $\mathbf{u}(0)$. From the definition of $\mathscr{L}_i$ it follows that the frequencies $\nu_1, \ldots, \nu_n$ are all different and hence that $\|\mathbf{p}_i\|^2$ is modulated in $\hat{u}(t)$ with a unique frequency. Consequently, analysis of $\hat{u}(t)$, e.g. using a Fourier transform, yields a static, i.e. non-temporal, translation-invariant ‘fingerprint’ of $\mathbf{u}(0)$.

Owing to the cellular structure of the MCNN, large images can be processed using a massively parallel computer. Since each cell requires only four connections, a simple connection structure, e.g. a matrix structure, will do. Moreover, when other non-digital solutions are considered, e.g. using analogue VLSI, optics, fluid mechanics, mechanical constructions, etc., it may be useful to realize that $\hat{u}(t)$ may be directly derived from the elastic energy of the system.

3.4. Simulation

An MCNN with a toroidal connection structure has been applied to generate translation-invariant representations of *handwritten signature* images. Three persons were asked to draw their signature four times at different positions within a 2×1 inch^2 area. The 12 different images are depicted in the left section of Figure 4. Each image may be classified as either the signature of person 1 (see Figures 4(a)–4(d)), person 2 (see Figures 4(e)–4(h)) or person 3 (see Figures 4(i)–4(l)). We emphasize that small differences between the signatures of a single person exist because they are drawn at different times.

The images were scanned at 25 dots per inch (dpi), resulting in 50×25 pixels at 256 grey levels, i.e. 1 byte per pixel. Each image is normalized and subsequently presented to a 50×25 MCNN with a toroidal connection structure as in the example described above. Simulations of the MCNN were performed using a fourth- and fifth-order Runge–Kutta–Fehlberg integration procedure with a variable step size and a tolerance of 10^{-5}.

In order to get an indication of the bandwidth of $\hat{u}(t)$ for a 50×25 MCNN, the lowest and the highest frequency can be determined by presenting a low- and a high-spatial-frequency pattern to the network respectively. In Figures 5(a) and 5(b) two low-spatial-frequency patterns are depicted with their corresponding signals. Figure 5(a) corresponds to 0·04 Hz, which is the lowest frequency in $\hat{u}(t)$. In Figure 5(c) a high-spatial-frequency pattern, i.e. a checkerboard, is presented, resulting in a frequency of 0·90 Hz, i.e. the highest frequency in $\hat{u}(t)$. We note that these frequencies can be scaled by multiplying $\mathbf{A}$ by a scalar, i.e. changing the parameter a in equation (3).

Network simulations for the signature images were performed for $t \in [0, 50]$, resulting in the curves $\hat{u}_1(t), \ldots, \hat{u}_{12}(t)$ shown in the middle section of Figure 4. Unlike the original images, the temporal signals are difficult to distinguish (for a human being) at first sight. Therefore we have also depicted in the right section of Figure 4 the deviation of each signal with respect to the average signal $\frac{1}{12} \sum_{i=1}^{12} \hat{u}_i(t)$ for all signatures taken. In order to process the output signals for further classification, $\hat{u}_i(t)$ will be sampled. The performance of the system depends on the number of samples used for classification. Ideally, this number is determined by the distribution of the harmonic components of $\hat{u}(t)$ in the frequency spectrum. When fewer samples are used, small frequency differences cannot be distinguished, i.e. it seems as if the network has smaller dimensions. When more samples are taken, classification performance will not increase significantly.

We have roughly estimated the number of required samples for the signals presented in Figure 4 as 250 by counting the number of distinct spatial frequencies that can exist without aliasing in the network. Hence each signal is sampled using $\Delta t = 0{\cdot}2$ s for $t \in [0, 50]$, resulting in 12 vectors $\hat{\mathbf{u}}_1, \ldots, \hat{\mathbf{u}}_{12}$ with $\hat{u}_{ij} = \hat{u}_i(j\,\Delta t)$. We have analysed the vectors using two different approaches.

First, it can be shown that two signals corresponding to the same person have a higher correlation than signals corresponding to different persons. Let $\mathbf{X}$ be the correlation matrix of the normalized vectors $\hat{\mathbf{u}}_1, \ldots, \hat{\mathbf{u}}_{12}$, i.e.

$$x_{ij} = \frac{\hat{\mathbf{u}}_i^{\mathrm{T}} \hat{\mathbf{u}}_j}{\| \hat{\mathbf{u}}_i \| \, \| \hat{\mathbf{u}}_j \|} \tag{21}$$

The columns of $\mathbf{X}$, i.e. $\mathbf{x}_1, \ldots, \mathbf{x}_{12}$, are reduced representations of $\hat{\mathbf{u}}_1, \ldots, \hat{\mathbf{u}}_{12}$ respectively. We have clustered $\mathbf{x}_1, \ldots, \mathbf{x}_{12}$ using a two-dimensional self-organizing feature map[4] consisting of 10×10 neurons. The resulting two-dimensional representation of the clusters is depicted in Figure 6. This representation shows that signals are corresponding to the same person, e.g. (a)–(d), have a stronger correlation than signals which correspond to the different persons, e.g. (a) and (k). We note in passing that pixel

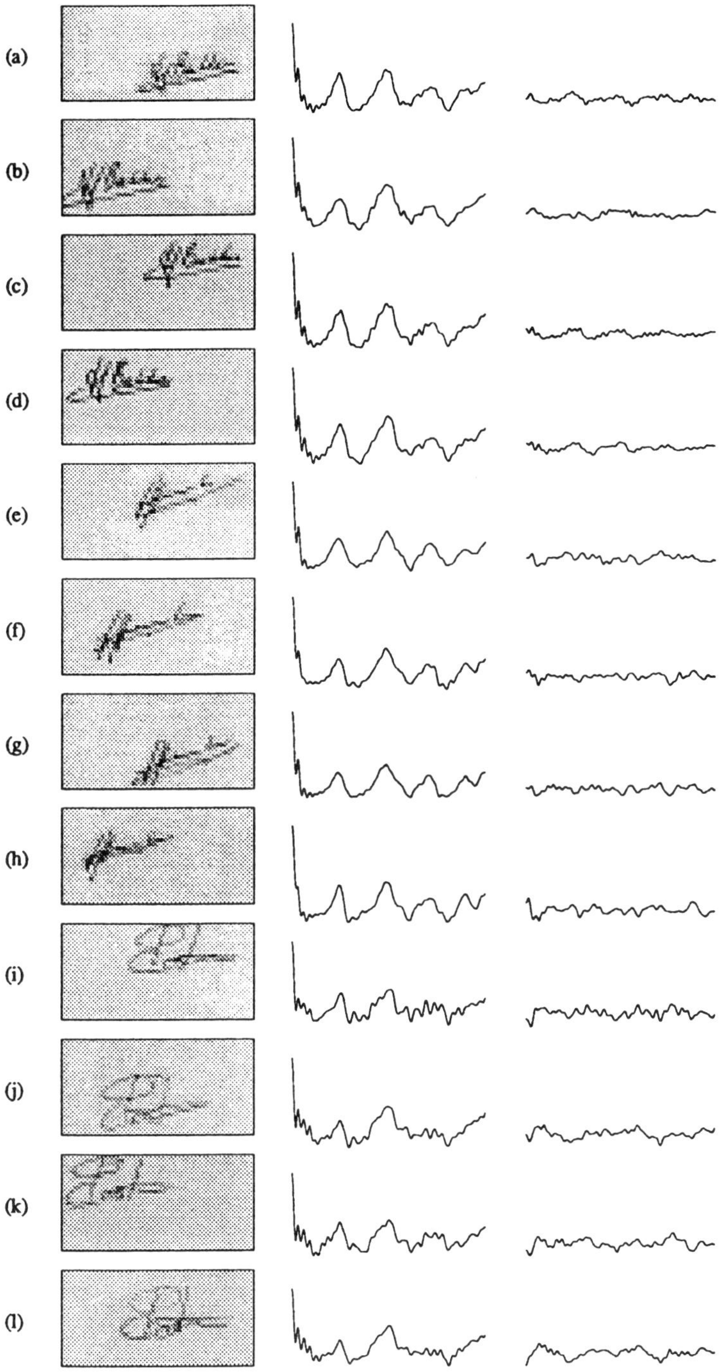

Figure 4. MCNN input and output signals. The left section shows the signature images used as initial deflection patterns in the simulations. The middle section shows the output $\hat{u}(t)$ for each image respectively. The right section shows the difference between $\hat{u}(t)$ and the average of the 12 signals shown in the middle section. Three persons were asked to draw their signature four times: (a)–(d) person 1; (e)–(h) person 2; (i)–(l) person 3

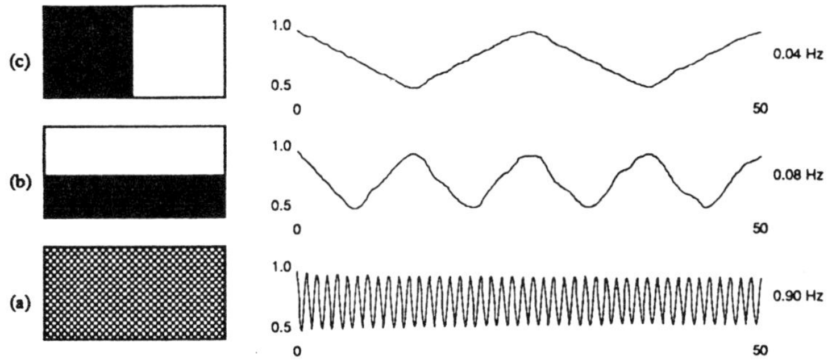

Figure 5. Three typical frequencies in $\hat{u}(t)$ of a 50×25 MCNN. Spatial frequency is related to temporal frequency: (a) lowest-spatial-frequency pattern in horizontal direction corresponds to lowest frequency 0·04 Hz; (b) lowest-spatial-frequency pattern in vertical direction corresponds to frequency 0·08 Hz; (c) highest-spatial-frequency pattern corresponds to highest frequency 0·90 Hz

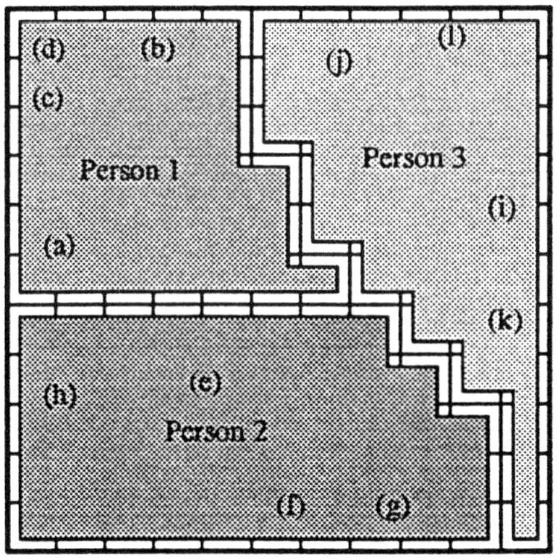

Figure 6. Clustering of the correlations between the output signals depicted in the middle section of Figure 4 using a two-dimensional 10×10 Kohonen self-organizing feature map (neighbourhood size is four, learning rate is 0·3)

correlation of the images themselves for the purpose of classification would be *quite meaningless* owing to translation of the signature within different images.

Secondly, we will show that a back-propagation[5,6] neural network (BPNN) is capable of *extracting position-invariant features* characterizing a signature. Vectors $\hat{\mathbf{u}}_1, \ldots, \hat{\mathbf{u}}_{12}$ are presented to a BPNN with 250 input neurons, 20 hidden neurons and three output neurons. The output neurons are used for classification of the input in three different classes, i.e. persons 1, 2 and 3 respectively. In order to examine the generality of the extracted features, a learn set and a test set were constructed. The signatures shown in Figures 4(a), 4(b), 4(e), 4(f), 4(i) and 4(j) are used as examples for the test set to see how well the network has generalized from the examples in the learn set, i.e. Figures 4(c), 4(d), 4(g), 4(h), 4(k) and 4(l). After approximately 8000 cycles (learning rate and momentum are both 0·1) the average of the squared errors on the learn set is approximately 0·0003 and on the test set it is approximately 0·007 (see Figure 7). The errors of the individual examples in the test set and the learn set are shown in Tables I(A) and I(B) respectively. We note that the average of the squared errors (see Figure 7) has been calculated by averaging the squared differences between the output and target values for all output neurons and for all examples in either the learning set (solid curve) or the test set (dashed curve).

Although the network has not previously 'seen' the examples in the test set, it can distinguish sufficiently between signatures of different persons while generalizing over the slight variations in a signature of the same person (see Table I). Apparently, the network has learned the essential position-

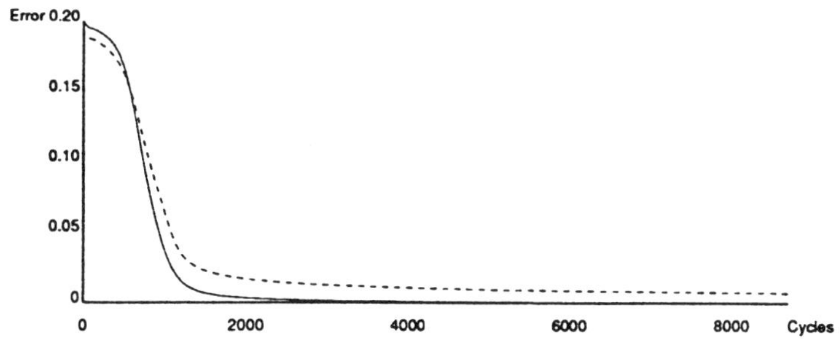

Figure 7. Curves for the average of squared errors for the learn set (solid curve) and the test set (dashed curve). Back-propagation learning was applied to a network with 250 input neurons, 20 hidden neurons and three output neurons (learning rate and momentum were both set to 0·1)

Table I. Results of back-propagation learning for a neural network consisting of three layers containing 250, 20 and three neurons respectively. The labels in the first column correspond to the labelling of the images in Figure 4. (A) shows the errors for the examples in the test set after 8000 cycles through the learn set. The errors for the examples in the learn set are presented in (B). The learning rate and momentum were set to 0·1

(A) Results of the test set

Image (Figure 4)	Person 1		Person 2		Person 3	
	Output	Target	Output	Target	Output	Target
(a)	0·72	1·00	0·15	0·00	0·02	0·00
(b)	0·89	1·00	0·03	0·00	0·04	0·00
(e)	0·07	0·00	0·88	1·00	0·01	0·00
(f)	0·04	0·00	0·96	1·00	0·00	0·00
(i)	0·01	0·00	0·02	0·00	0·98	1·00
(j)	0·05	0·00	0·02	0·00	0·92	1·00

(B) Results of the learning set

Image (Figure 4)	Person 1		Person 2		Person 3	
	Output	Target	Output	Target	Output	Target
(c)	0·97	1·00	0·02	0·00	0·01	0·00
(d)	0·98	1·00	0·02	0·00	0·01	0·00
(g)	0·01	0·00	0·97	1·00	0·03	0·00
(h)	0·02	0·00	0·99	1·00	0·00	0·00
(k)	0·02	0·00	0·01	0·00	0·97	1·00
(l)	0·01	0·00	0·02	0·00	0·98	1·00

invariant features which distinguish the signature of one person from any other person. We note that normally position invariance is difficult to learn with the back-propagation scheme.

4. CONCLUSIONS

In this paper we have presented the Membrain model describing neural networks whose functional architecture is based on a network of linearly coupled oscillators. The analytical solution shows that the network behaviour is a multimode oscillation with the appearance of transversally travelling waves, e.g. the waves appearing on a vibrating membrane. Membrain neural networks may be realized particularly well when they have a cellular connection structure similar to cellular neural networks (CNNs). Moreover, we show that such a Membrain CNN (MCNN) is actually equivalent to a second-order CNN. The MCNN may be applied to image processing, interpreting the network *initial state* as a grey-scale image. As an example, we presented an MCNN with a toroidal connection structure which may be used for generating translation-invariant image representations. It is shown analytically that the power spectrum of the temporal wave pattern can be interpreted as a translation-invariant 'fingerprint' image. Owing to the symmetric connection structure, a one-dimensional modulation of the power spectrum can be easily obtained by summing the squared neuron states, i.e. membrane deflections. Finally, simulations were performed confirming that the temporal signal is useful for (1) position-invariant correlation of images and (2) extraction of position-invariant features for image classification.

5. FUTURE RESEARCH

Concerning the application of a toroidal MCNN to signature classification, we consider taking into account the writing velocity and pressure for signature classification. This can be accomplished by applying a point force to the network with varying strength at a varying position. Depending on the drawing speed and the pressure applied to the pen, a characteristic wave pattern emerges. Since these properties are characteristic of a person's signature, this method will enhance the ability of the system to reject false signatures. We note that this application requires a different form of network *input* i.e. using the external force $\mathbf{f}(t)$ rather than using the initial deflection pattern.

Recently, Pribram[9] has described a neural model in which brain function is modelled as wave propagation similar to the propagation of transversal waves in the MCNN. In Pribram's model, however, input is presented to the network as a *perturbation* to the system's stationary wave state. In the Membrain model, input may also be presented as a perturbation by using $f_{ij}(t)u_{ij}(t)$ instead of $f_{ij}(t)$. We note that this actually results in a piecewise addition of the image components $f_{ij}(t)$ to the elements on the main diagonal of the connection matrix. Hence a perturbation will change the eigenvalues of $\mathbf{A}$ and consequently $\hat{u}(t)$ will be a *frequency modulation* of the 'fingerprint' rather than an amplitude modulation.

The MCNN model has many more potential application areas. For example, other, possibly non-homogeneous, connection structures may be used to enforce other types of invariance corresponding to particular feature spaces. Ideally, such connection structures should be adapted automatically, e.g. using Hebbian learning.

In any case, the quite popular diffusion model of neural activity can be replaced by an equally fruitful oscillator model. This shows once again that the magic of neural networks is related to their collective many-body behaviour rather than to their processing model for individual neurons.

ACKNOWLEDGEMENTS

We would like to thank Eric Postma and Piet Boekhoudt for their many fruitful discussions about the work presented here. Furthermore, we thank IBM for their hardware support under the Joint Study Agreement DAEDALOS (#289651).

APPENDIX: LAPLACE TRANSFORMS

The system of uncoupled second-order linear differential equations described in equation (11) may be solved analytically using a Laplace transform. Let $x_n(t)$ and $y_n(t)$ denote the nth component of $\mathbf{x}(t)$ and $\mathbf{y}(t)$ respectively. The typical differential equation for $x_n(t)$ looks as follows:

$$\frac{\mathrm{d}^2 x_n(t)}{\mathrm{d}t^2} = \gamma_n x_n(t) - \alpha \frac{\mathrm{d}x_n(t)}{\mathrm{d}t} + y_n(t) \tag{22}$$

Let $X(s)$ denote the Laplace transform of $x_n(t)$ and $Y(s)$ the Laplace transform of $y_n(t)$; then

$$s^2X(s) - sx_n(0) - \dot{x}_n(0) = \gamma_n X(s) - \alpha(sX(s) - x_n(0)) + Y(s) \tag{23}$$

From this it follows that

$$X(s) = \frac{1}{s^2 - \gamma_n + \alpha s} Y(s) + \frac{1}{s^2 - \gamma_n + \alpha s} \dot{x}_n(0) + \frac{s+\alpha}{s^2 - \gamma_n + \alpha s} x_n(0) \tag{24}$$

The solution of equation (11) can be determined using an inverse Laplace transform, yielding

$$\mathscr{L}^{-1}\left(\frac{1}{s^2 - \gamma_n + \alpha s}\right) = \mathscr{L}^{-1}\left(\frac{1}{(s+\alpha/2)^2 - \gamma_n - \alpha^2/4}\right)$$

$$= \begin{cases} (1/\omega_n)\mathrm{e}^{-(\alpha/2)t}\sin(\omega_n t), \quad \omega_n^2 = -\gamma_n - \alpha^2/4 & \text{if } \gamma_n < -\alpha^2/4 \\ t\mathrm{e}^{-(\alpha/2)t} & \text{if } \gamma_n = -\alpha^2/4 \\ (1/\omega_n)\mathrm{e}^{-(\alpha/2)t}\sinh(\omega_n t), \quad \omega_n^2 = \gamma_n + \alpha^2/4 & \text{if } \gamma_n > -\alpha^2/4 \end{cases} \tag{25}$$

and

$$\mathscr{L}^{-1}\left(\frac{s+\alpha}{s^2 - \gamma_n + \alpha s}\right) = \mathscr{L}^{-1}\left(\frac{s+\alpha/2}{(s+\alpha/2)^2 - \gamma_n - \alpha^2/4}\right) + \mathscr{L}^{-1}\left(\frac{\alpha/2}{(s+\alpha/2)^2 - \gamma_n - s\alpha^2/4}\right)$$

$$= \begin{cases} \mathrm{e}^{-(\alpha/2)t}\cos(\omega_n t) + (\alpha/2\omega_n)\mathrm{e}^{-(\alpha/2)t}\sin(\omega_n t), \quad \omega_n^2 = -\gamma_n - \alpha^2/4 & \text{if } \gamma_n < -\alpha^2/4 \\ \mathrm{e}^{-(\alpha/2)t} + (\alpha/2)t\mathrm{e}^{-(\alpha/2)t} & \text{if } \gamma_n = -\alpha^2/4 \\ \mathrm{e}^{-(\alpha/2)t}\cosh(\omega_n t) + (\alpha/2\omega_n)\mathrm{e}^{-(\alpha/2)t}\sinh(\omega_n t), \quad \omega_n^2 = \gamma_n + \alpha^2/4 & \text{if } \gamma_n > -\alpha^2/4 \end{cases}$$

Given these Laplace transforms, it is possible to determine the solutions for $x(t)$.

(I) $\gamma_n < -\alpha^2/4$ ($\omega_n^2 = -\gamma_n - \alpha^2/4$):

$$x_n(t) = \frac{1}{\omega_n}\int_0^t y(\tau)\mathrm{e}^{-(\alpha/2)(t-\tau)}\sin[(t-\tau)\omega_n]\,\mathrm{d}\tau + \frac{\mathrm{e}^{-(\alpha/2)t}\sin(\omega_n t)}{\omega_n}\dot{x}_n(0) + \left(\mathrm{e}^{-(\alpha/2)t}\cos(\omega_n t) + \frac{\alpha}{2\omega_n}\mathrm{e}^{-(\alpha/2)t}\sin(\omega_n t)\right)x_n(0) \tag{27}$$

(II) $\gamma_n = -\alpha^2/4$:

$$x_n(t) = \int_0^t y(\tau)(t-\tau)\mathrm{e}^{-(\alpha/2)(t-\tau)}\,\mathrm{d}\tau + t\mathrm{e}^{-(\alpha/2)t}\dot{x}_n(0) + \left(\mathrm{e}^{-(\alpha/2)t} + \frac{\alpha}{2}t\mathrm{e}^{(\alpha/2)t}\right)x_n(0) \tag{28}$$

(III) $\gamma_n > -\alpha^2/4$ ($\omega_n^2 = \gamma_n + \alpha^2/4$):

$$x_n(t) = \frac{1}{\omega_n}\int_0^t y(\tau)\mathrm{e}^{-(\alpha/2)(t-\tau)}\sinh[(t-\tau)\omega_n]\,\mathrm{d}\tau + \frac{\mathrm{e}^{-(\alpha/2)t}\sinh(\omega_n t)}{\omega_n}\dot{x}_n(0) + \left(\mathrm{e}^{-(\alpha/2)t}\cosh(\omega_n t) + \frac{\alpha}{2\omega_n}\mathrm{e}^{-(\alpha/2)t}\sinh(\omega_n t)\right)x_n(0) \tag{29}$$

REFERENCES

1. W. E. Boyce and R. C. DiPrima, *Elementary Differential Equations and Boundary Value Problems*, 3rd edn, Wiley, New York, 1977.
2. L. O. Chua and L. Yang, 'Cellular neural networks: theory', *IEEE Trans. Circuits and Systems*, **CAS-35**, 1257–1272 (1988).
3. T. Roska and L. O. Chua, 'Cellular neural networks with nonlinear and delay-type template elements', *Proc. IEEE Int. Workshop on Cellular Neural Networks and Their Applications, CNNA-90*, Budapest, 1990, IEEE, New York, 1991, pp. 12–25.

4. T. Kohonen, *Self-organisation and Associative Memory*, 2nd revised edn, Springer, Berlin/Heidelberg/New York, 1988.
5. D. E. Rumelhart, G. E. Hinton and R. J. Williams, 'Learning internal representations by error propagation', in *Parallel Distributed Processing. Volume 1: Foundations*, MIT Press, Cambridge, 1986, pp. 318–362.
6. J. Henseler and P. J. Braspenning, 'Training complex multi-layer neural networks', *Proc. Latvian Signal Processing Int. Conf.*, I. Bilinsky (ed.), Riga, 1990, Vol. 2, pp. 301–305.
7. J. Henseler and P. J. Braspenning, 'A cortex-like architecture of a cellular neural network', *Proc. IEEE Workshop on Cellular Neural Networks and Their Applications, CNNA-90*, Budapest, 1990, IEEE, New York, 1991, pp. 244–253.
8. R. Courant and D. Hilbert, *Methods of Mathematical Physics*, 1st English edn, Interscience, New York, 1953, pp. 275–396.
9. K. Pribram, *Brain and Perception: Holonomy and Structure in Figural Processing*, Lawrence Erlbaum, Hillsdale, NJ, 1991.

ON THE UNIVERSE OF STABLE CELLULAR NEURAL NETWORKS†

LEON O. CHUA AND CHAI WAH WU

Electronics Research Laboratory, Department of Electrical Engineering and Computer Sciences, University of California, Berkeley, CA 94720, U.S.A.

SUMMARY

A cellular neural network (CNN) is a novel analogue circuit architecture with many desirable features. This paper extends previous stability results of CNNs to include classes of strictly sign-symmetric and acyclic templates. We show that most of the 3×3 strictly sign-symmetric templates are stable almost everywhere, with the unknown templates reduced to three classes. We also introduce template graphs and CNN graphs and utilize them to obtain results concerning stability and irreducibility of CNN templates.

1. INTRODUCTION

The cellular neural network (CNN), a new type of analogue circuit, was invented recently and shown to have many desirable properties.[1] CNNs consist of many parallel analogue processors computing in real time. One desirable feature is that these processors, arranged in a two-dimensional grid, only have local connections, which lend themselves easily to VLSI implementations. The connections between these processors are determined by a cloning template, which describes the strength of nearest-neighbour interconnections. The cloning templates we will consider are all space-invariant, meaning that all the processors have the same relative connections.

Some important applications of CNNs are in the area of image processing, where several templates have been discovered for edge detection, noise removal, image thinning, motion detection and other image-processing functions.[2–4] For proper operation of such templates a CNN must be completely stable, i.e. for all initial conditions the states converge to some DC equilibrium point. Generally, different initial states will converge to different equilibrium points. In the original paper[1] complete stability has been proved for CNNs with symmetric templates.

CNNs form a special subclass of systems of differential equations, and by using results obtained for competitive–co-operative systems, a larger class of templates can be shown to be stable. In Reference 5 a class of non-symmetric templates has been shown to be stable almost everywhere using results in Reference 6. Asymmetric templates have also been studied in Reference 7.

This paper extends the results in Referencc 5 and shows that most 3×3 strictly sign-symmetric templates are stable almost everywhere. We also show that acyclic templates are completely stable. To facilitate analysis of CNN templates and visualization of the connections between CNN cells, we introduce the use of template graphs and CNN graphs and show several results relating to these graphs.

† Part of this research has been reported in the Proceedings of the 1990 International Workshop on Cellular Neural Networks and Their Applications held in Budapest.

2. PRELIMINARY DEFINITIONS

An $M \times N$ CNN is governed by the following set of differential equations:[1]

$$C\frac{\mathrm{d}v_{x_{ij}}(t)}{\mathrm{d}t} = -\frac{1}{R_x}v_{x_{ij}}(t) + \sum_{C_{kl}\in N_r(ij)} A(i,j;k,l)v_{y_{kl}}(t) + \sum_{C_{kl}\in N_r(ij)} B(i,j;k,l)v_{u_{kl}}(t) + I \tag{1}$$

$$N_r(ij) = \{C_{kl}:\max(|k-i|, |l-j|) \leqslant r\} \quad \text{for some positive integer } r \tag{2}$$

where $v_{x_{ij}}$, $v_{u_{ij}}$ and $v_{y_{ij}}$ denote the state voltage, input voltage and output voltage of a cell respectively. We assume that the input is constant and that the A-template is space-invariant, i.e. for all i, j, k, l, m and n such that $1 \leqslant i, k, i+m, k+m \leqslant M$ and $1 \leqslant j, l, j+n, l+n \leqslant N$, $A(i,j;k,l) = A(i+m, j+n; k+m, l+n)$.

Also, $v_y = \hat{f}(v_x) = \frac{1}{2}[|v_x+1| - |v_x-1|]$. We will often use a continuously differentiable (C^1) approximation of this piecewise linear function in our stability analysis.

Definition 1

A dynamical system $\dot{\mathbf{x}} = \mathbf{F}(\mathbf{x})$, $\mathbf{F} \in C^1$ is co-operative if for all $i \neq j$

$$\frac{\partial \mathbf{F}_i(\mathbf{x})}{\partial \mathbf{x}_j} \geqslant 0 \tag{3}$$

Definition 2

An $n \times n$ matrix $\mathbf{A} = [\mathbf{A}_{ij}]$ is irreducible if for all $i \neq j$ there exists a chain of indices $i = k_0, \ldots, k_m = j$ such that

$$\mathbf{A}_{k_r k_{r-1}} \neq 0, \quad r = 1, 2, \ldots, m \tag{4}$$

Definition 3

A dynamical system $\dot{\mathbf{x}} = \mathbf{F}(\mathbf{x})$, $\mathbf{F} \in C^1$ is irreducible if $D\mathbf{F}(\mathbf{x})$ is irreducible for all $\mathbf{x}$.

Definition 4

A dynamical system is stable almost everywhere if the set of initial values in which the system does not converge to a constant equilibrium point has Lebesque measure zero.†

A dynamical system satisfying Definition 4 is *not* necessarily *completely stable* since it may contain unstable limit cycles or other exotic trajectories. This weaker form of stability is easier to prove than complete stability and is all that we need in practice.

Definition 5

A cloning template $A(i,j;\ k,l)$ is positive if $A(i,j;\ k,l) \geqslant 0$ for all $(i,j) \neq (k,l)$.

3. TEMPLATE GRAPH Γ_A AND CNN GRAPH Γ

A distinguishing property of CNNs is that the templates are local; there exists a positive integer r such that $A(i,j;k,l) = 0$ for all $C_{kl} \notin N_r(ij)$. Space invariance implies that r can be made independent of i and

† This definition of 'stable (almost everywhere)' is used in the sense of Hirsch,[8] who called it '(almost) convergent'.

Table I. Table of templates and their template graphs and CNN graphs

Name	A-template	Template graph	CNN graph
Hole filler (a)	0·0 1·0 0·0 1·0 2·0 1·0 0·0 1·0 0·0	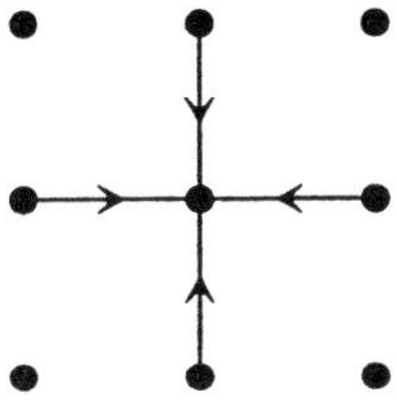	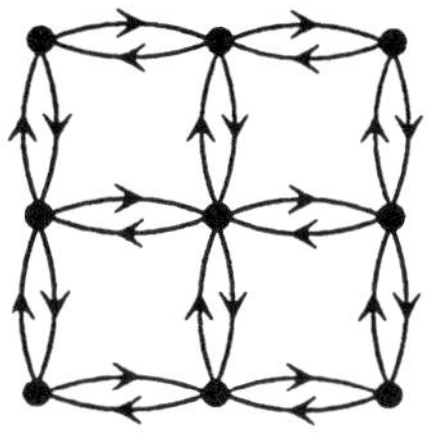
Edge detection (b)	0·0 −0·5 0·0 −0·5 2·0 −0·5 0·0 −0·5 0·0	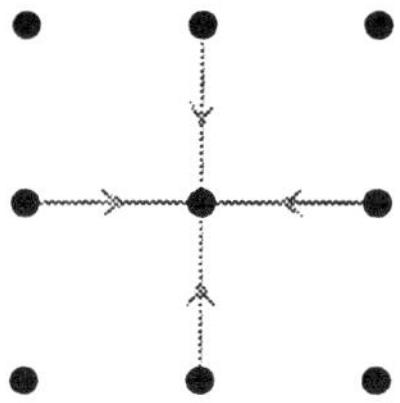	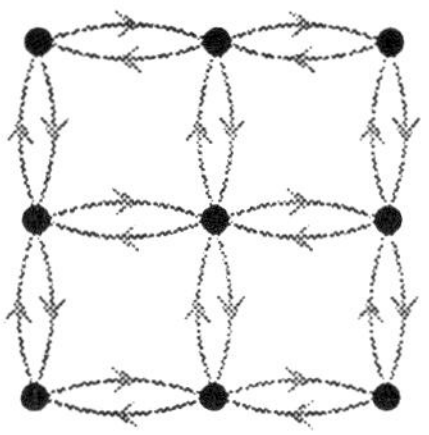
Horizontal shadow detector (c)	0·0 0·0 0·0 0·0 2·0 2·0 0·0 0·0 0·0	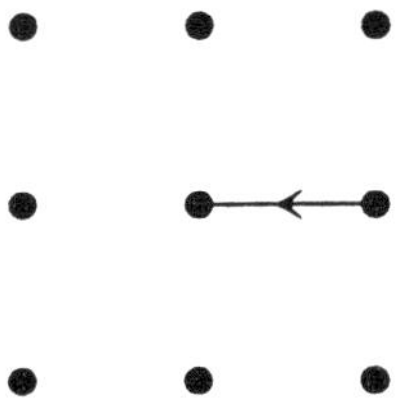	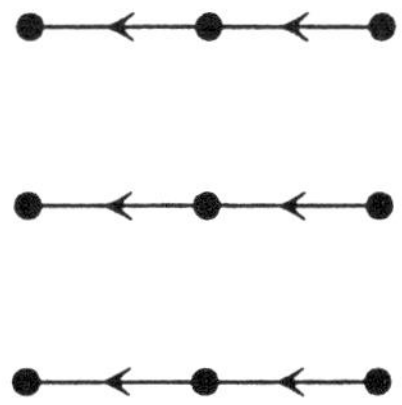
Connected component detector (d)	0·0 0·0 0·0 1·0 2·0 −1·0 0·0 0·0 0·0	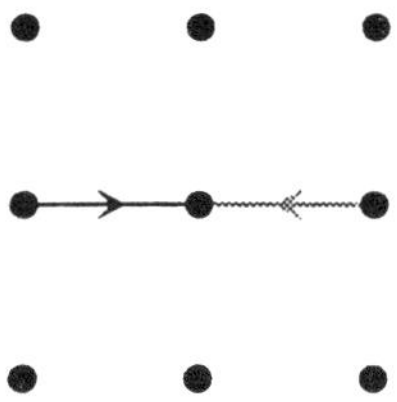	
(e)	1·0 0·0 2·5 0·0 2·0 0·0 1·0 0·0 1·5	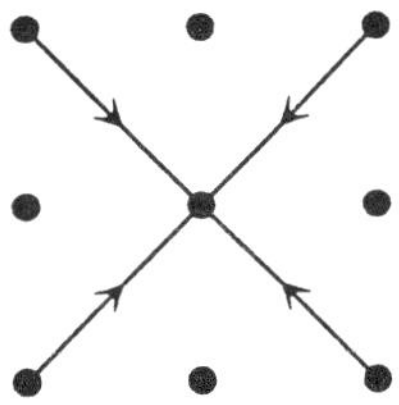	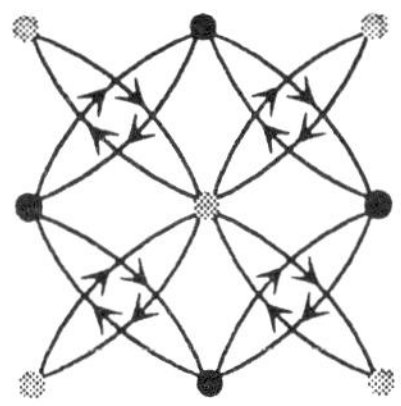
(f)	0·0 0·0 0·0 0·0 2·0 2·5 0·0 −1·5 0·0	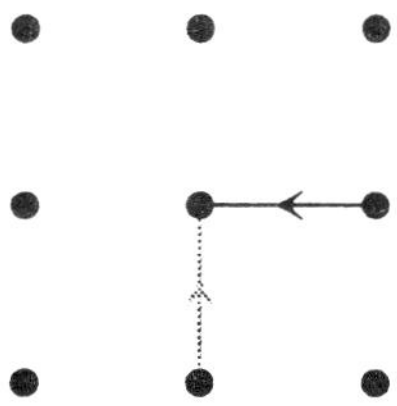	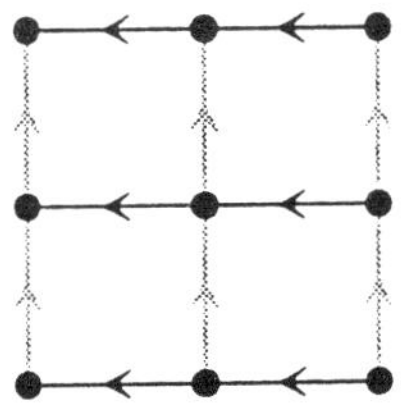

j and we call the least such r the radius of the A-template.† The radius of a B-template is defined in the same way.

We associate with each A-template a labelled (+ or −) directed graph Γ_A. The $(2r+1)^2$ nodes of Γ_A will be the cells in $N_r(ij)$ for some i and some j, with C_{ij} fixed as the centre cell. For all cells $C_{kl} \in N_r(ij)$ we have a directed edge from cell C_{kl} to cell C_{ij} if and only if $A(i,j;k,l) \neq 0$. The label associated with this edge will be the sign of $A(i,j;k,l)$. We will also call this edge positive or negative depending on the sign. Because of space invariance, the choice of C_{ij} is irrelevant. We call Γ_A the *template graph*. This graph allows us to see how the cells in a CNN are connected to each other locally (i.e. to neighbouring cells) through the template.

Similarly, for each $M \times N$ CNN we associate a labelled directed graph Γ. The MN nodes in Γ will be the cells in the CNN. For all cells C_{ij} and C_{kl} in the graph there is an edge from C_{kl} to C_{ij} if and only if $A(i,j;k,l) \neq 0$. We call Γ the *CNN graph*.‡ This graph shows how the cells in a CNN are connected to each other locally and globally (i.e. to other cells that affect the dynamics.)

We can generate Γ from the template graph Γ_A of the A-template by considering Γ_A as a subgraph of Γ and repeating Γ_A periodically over each node of Γ.

Table I lists some templates along with their template graphs and CNN graphs.§ [2,10–12] Each branch of a template graph or a CNN graph has a *direction* (arrowhead) and a 'sign' label (+ or −). In Table I, instead of attaching labels to edges, positive edges are shown black and negative edges are shown grey.

If we look at the CNN graph of the shadow detector (Table I, (c)), we see that the rows are disconnected. This means that the state of a cell in a certain row cannot affect or depend on the state of a cell in another row. Thus we can consider each row as an independent one-dimensional CNN and analyse its behaviour separately. Furthermore, all the edges point to the left, meaning that the cells can only influence the state of cells to their left and not vice versa. All this is consistent with the operation of a horizontal shadow detector.

Definition 6

A directed graph is irreducible if for all $i \neq j$ there is a directed path from node i to node j.

Definition 7

An $n \times n$ A-template is cell-linking if the CNN graph of the associated $n \times n$ CNN is irreducible.

This definition of a cell-linking template is clearer than the one given in Reference 5, since the irreducibility of a CNN graph depends on the size of the CNN. In practice, the size of the CNN is usually larger than the size of the template and the difference in the two definitions is not very important.

A cell-linking template corresponds to an irreducible CNN graph Γ of the same size. The hole filler (Table I, (a)) is a cell-linking template and the corresponding Γ of a 3×3 CNN is irreducible. This means that every cell can influence every other cell. This is consistent with the fact that hole filling is a global property; whether a cell is part of a hole or not depends on cells around it, possibly quite a distance away.

A disconnected directed graph cannot be irreducible. Hence the CNN graphs of (c), (d) and (e) in Table I are reducible. An example of a *connected* CNN graph that is reducible is shown in (f) of Table I.

Definition 8

A directed graph is symmetric if an edge from node i to node j implies an edge exists from node j to node i.

---

† We use the notation 'a template of radius r' or 'a $2r+1$ by $2r+1$ template'.

‡ In References 8 and 9, CNN graphs are called 'labelled directed interaction graphs' and 'signed interaction graphs' respectively.

§ The CNN graphs are shown for a 3×3 CNN.

By induction, a directed path from node i to node j in a symmetric digraph implies the existence of a directed path from node j to node i.

Definition 9

E, a subgraph of a digraph Γ, is said to be maximally irreducible if any irreducible subgraph of Γ containing E is equal to E.

A digraph can be partitioned into maximally irreducible subgraphs. A symmetric digraph can be partitioned into one or more maximally irreducible subgraphs which are *disconnected* from each other. To prove this assertion, let C_i be a cell in maximally irreducible subgraph 1 and C_j be a cell in maximally irreducible subgraph 2. If there is a directed path from C_i to C_j, i.e. if the two subgraphs are connected, then by the symmetry property there must be a directed path from C_j to C_i. Every element in subgraph 1 is either C_i or has a directed path to C_i, hence to C_j, hence to every element in subgraph 2 and vice versa. This means that the union of subgraph 1 and subgraph 2 is irreducible. Since subgraph 1 and subgraph 2 are maximally irreducible, this implies that subgraph 1 = subgraph 1 $\cup$ subgraph 2 = subgraph 2. Therefore the maximally irreducible subgraphs of Γ are not connected to each other.

Consider graph (a) in Table I. This graph is symmetric and consists of one maximally irreducible subgraph; namely, itself. Graph (e) in Table I is symmetric and can be separated into two maximally irreducible subgraphs which are disconnected. In Table I, (e) nodes corresponding to the two maximally irreducible subgraphs are coloured black and grey respectively. This CNN is equivalent to two independent CNNs.

Definition 10

A directed graph is called acyclic if it has no cycles, i.e. there are no nodes $v_1, \ldots, v_j, v_{j+1}$, with $v_{j+1} = v_1$, such that there is an edge from v_i to v_{i+1}, $i = 1, \ldots, j$.

If a graph is acyclic, then a directed path from node i to node j implies no directed paths from node j to node i. Intuitively, this means that the dependence of cells is only in one direction. There is no indirect feedback through other cells.

The stability results obtained in Reference 5 and in this paper are dependent mostly on the CNN graph. This is useful since it allows us to determine the stability of some CNNs by analysing solely their CNN graph Γ.

4. TEST FOR IRREDUCIBILITY OF Γ

The results in Reference 6 that we will use require that the systems be irreducible. In the case of CNNs this corresponds to checking whether the CNN graph is irreducible. One way to do this is to start with any node of Γ and check whether the maximally irreducible subgraph containing that node is equal to Γ itself. Because Γ is generated by a space-invariant template, we can give simpler conditions for Γ being irreducible.

We call cell C_{kl} the the *conjugate* of cell C_{ij} if C_{kl} is the cell symmetrical to C_{ij} with respect to the centre of the cell array. We will denote this as follows: $C_{ij}^* = C_{kl}$. For example, cell a and cell a^* in Figure 1 are conjugate cells. From the regularity of our CNN cell array, each cell's conjugate cell is also a cell in the array.

Proposition 1

There exists a directed path of length n in Γ from cell a to cell b if and only if there exists a directed path of length n in Γ from cell b^* to cell a^*.

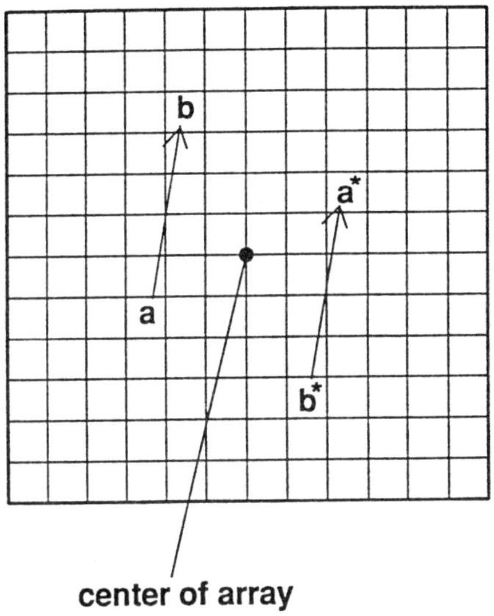

Figure 1. Cell a^* is the conjugate cell of cell a. There exists a directed edge from cell a to cell b if and only if there exists a directed edge from cell b^* to cell a^*

Proof 'Only if': by induction on the length of the directed path. It is clear from Figure 1 that if there is an edge from a to b, then there is an edge from b^* to a^* by space invariance of the templates.

Suppose a directed path of length k from a to b implies that a directed path of length k exists from b^* to a^*. A directed path of length $k+1$ from a to b consists of a directed path of length k from a to some cell c and an edge from c to b. There is an edge from b^* to c^* and by hypothesis there is a directed path of length k from c^* to a^*. Therefore there is a directed path of length $k+1$ from b^* to a^* and the result follows by induction. The 'if' part of the proposition follows from the fact that $a^{**} = a$. ∎

We can now give the following condition for irreducibility.

Theorem 1

A CNN graph Γ is irreducible if and only if some cell in Γ and its conjugate cell both have directed paths to every cell besides itself in the CNN array.

Proof The 'only-if' part is obvious. To prove the 'if' part, let cells a and a^* both have directed paths to every other cell in the array. Consider cell c different from a. Then cell c^* is a cell in the array different from a^*. Therefore there is a directed path from a^* to c^*. By Proposition 1 there is a directed path from c to a and hence there is a directed path from c to every cell in the array. Therefore Γ is irreducible. ∎

For $M = N = 3$ we have the following corollary.

Corollary 1

The CNN graph Γ of a 3×3 CNN is irreducible if and only if the centre cell in Γ has directed paths to the other eight cells.

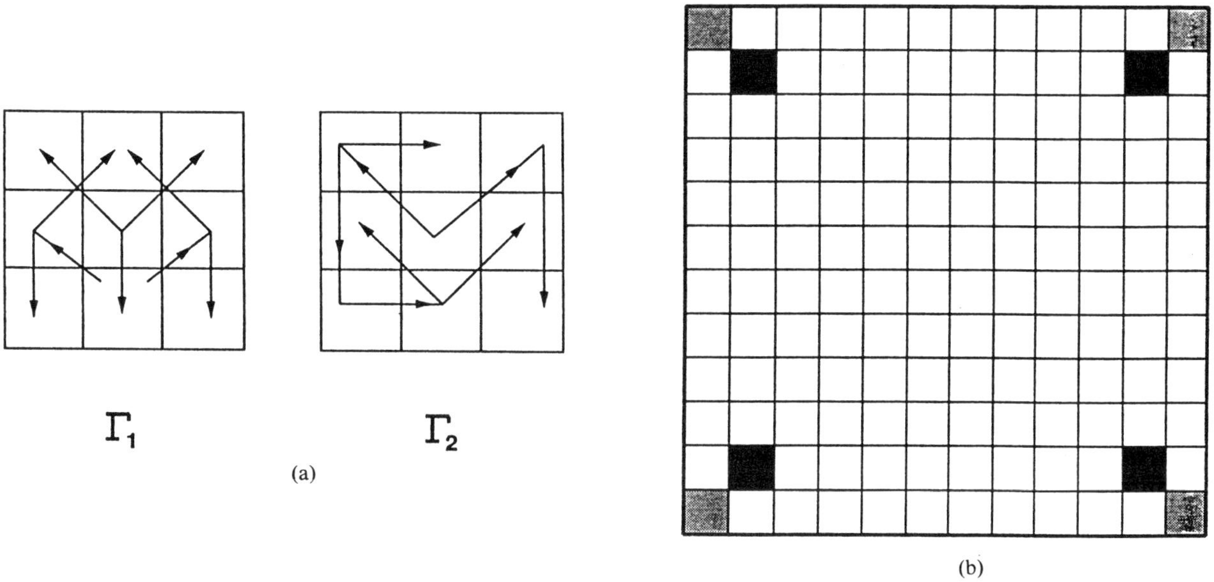

Figure 2. (a) The requirement for irreducibility is satisfied for the 3 × 3 CNN graphs generated by templates A_1 and A_2 (b) Paths from four of the eight cells shown in black and grey, as specified in Theorem 2, provide sufficient conditions for Γ being irreducible

Let us analyse the two A-templates described by

0	1	0
0	A	0
1	0	1

A_1

0	0	1	0	0
0	0	0	0	0
0	1	A	0	0
0	1	0	1	0
0	0	0	0	0

A_2

As shown in Figure 2(a), the CNN graphs Γ_1 and Γ_2 of a 3 × 3 CNN, corresponding to templates A_1 and A_2 respectively, satisfy the requirements stated in Corollary 1.

Therefore the CNN graphs generated by A_1 and A_2 are irreducible for any 3 × 3 CNN and both templates are cell-linking.

Remark 1 The CNN graph of a 3 × 3 CNN is identical to *any* 3 × 3 subgraph of an $M \times N$ CNN graph ($M \geqslant 3$, $N \geqslant 3$) utilizing the same A-template. From this it follows that if the CNN graph is irreducible for a 3 × 3 CNN, then the CNN graph for a larger CNN is also irreducible. Corollary 1 can then be restated as saying that a CNN graph Γ is irreducible if the centre cell of a 3 × 3 subgraph of Γ has paths *in the subgraph* to the other eight cells in the subgraph. For example, we looked at a 3 × 3 CNN graph (or a 3 × 3 subgraph of the CNN graph) even though template A_2 is a 5 × 5 template.

In the following theorem, to show that a CNN graph Γ is irreducible, we need to check the paths in the CNN graph Γ′ of a smaller CNN or equivalently we need to check the paths in a subgraph of Γ. More specifically, we need to check the paths emanating from two of the four 'corner' cells in the boundary of the CNN (shown in grey in Figure 2(b)) as well as those of two of the four 'near-corner' cells adjacent to the corner cells along the diagonals (shown in black in Figure 2(b)).

Theorem 2

Given an A-template and the CNN graph Γ associated with an $M \times N$ ($M \geqslant 3$, $N \geqslant 3$) CNN utilizing that template, if the upper and lower left 'near-corner' cells in Γ each has directed paths going to all of its eight nearest neighbours and if the upper and lower left 'corner' cells in Γ each has a directed path to its corresponding 'near-corner' cell, then a CNN of at least $(2M+2) \times (2N+2)$ cells utilizing the same A-template will generate an irreducible CNN graph.

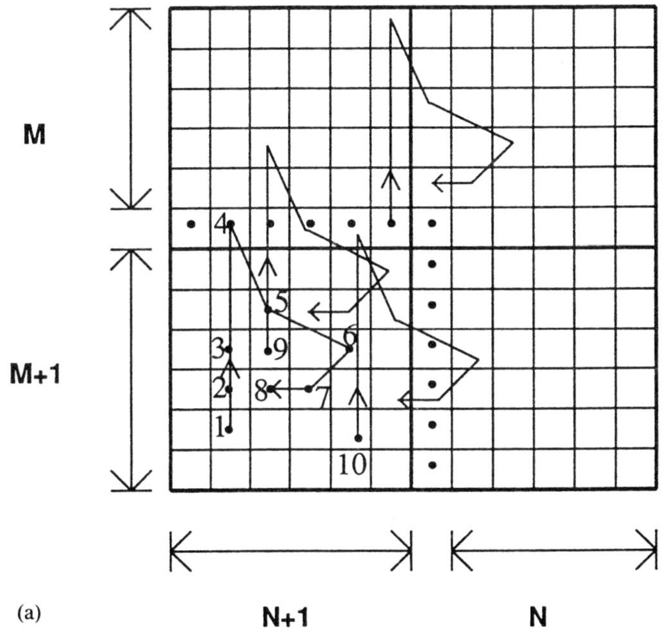

(a)

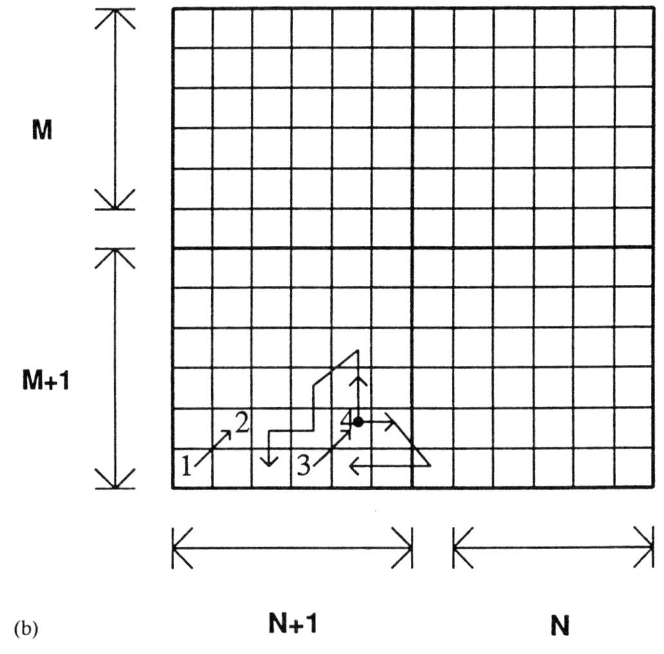

(b)

Figure 3. (a) Each cell in the augmented quadrant that is not a boundary cell has paths going to its eight nearest neighbours. (b) A cell on the boundary of the quadrant has paths to its nearest neighbours. There are five nearest neighbours for a non-corner boundary cell and three for a corner cell

Proof First we note that a CNN graph on a regular grid (where a node is located at the centre of each square cell in Figure 3(a)) is irreducible if and only if each node has directed paths to all its nearest neighbours. In a CNN graph there are eight nearest neighbours for an internal cell, five for a boundary cell and three for a corner cell. Consider a CNN of $(2M+2)\times(2N+2)$ cells and its CNN graph Γ divided into four quadrants as shown in Figure 3(a). We need to show that every node in Γ has directed paths to all its nearest neighbours. Consider first the lower left quadrant. We define an augmented quadrant as a quadrant augmented with one more layer of cells. For example, the cells outside the lower left quadrant in Figure 3(a) with a dot in the centre together with the lower left quadrant form the corresponding augmented quadrant. By a boundary cell we mean a cell on the boundary of the $(2M+2)\times(2N+2)$ array. A directed path from the 'near-corner' cell to one of its nearest neighbours for some Γ is shown as connecting nodes 1–8 in Figure 3(a). Because of space invariance, any directed path in Γ can be shifted to start at another node and still remain a valid directed path in Γ as long as the entire path stays within Γ. Because the augmented quadrant is $(M+2)\times(N+2)$, a path from the 'near-corner' cell to its neighbours lies entirely within the augmented quadrant. Consider a directed path from the 'near-corner' cell to its nearest neighbours shifted so that it starts at another cell in the augmented quadrant. Because the augmented quadrant is $(M+2)\times(N+2)$, the maximum amount of shift will be a distance of M up and N to the right and the path will remain within Γ. For example, the two directed paths in Figure 3(a) starting from nodes 9 and 10 are just shifted versions of the path from node 1 to node 8. They lie completely within Γ and are therefore valid paths in Γ. It is clear from Figure 3(a) that every cell in that augmented quadrant that is not a boundary cell has directed paths to its nearest neighbours. They also have directed paths to every cell in the augmented quadrant.

By assumption, the lower left corner cell has a directed path, call it p_1, towards its upper right neighbour, lying completely inside the quadrant. Consider a cell on the bottom boundary of the lower left quadrant. The distance between this cell and the lower left corner cell is at most N. Therefore the path p_1 shifted so that it starts at this cell will still lie entirely within Γ. The shifted path will end at the cell's upper right neighbour, which is in the augmented quadrant and is not a boundary cell. Since a cell in the augmented quadrant not on the boundary has directed paths to every cell in the augmented quadrant, this cell on the bottom boundary also has directed paths to every cell in the augmented quadrant, in particular its nearest neighbours. In Figure 3(b), node 3 has a directed path to node 4 since node 1 has a directed path to node 2. Node 4 is not a boundary cell, so node 4 has directed paths to every other cell in the augmented quadrant. Therefore node 3 also has directed paths to every cell in the augmented quadrant.

A similar argument holds for a cell on the left boundary in the lower left quadrant. Therefore every cell in the lower left quadrant has directed paths to its nearest neighbours. The cells in the upper left quadrant follow a similar reasoning. By Proposition 1 there is a directed path from both the upper and lower right 'corner' cell towards its corresponding 'near-corner' cell lying completely in the corresponding quadrant. Because $M \geqslant 3$ and $N \geqslant 3$, the upper or lower left 'near-corner' cell's neighbours are within the quadrant and thus have a directed path to the corresponding 'near-corner' cell. Because the augmented quadrant is $(M+2)\times(N+2)$, it can be shown that these paths lie completely within the corresponding augmented quadrant. By Proposition 1 the upper and lower right 'near-corner' cells have directed paths towards their neighbours lying completely within the corresponding augmented quadrant. We can then use the same analysis for the upper and lower right quadrants as well. It follows that every cell has directed paths to its nearest neighbours and thus Γ is irreducible. From Remark 1 it is clear that the CNN graphs of even larger CNNs (using the same A-template) will also be irreducible.† ■

5. STATE TRANSFORMATIONS

Our goal in this section is to find state transformations that produce a transformed system similar in form to a CNN with positive cell-linking templates. Since the *stability* of a dynamical system is preserved by such transformations, we can greatly enlarge the class of stable CNNs.

† It can be shown that the CNN graph of the $2M\times 2N$ CNN using this template is irreducible as well.

Consider the state equation of an $M \times N$ CNN (1). As in Reference 5, let us repack the state variables $v_{x_{ij}}$ of the CNN into a vector $\tilde{\mathbf{x}}$ of size MN. Similarly, let us repack the input and output variables $v_{u_{ij}}$ and $v_{y_{ij}}$ into $\tilde{\mathbf{u}}$ and $\tilde{\mathbf{y}}$ respectively (using the same ordering as that for $\tilde{\mathbf{x}}$). The invertible ordering will be called σ, i.e. $\tilde{\mathbf{x}}_{\sigma(i,j)} = v_{x_{ij}}$. Without loss of generality let $R_x = 1$, $C = 1$ and the state equations assume the form

$$\dot{\tilde{\mathbf{x}}} = \mathbf{F}(\tilde{\mathbf{x}}) = -\tilde{\mathbf{x}} + \tilde{\mathbf{A}}\mathbf{f}(\tilde{\mathbf{x}}) + \tilde{\mathbf{B}}\tilde{\mathbf{u}} + \tilde{\mathbf{I}} \tag{5}$$

where $\tilde{\mathbf{A}}$ and $\tilde{\mathbf{B}}$ are $MN \times MN$ matrices containing the coefficients $A(i,j;k,l)$ and $B(i,j;k,l)$ respectively. $\tilde{\mathbf{I}}$ is a constant vector of length MN and $\mathbf{f}_i(\mathbf{x}) = \hat{f}(\mathbf{x}_i) \approx \frac{1}{2}[|\mathbf{x}_i + 1| - |\mathbf{x}_i - 1|]$ for $i = 1, \ldots, MN$.† By our choice of $\mathbf{f}(\mathbf{x})$, $\partial \mathbf{f}_i(\mathbf{x})/\partial x_i > 0$, and for $i \neq j$

$$\frac{\partial \mathbf{F}_i(\tilde{\mathbf{x}})}{\partial \tilde{\mathbf{x}}_j} = \tilde{\mathbf{A}}_{ij} \frac{\partial \mathbf{f}_j(\tilde{\mathbf{x}})}{\partial \tilde{\mathbf{x}}_j} > 0 \Leftrightarrow \tilde{\mathbf{A}}_{ij} > 0 \tag{6}$$

The system is co-operative if and only if $\tilde{\mathbf{A}}$ has only non-negative off-diagonal elements. CNNs whose templates are not positive have corresponding $\tilde{\mathbf{A}}$-matrices that contain some *negative* off-diagonal elements. Our goal is to find a suitable transformation that produces an equivalent dynamical system which is co-operative and irreducible. We will consider a type of transformation where only the signs of some strategically chosen states are changed. Changing the signs of state variables amounts to multiplying the state vector by $\mathbf{J} = \mathrm{diag}\{(-1)^{n_1}, (-1)^{n_2}, \ldots, (-1)^{n_k}\}$, where $k = MN$ and $n_i \in \{0, 1\}$. Setting $\mathbf{J}\tilde{\mathbf{x}}' = \tilde{\mathbf{x}}$, the state equation transforms as

$$\dot{\tilde{\mathbf{x}}} = \mathbf{F}(\tilde{\mathbf{x}}) = -\tilde{\mathbf{x}} + \tilde{\mathbf{A}}\mathbf{f}(\tilde{\mathbf{x}}) + \tilde{\mathbf{B}}\tilde{\mathbf{u}} + \tilde{\mathbf{I}} \Leftrightarrow \mathbf{J}\dot{\tilde{\mathbf{x}}}' = -\mathbf{J}\tilde{\mathbf{x}}' + \tilde{\mathbf{A}}\mathbf{f}(\mathbf{J}\tilde{\mathbf{x}}') + \tilde{\mathbf{B}}\tilde{\mathbf{u}} + \tilde{\mathbf{I}} \tag{7}$$

$$\dot{\tilde{\mathbf{x}}}' = -\tilde{\mathbf{x}}' + \mathbf{J}^{-1}\tilde{\mathbf{A}}\mathbf{f}(\mathbf{J}\tilde{\mathbf{x}}') + \mathbf{J}^{-1}\tilde{\mathbf{B}}\tilde{\mathbf{u}} + \mathbf{J}^{-1}\tilde{\mathbf{I}} + \mathbf{F}'(\tilde{\mathbf{x}}') \tag{8}$$

The Jacobian matrix of $\mathbf{F}'$ is

$$D\mathbf{F}'(\tilde{\mathbf{x}}') = \mathbf{J}^{-1}\tilde{\mathbf{A}}D\mathbf{f}(\mathbf{J}\tilde{\mathbf{x}}')\mathbf{J} - \mathbf{U} \tag{9}$$

where $\mathbf{U}$ denotes the identity matrix. $D\mathbf{f}$ and $\mathbf{J}$ are both diagonal matrices and thus they commute. Noting that $\mathbf{J}^{-1} = \mathbf{J}$, we have

$$D\mathbf{F}'(\tilde{\mathbf{x}}') = \mathbf{J}\tilde{\mathbf{A}}\mathbf{J}D\mathbf{f}(\mathbf{J}\tilde{\mathbf{x}}') - \mathbf{U} \tag{10}$$

For $i \neq j$

$$\frac{\partial \mathbf{F}_i'(\tilde{\mathbf{x}}')}{\partial \tilde{\mathbf{x}}_j'} = (\mathbf{J}\tilde{\mathbf{A}}\mathbf{J})_{ij} \left.\frac{\mathrm{d}\hat{f}(v)}{\mathrm{d}v}\right|_{(\mathbf{J}\tilde{\mathbf{x}}')_j} \tag{11}$$

Equation (11) implies that the system in (8) is co-operative and irreducible if and only if $\tilde{\mathbf{A}}' = \mathbf{J}\tilde{\mathbf{A}}\mathbf{J}$ is non-negative in the off-diagonal elements and is irreducible.

Theorem 3

The system (8) is stable almost everywhere if the equilibria are isolated and $\tilde{\mathbf{A}}'$ is non-negative in the off-diagonal elements and is irreducible.

Proof The proof is similar to that of Lemma 1 in Reference 5. First we note that from any fixed $\tilde{\mathbf{x}}(0)$ the state vector $\tilde{\mathbf{x}}$ in (5) has a bounded forward orbit by Theorem 1 in Reference 1 and therefore $\tilde{\mathbf{x}}'$ has a bounded forward orbit. The vector $\tilde{\mathbf{x}}'$ is an equilibrium point of (8) if and only if $\mathbf{J}\tilde{\mathbf{x}}'$ is an equilibrium point of (5). Since the system (8) is co-operative and irreducible, Theorem 3 follows from Reference 6. ■

† We use the symbol '≈' since in the following proofs we will assume $\hat{f}(x_i)$ to be any continuously differentiable and strictly increasing approximation of the piecewise linear function. To simplify the proofs, we also assume $\hat{f}$ to be odd ($\hat{f}(-x) = -\hat{f}(x)$).

Remark 2 If the equilibria are not isolated, then we can only conclude that almost all initial states *approach* the set of equilibrium points. For $A(i, j; i, j) > 1/R_x$ the stable equilibria are the equilibria in the saturation region and are therefore isolated.

Remark 3 The condition that the CNN is irreducible is important. A reducible co-operative CNN can always be decomposed as a cascade of co-operative and irreducible systems. However, it is not true that a cascade of stable systems is necessarily stable as a whole (see Reference 8, p. 342).

If the transformed system is stable almost everywhere, then the original system is also stable almost everywhere. It is clear that $\mathbf{J\tilde{A}J}$ is irreducible if and only if $\tilde{\mathbf{A}}$ is irreducible. $\tilde{\mathbf{A}}$ being irreducible corresponds to the associated CNN graph being irreducible and we can use the results in Section 4 to check whether $\tilde{\mathbf{A}}$ is irreducible.

Our objective is to find a $\mathbf{J}$ in the diagonal form described above such that $\mathbf{J\tilde{A}J}$ is an off-diagonally non-negative matrix:

$$(\mathbf{J\tilde{A}J})_{ij} = (-1)^{n_i}(\tilde{\mathbf{A}}\mathbf{J})_{ij} = (-1)^{n_i}\tilde{\mathbf{A}}_{ij}(-1)^{n_j} = (-1)^{n_i+n_j}\tilde{\mathbf{A}}_{ij} \tag{12}$$

Recall that the matrix $\tilde{\mathbf{A}}$ results from a space-invariant template. Assume that for all $i \neq j$, $\tilde{\mathbf{A}}_{ij} \neq 0$. Given this assumption and the requirement that $\mathbf{J\tilde{A}J}$ has non-negative off-diagonal elements, some properties of $\mathbf{J}$ can be obtained. $\tilde{\mathbf{A}}_{ij} \neq 0$ and $(\mathbf{J\tilde{A}J})_{ij} \geqslant 0$ for all $i \neq j$ imply that

$$(-1)^{n_i+n_j}\tilde{\mathbf{A}}_{ij} = |\tilde{\mathbf{A}}_{ij}| \neq 0 \tag{13}$$

Let $i = \sigma(a, b)$, $j = \sigma(a+l, b)$ and $k = \sigma(a+2l, b)$ for some $1 \leqslant a \leqslant M-2l$, $1 \leqslant b \leqslant N$ and $l = 1, \ldots, \lfloor (M-1)/2 \rfloor$. Because of space invariance, $\tilde{\mathbf{A}}_{ij} = \tilde{\mathbf{A}}_{jk}$ and we have

$$(-1)^{n_i+n_j}\tilde{\mathbf{A}}_{ij} = |\tilde{\mathbf{A}}_{ij}| = |\tilde{\mathbf{A}}_{jk}| = (-1)^{n_j+n_k}\tilde{\mathbf{A}}_{jk} = (-1)^{n_j+n_k}\tilde{\mathbf{A}}_{ij} \tag{14}$$

Consequently

$$(-1)^{n_i} = (-1)^{n_k} \Rightarrow n_i = n_k \tag{15}$$

The same conclusion follows if we take $i = \sigma(a, b)$, $j = \sigma(a, b+l)$ and $k = \sigma(a, b+2l)$ for some $1 \leqslant a \leqslant M$, $1 \leqslant b \leqslant N-2l$ and $l = 1, \ldots, \lfloor (N-1)/2 \rfloor$. Therefore we have the following constraints on $\mathbf{J}$:

$$n_{\sigma(a,b)} = n_{\sigma(a+2l,b)}, \quad 1 \leqslant a \leqslant M-2l, \quad 1 \leqslant b \leqslant N, \quad l = 1, \ldots, \lfloor (M-1)/2 \rfloor \tag{16}$$

$$n_{\sigma(a,b)} = n_{\sigma(a,b+2l)}, \quad 1 \leqslant a \leqslant M, \quad 1 \leqslant b \leqslant N-2l, \quad l = 1, \ldots, \lfloor (N-1)/2 \rfloor \tag{17}$$

This gives us four possible choices for $\mathbf{J}$.†

From all the sign-changing transformations, only these four have any chance of generating a co-operative system. One of them, $\mathbf{J} = \mathbf{U}$, gives $\tilde{\mathbf{A}}' = \tilde{\mathbf{A}}$. The other three will be discussed in more detail.

Remark 4 A transformation that changes all the states to their negatives amounts to $\mathbf{J}$ being equal to $-\mathbf{U}$ and therefore $\tilde{\mathbf{A}}' = \tilde{\mathbf{A}}$, which does not help us in determining the stability of the system.

Remark 5 The four choices for $\mathbf{J}$ were found from the two sets of constraints (16) and (17), which in turn were derived from the space invariance of the templates and the absence of zeros in the off-diagonal elements of $\tilde{\mathbf{A}}$. Clearly these $\mathbf{J}$s will also yield correct results (i.e. $\mathbf{J\tilde{A}J}$ has non-negative off-diagonal elements) if there are some zero elements in the off-diagonal elements of $\tilde{\mathbf{A}}$. If we are restricting ourselves to considering only templates with zero elements in certain locations ($\tilde{\mathbf{A}}$ has zeros in some off-diagonal entries), then some of the constraints might not be necessary and more choices for $\mathbf{J}$ can then be found. We will demonstrate this later in the paper.

Remark 6 $\tilde{\mathbf{I}}$ is assumed to be constant and $\tilde{\mathbf{B}}$ is assumed to be generated by a space-invariant B-template, since they correspond to a CNN with space-invariant templates. For an arbitrary sign-changing

† There are actually eight choices for $\mathbf{J}$. We can group them into four pairs, with each pair consisting of $+\mathbf{J}$ and $-\mathbf{J}$. The effects of $+\mathbf{J}$ and $-\mathbf{J}$ on the cloning templates are identical, so we need only consider four choices for $\mathbf{J}$.

transformation matrix $\mathbf{J}$, $\mathbf{J}^{-1}\tilde{\mathbf{B}}$ need not correspond to a space-invariant B-template, $\mathbf{J}^{-1}\tilde{\mathbf{I}}$ need not be a constant vector and therefore (8) need not correspond to the repacked equation of a space-invariant CNN.

Remark 7 There are other constraints on $\mathbf{J}$ of the form

$$n_{\sigma(a,b)} = n_{\sigma(a \pm 2l, b \pm 2m)}, \quad 1 \leqslant a \leqslant M-2l, \quad 1 \leqslant b \leqslant N-2m,$$
$$l = 1, \ldots, \lfloor (M-1)/2 \rfloor, \quad m = 1, \ldots, \lfloor (N-1)/2 \rfloor \tag{18}$$

but they are equivalent to (16) and (17).

$\tilde{\mathbf{A}}'$ generated by the three transformations is space-invariant in the sense that there exists a space-invariant cloning template $A'(i,j;k,l)$ which generates $\tilde{\mathbf{A}}'$ through σ. This is made clear in the following proof. To simplify notations, let us return to state equation (1). The transformation will now be

$$v_{x'_{ij}} = v_{x_{ij}}(-1)^{n_{\sigma(i,j)}} \tag{19}$$

Let us denote the output variables after the transformation by $v_{y'_{ij}} = \hat{f}(v_{x'_{ij}})$. The state equation after the transformation is

$$C' \frac{\mathrm{d}v_{x'_{ij}}(t)}{\mathrm{d}t} = -\frac{1}{R'_x} v_{x'_{ij}}(t) + \sum_{C_{kl} \in N_r(ij)} A'(i,j;k,l) v_{y'_{kl}}(t) + \sum_{C_{kl} \in N_r(ij)} B'(i,j;k,l) v_{u_{kl}}(t) + I'_{ij} \tag{20}$$

where $R'_x = R_x = 1$ and $C' = C = 1$. For now, let us only consider 3×3 templates.

Transformation 1

$v_{x'_{ij}}(t) = v_{x_{ij}}(t)(-1)^j$, i.e. $v_{x'_{ij}}(t) = v_{x_{ij}}(t)$ if j is even, $v_{x'_{ij}}(t) = -v_{x_{ij}}(t)$ if j is odd. The cells with their states changed to their negatives form *alternating columns*. The effect on the A-templates is as follows:

$+a$	$+b$	$+c$		$-a$	$+b$	$-c$
$+d$	$+e$	$+f$	$\rightarrow$	$-d$	$+e$	$-f$
$+g$	$+h$	$+i$		$-g$	$+h$	$-i$

A $\qquad$ A'

Proof Recall that $v_{y_{ij}}(t) = \hat{f}(v_{x_{ij}}(t))$ and $\hat{f}$ is odd. Hence, if $v_{x'_{ij}}(t) = -v_{x_{ij}}(t)$, then $v_{y'_{ij}}(t) = -v_{y_{ij}}(t)$.

1. Suppose j is even; then $v_{x'_{ij}}(t) = v_{x_{ij}}(t)$:

$$\frac{\mathrm{d}v_{x'_{ij}}(t)}{\mathrm{d}t} = \frac{\mathrm{d}v_{x_{ij}}(t)}{\mathrm{d}t} = A(i,j;i+1,j+1)v_{y_{(i+1)(j+1)}}(t) + \cdots$$
$$= A(i,j;i+1,j+1)(-v_{y'_{(i+1)(j+1)}}(t)) + \cdots \quad \text{(note that } j+1 \text{ is odd)}$$
$$= A'(i,j;i+1,j+1)v_{y'_{(i+1)(j+1)}}(t) + \cdots \tag{21}$$

where the symbol '$\cdots$' indicates the *remaining terms* which are not relevant to this part of the proof. Therefore $A'(i,j;i+1,j+1) = -A(i,j;i+1,j+1)$. The same argument is valid when $i+1$ is replaced by i or $i-1$ and when $j+1$ is replaced by $j-1$ ($j+1$ and $j-1$ are both odd):

$$\frac{\mathrm{d}v_{x'_{ij}}(t)}{\mathrm{d}t} = \frac{\mathrm{d}v_{x_{ij}}(t)}{\mathrm{d}t} = A(i,j;i+1,j)v_{y_{(i+1)j}}(t) + \cdots$$
$$= A(i,j;i+1,j)v_{y'_{(i+1)j}}(t) + \cdots$$
$$= A'(i,j;i+1,j)v_{y'_{(i+1)j}}(t) + \cdots \tag{22}$$

Therefore $A'(i,j;i+1,j) = A(i,j;i+1,j)$. The same argument is valid when $i+1$ is replaced by $i-1$ or i.

2. Suppose j is odd; then $v_{x'_{ij}}(t) = -v_{x_{ij}}(t)$:

$$\begin{aligned}\frac{\mathrm{d}v_{x'_{ij}}(t)}{\mathrm{d}t} = -\frac{\mathrm{d}v_{x_{ij}}(t)}{\mathrm{d}t} &= -A(i,j;i+1,j+1)v_{y_{(i+1)(j+1)}}(t) + \cdots \\ &= -A(i,j;i+1,j+1)v_{y'_{(i+1)(j+1)}}(t) + \cdots \quad \text{(note that } j+1 \text{ is even)} \\ &= A'(i,j;i+1,j+1)v_{y'_{(i+1)(j+1)}}(t) + \cdots \end{aligned} \tag{23}$$

Therefore $A'(i,j;i+1,j+1) = -A(i,j;i+1,j+1)$. The same argument is valid when $i+1$ is replaced by i or $i-1$ and when $j+1$ is replaced by $j-1$ ($j+1$ and $j-1$ are both even):

$$\begin{aligned}\frac{\mathrm{d}v_{x'_{ij}}(t)}{\mathrm{d}t} = -\frac{\mathrm{d}v_{x_{ij}}(t)}{\mathrm{d}t} &= -A(i,j;i+1,j)v_{y_{(i+1)j}}(t) + \cdots \\ &= -A(i,j;i+1,j)(-v_{y'_{(i+1)j}}(t)) + \cdots \\ &= A'(i,j;i+1,j)v_{y'_{(i+1)j}}(t) + \cdots \end{aligned} \tag{24}$$

Therefore $A'(i,j;i+1,j) = A(i,j;i+1,j)$. The same argument is valid when $i+1$ is replaced by $i-1$ or i. ■

We see that $\tilde{\mathbf{A}}'$ corresponds to a space-invariant cloning template $A'(i,j;k,l)$.

Transformation 2

$v_{x'_{ij}}(t) = v_{x_{ij}}(t)(-1)^i$, i.e. $v_{x'_{ij}}(t) = v_{x_{ij}}(t)$ if i is even, $v_{x'_{ij}}(t) = -v_{x_{ij}}(t)$ if i is odd. The cells with their states changed to their negatives form *alternating rows*. The effect on the A-templates is depicted below:

$+a$	$+b$	$+c$		$-a$	$-b$	$-c$
$+d$	$+e$	$+f$	→	$+d$	$+e$	$+f$
$+g$	$+h$	$+i$		$-g$	$-h$	$-i$
	A				A'	

The proof is similar to that of the previous transformation. In essence, this is Transformation 1 rotated through 90°.

Transformation 3

$v_{x'_{ij}}(t) = v_{x_{ij}}(t)(-1)^{i+j}$, i.e. $v_{x'_{ij}}(t) = v_{x_{ij}}(t)$ if $i+j$ is even, $v_{x'_{ij}}(t) = -v_{x_{ij}}(t)$ if $i+j$ is odd. In this transformation the cells with their states changed to their negatives form a *checkerboard pattern*. The effect on the A-templates is depicted as follows:

$+a$	$+b$	$+c$		$+a$	$-b$	$+c$
$+d$	$+e$	$+f$	→	$-d$	$+e$	$-f$
$+g$	$+h$	$+i$		$+g$	$-h$	$+i$
	A				A'	

Observe that Transformation 3 is equivalent to an application of Transformation 1 followed by Transformation 2.

We can now summarize the above results in the form of a theorem. Note that the transformed $n \times n$ template $A'(i,j;k,l)$ is positive cell-linking if and only if $\tilde{\mathbf{A}}'$ is off-diagonally non-negative and irreducible for an $n \times n$ CNN.

Theorem 4

If through one of the above three transformations on an $n \times n$ template $A(i, j; k, l)$ the resulting $A'(i, j; k, l)$ is positive cell-linking and all the equilibria are isolated, then the $n \times n$ (or larger) CNN with the template $A(i, j; k, l)$ is stable almost everywhere.

In Reference 13, Seiler used Transformation 1 to prove the stability of one-dimensional templates.

Let us look more closely at how the transformation **J** and the change of the A-templates are related. The transformation either changes the state of a cell to its negative or it does not change the state of a cell. Consider two cells C_{ij} and C_{kl} connected by a template element $A(i, j; k, l)$. The corresponding nodes in Γ are connected by an edge. If the states of both cells are changed to their negatives by the transformation, then the connection between these two cells remains the same after the transformation ($A'(i, j; k, l) = A(i, j; k, l)$). The same result occurs if the states of both cells remain unchanged. If one cell's state is changed to its negative while the other cell's state remains unchanged, then the connection between these two cells is changed to its negative ($A'(i, j; k, l) = -A(i, j; k, l)$). By looking at how a cell is changed by the transformation relative to the surrounding cells, we can determine the transformation's effect on the template $A(i, j; k, l)$.

For example, let us consider Transformation 1. In Figure 4(a) the black cells correspond to cells that have their state variables changed to their negatives and the white cells correspond to cells that have their state variables left alone.

Consider first an arbitrary white cell, which we shall call the centre cell, and its nearest neighbours as they would appear in Γ. (In Figure 4(b) the labels have been omitted and only the edges pointing towards the centre cell are shown.)

The transformation does not change the state of the centre cell. The cells nearest to the left and to the right of the centre cell have their states changed, so all the edges pointing towards the white cell from its nearest right and left neighbours have changed sign. The right and left (with respect to the centre) elements on the A-template of the centre cell correspond to these edges and are also changed to their negatives. The cells above and below the centre cell remain the same and thus the elements above and below the centre on the A-template of the centre cell do not change. Similarly, the diagonal elements of the A-template are changed to their negatives. The effect of this transformation on the A-template can be read off directly by looking at the difference in colour (black or white) between the centre cell and its nearest neighbours in Figure 4(b).

Consider next an arbitrary black cell (which we call the centre cell now) and its nearest neighbours as they would appear in Γ (Figure 4(c)).

The centre cell has its state changed to its negative while the cells to the left and to the right of the centre cell are left alone. Therefore the left elements and the right elements (with respect to the centre) of the A-template of the centre cell are changed to their negatives. The cells immediately above and below the centre cell are also changed to their negatives along with the centre cell, so the elements above and below the centre on the A-template of the centre cell do not change. Again, the effect on the A-template is apparent from the colour pattern of Figure 4(c).

The effect of the transformation on the A-template is the same in both cases. This confirms the fact that these transformations preserve space invariance of the A-templates. For a numerical example of these transformations see Reference 14.

The effects of these transformations on larger-size cloning templates are analogous. For example, Transformation 1 causes alternate columns in the cloning template to change sign while the other columns, which include the middle column, remain unaffected.

The same technique can be applied to other network topologies as well. For example, in a hexagonal network we would have a transformation that changes a template as in Figure 5. By rotating this transformation through 60° and 120°, we obtain two other transformations.

Remark 8 These simple transformations result from the special structure of the CNN. For state transformations applicable to more general dynamical systems see References 15 and 8.

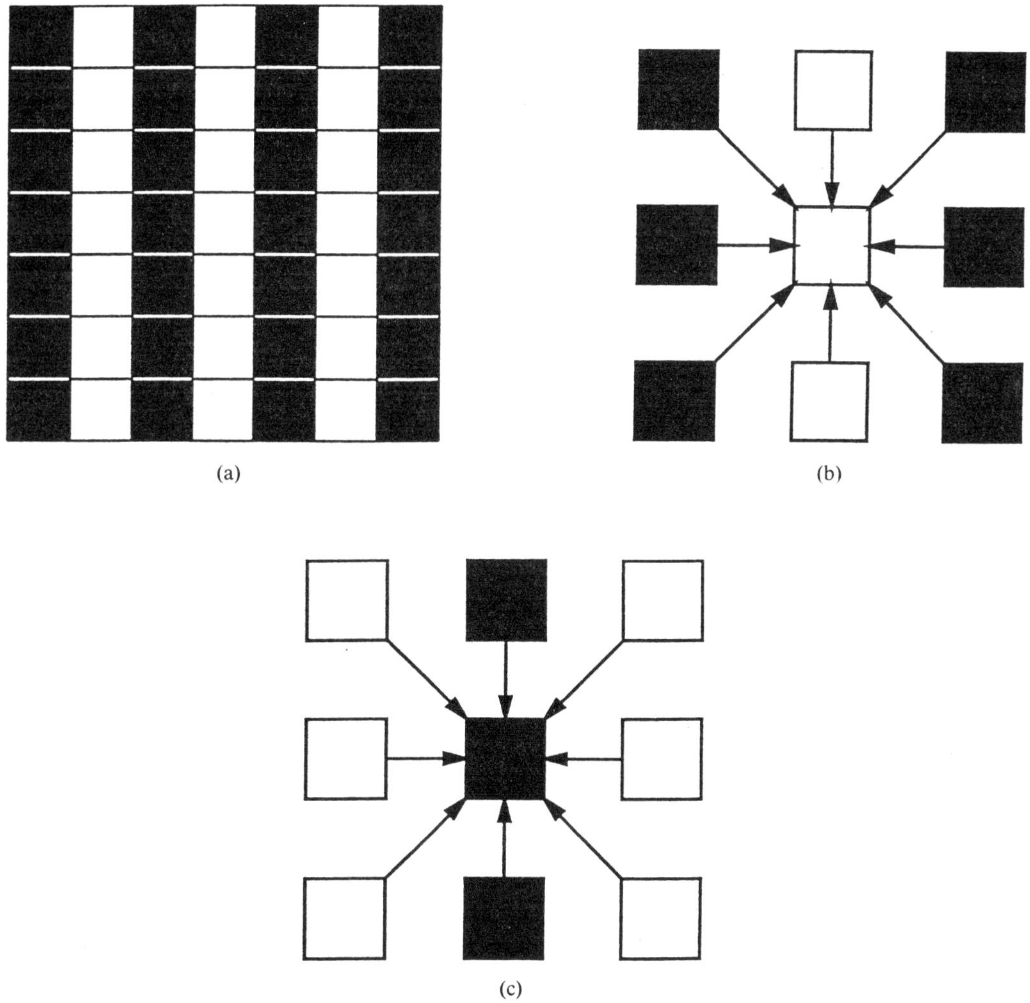

Figure 4. (a) Under Transformation 1, black cells have their states reversed and white cells remain unchanged. (b) Cells to the left and right of the centre cells have their states changed to their negatives. (c) Cells in the centre column have their states changed while the states of cells to the left and right have not changed

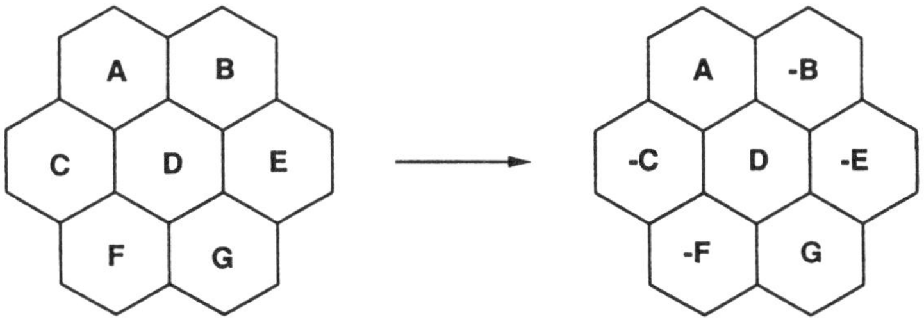

Figure 5. Effect of a transformation on a hexagonal template

6. STRICTLY SIGN-SYMMETRIC TEMPLATES

All the stability results we have obtained so far are for cell-linking templates. We will now introduce a class of templates that are not necessarily cell-linking.

Definition 11

A template is called strictly sign-symmetric if whenever $A(i,j;k,l) \neq 0$, $A(i,j;k,l)A(k,l;i,j) > 0$.†

In view of the repacked equation (5), this is equivalent to saying that $\tilde{\mathbf{A}}_{ij}\tilde{\mathbf{A}}_{ji} > 0$ if $\tilde{\mathbf{A}}_{ij} \neq 0$. Because of space invariance, $A(i,j;i+m,j+n)A(i,j;i-m,j-n) > 0$ if $A(i,j;i+m,j+n) \neq 0$. For such a cloning template the sign of each *non-zero* element is identical to the sign of its corresponding symmetric (with respect to the centre element) element. For example, in Table I only templates (a), (b) and (e) are strictly sign-symmetric.

In our next lemma we say that a CNN is composed of *independent* subsystems if the state variables can be partitioned into n mutually exclusive subsets of state variables $S_1, \ldots, S_n$ such that $A(i,j;k,l) = 0$ for $v_{x_{ij}} \in S_a$, $v_{x_{kl}} \in S_b$ and $S_a \neq S_b$. In other words, the state of the cells in one subset in the partition is independent of the state of the cells in the other subsets in the partition.

Lemma 1

A CNN composed of independent subsystems is stable almost everywhere if each subsystem is stable almost everywhere.

Proof Consider first a system composed of two independent subsystems. The states are bounded[1] and we denote the state space of the system by $[a, a+b]^n$, the cartesian product of closed intervals $[a, a+b]$, $b > 0$. Similarly, let the state space of subsystem 1 be $[a, a+b]^k$ and that of subsystem 2 be $[a, a+b]^{n-k}$, $b > 0$. Assume that the state vector of the whole system are composed of first the state variables of subsystem 1, then the state variables of subsystem 2. Let E_i be the set of initial values in which subsystem i is not stable, $i \in \{1, 2\}$. Let E be the set of initial values in which the entire system is not stable. An initial state x is not stable for the entire system if the subvector of the first k co-ordinates belongs to E_1 or to E_1 or the subvector of the last $n-k$ co-ordinates belongs to E_2. Let $\alpha \subseteq [a, a+b]^k$ be a covering of E_1 with Lebesque measure $\varepsilon > 0$; then $\alpha' = \{(x_1, \ldots, x_n) : (x_1, \ldots, x_k) \in \alpha\}$ has Lebesque measure $b^{n-k}\varepsilon$. Similarly, let $\beta \subseteq [a, a+b]^{n-k}$ be a covering of E_2 with Lebesque measure ε, then $\beta' = \{(x_1, \ldots, x_n) : (x_{k+1}, \ldots, x_n) \in \beta\}$ has Lebesque measure $b^k\varepsilon$. Now $\alpha' \cup \beta'$ is a covering of E with Lebesque measure $\leqslant (b^{n-k} + b^k)\varepsilon$. Since ε can be made as small as possible, E has Lebesque measure zero. This shows that the whole system is stable almost everywhere. The proof for a finite number of subsystems is similar. ■

Theorem 5

A CNN utilizing positive strictly sign-symmetric templates with isolated equilibria is stable almost everywhere.

Proof A strictly sign-symmetric template generates a symmetric CNN graph. As shown in Section 3, a symmetric graph can be decomposed into maximally irreducible subgraphs which are disconnected from each other. Because all the edges have positive labels, the CNN can be decomposed into independent positive irreducible subnetworks and each subnetwork has been shown to be stable almost everywhere.[5] By Lemma 1 the whole CNN is stable almost everywhere. ■

For example, template (e) in Table I is a positive strictly sign-symmetric template and a CNN utilizing this template is stable almost everywhere.

---

† This definition of '*strict* sign symmetry' is more restrictive than that of 'sign symmetry' defined in Reference 8.

From Theorem 5 it follows that Theorems 3 and 4 are still true when the words 'irreducible' and 'cell-linking' are replaced by 'strictly sign-symmetric' and we can utilize the three transformations in Section 5 to show that other strictly sign-symmetric templates are also stable almost everywhere. Note that the 'cell-linking' and 'strictly sign-symmetric' properties are topological properties of the templates and are preserved under the three sign-changing transformations. Using these transformations, only cell-linking templates can be transformed into a positive cell-linking template and only strictly sign-symmetric templates can be transformed into a positive strictly sign-symmetric template.

Using the three sign-changing transformations, we find that the following classes of 3×3 strictly sign-symmetric templates are stable almost everywhere.† In the rest of the paper we assume that all the equilibrium points of the CNN are isolated.

In the following descriptions the elements refer to the elements of the templates, not including the centre. In the figures 'N' indicates that the corresponding weight is non-positive and 'P' indicates that the corresponding weight is non-negative; '+' and '−' indicate positive and negative weights respectively. The choice of the weights must be kept consistent with the definition of strict sign symmetry.

1. Corner elements are zero:

0	P	0
N	A	N
0	P	0

0	N	0
P	A	P
0	N	0

0	N	0
N	A	N
0	N	0

2. Corner elements are non-negative while the rest are non-positive:

P	N	P
N	A	N
P	N	P

3. Every element is non-positive except for two non-negative elements in the middle row or column:

N	P	N
N	A	N
N	P	N

N	N	N
P	A	P
N	N	N

The following templates are also stable almost everywhere, although they cannot be transformed into positive templates by the three transformations above:

−	0	+
0	A	0
+	0	−

+	0	−
0	A	0
−	0	+

One way to show that they are stable almost everywhere is by constructing Γ of the first template.

By separating Γ into two maximally irreducible subgraphs (Figure 6), we see that both components are topologically equivalent to the CNN graph of a strictly sign-symmetric template with the corner elements equal to zero:

0	+	0
−	A	−
0	+	0

which we know is stable almost everywhere.

† Most of these 3×3 strictly sign-symmetric templates are cell-linking and the stability can be deduced without using Theorem 5. For bigger templates, strictly sign-symmetric templates are not necessarily cell-linking.

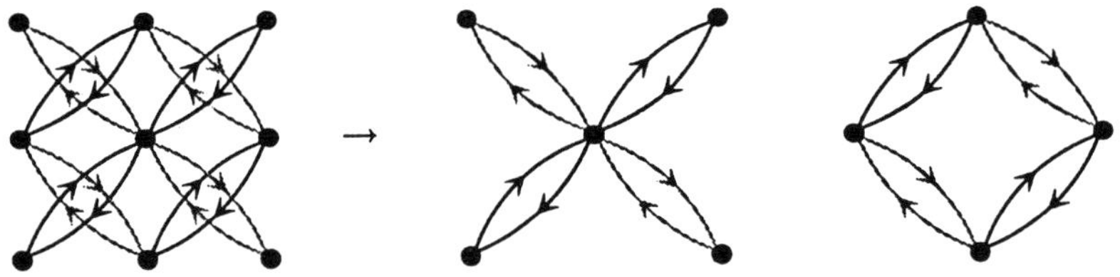

Figure 6. The CNN graph separated into two maximally irreducible subgraphs

Since the stability results we obtained are solely dependent on the template $A(i, j; k, l)$ or equivalently on Γ, the systems corresponding to the two components are stable almost everywhere and so is the whole system by Lemma 1. By way of any of the three transformations the second template can be shown to be stable almost everywhere as well.

There are only four state transformations (one of them trivial) of the type discussed in Section 5 such that Theorem 4 can be used on all types of templates. Other transformations can exist if we restrict ourselves to templates with zero elements in certain locations. To illustrate this, we will give an alternative proof of the stability of the first template in the pair shown above.

Because the template has some zero elements, the following state transformation will work in Theorem 4.

Transformation 4

$v_{x'_{ij}}(t) = -v_{x_{ij}}(t)$ if $i + j \equiv 0 \bmod 4$ or $i + j \equiv 1 \bmod 4$, $v_{x'_{ij}}(t) = v_{x_{ij}}(t)$ otherwise. The cells that have their states changed to their negatives are shown in black in Figure 7(a).

The effect on the A-template is

$+a$	0	$+c$		$-a$	0	$+c$
0	$+e$	0	$\rightarrow$	0	$+e$	0
$+g$	0	$+i$		$+g$	0	$-i$
	A				A'	

This transformation changes the A-template in question (with $a < 0$ and $i < 0$) into a positive strictly sign-symmetric template which we know is stable almost everywhere.

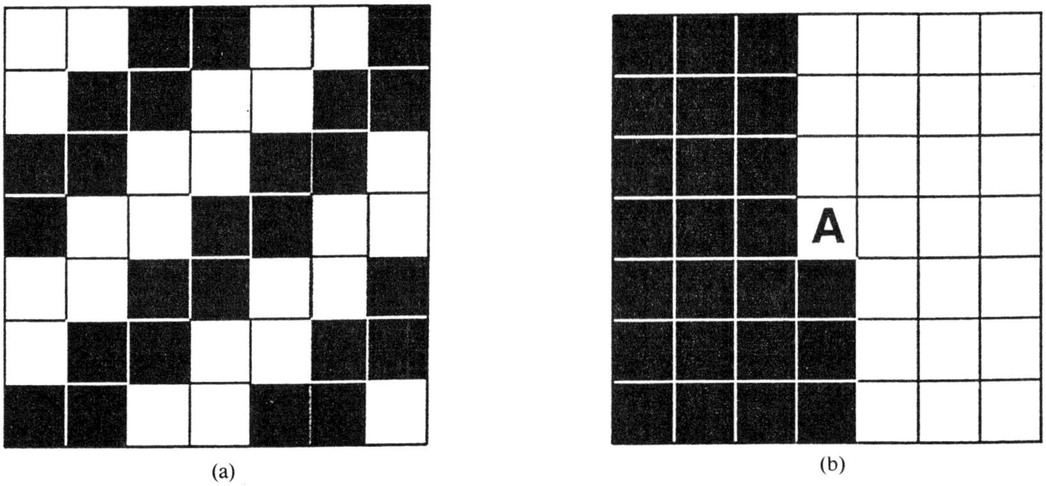

Figure 7. (a) Pattern of how the cells are transformed using Transformation 4. Black cells have their states changed to their negatives and white cells are left unchanged. (b) The white portion of the template (excluding the centre) having only zero elements is a sufficient condition for the template to generate an acyclic CNN graph

7. ACYCLIC Γ

Theorem 6

A CNN with an acyclic Γ and $A(i, j; i, j) > R_x^{-1}$ is completely stable.

Proof Because Γ is acyclic, there exists a set of cells (nodes in Γ) with indegree zero. Since they receive no input from other cells and the external input is constant, these cells converge to an equilibrium. As shown in Reference 1, $A(i, j; i, j) > R_x^{-1}$ implies that stable equilibria have magnitude greater than unity. The states are bounded and the output of these cells will not change after a finite time.† Delete these cells and the edges coming out of them and the resulting graph will have another set of cells with indegree zero. The only varying influence or input acting on these cells is from cells whose output will not change after a finite time. Thus these cells also converge to a constant output after a finite time. This argument can be repeated until all the cells have been shown to converge. ∎

A sufficient condition for a template to generate an acyclic Γ is the following.

Theorem 7

A template such that

$$A(i, j; k, l) = \begin{cases} 0 & l > j \\ 0 & l = j \quad \text{and} \quad k > i \\ \text{arbitrary} & \text{otherwise} \end{cases}$$

generates a CNN with an acyclic Γ.

For example, in a 7×7 template Theorem 7 requires that the white portion in Figure 7(b) (excluding the centre) has only zero elements for the template to generate an acyclic Γ.

Proof We need to show that there are no cycles in Γ. Imagine that the cells are arranged on a regular rectangular grid in $\mathbb{R}^2$ with the edges of Γ being vectors connecting the nodes. Then the edges have a certain direction that we can decompose into a horizontal and a vertical component. The hypothesis in Theorem 7 implies that the edges in Γ have a horizontal direction $\geqslant 0$. A cycle must include an edge with horizontal direction >0, because otherwise the cycle must consist solely of edges with horizontal direction $= 0$. The only edge that has no horizontal direction is an edge with a positive vertical direction and such a path can never be a cycle. Thus a cycle must include edges with horizontal direction >0. However, there are no edges with horizontal direction <0. Therefore there are no cycles in the graph.

For simplicity we call such a template acyclic. Obviously an acyclic template is still acyclic after rotation or reflection. An example of a template which satisfies the hypothesis of Theorem 7 (after reflection) is the horizontal shadow detector (Table I, (c)).

Acyclic templates can neither be strictly sign-symmetric (unless it is the zero template) nor cell-linking.

By considering an L-layer $M \times N$ CNN as an $L \times M \times N$ array of cells, the definition of a template graph and a CNN graph can be extended to a multiple-layer CNN. It is easy to show that the CNN graph of a feedforward multiple-layer CNN consisting of linear threshold layers[16] is acyclic.

8. CONCLUDING REMARKS

In this paper we have extended the stability results in Reference 5 to a class of CNN templates which are not necessarily positive cell-linking. Among them is a subclass of strictly sign-symmetric templates. Some

† We use the piecewise linear function $\frac{1}{2}[|x+1| - |x-1|]$ instead of an approximation for $\hat{f}(x)$ in this proof.

strictly sign-symmetric templates whose stability we do not know are the following:

−	−	−
−	A	−
−	−	−

+	+	+
−	A	−
+	+	+

+	−	+
+	A	+
+	−	+

−	+	−
+	A	+
−	+	−

These four templates are equivalent and belong to the same equivalence class in the sense that they can be transformed into each other through rotations, reflections or the three transformations defined in Section 5. If one of them can be shown to be stable almost everywhere (or unstable), then so are the other three.

Similarly, the following template and its equivalent templates form another class of strictly sign-symmetric templates whose stability we do not know at present:

−	−	P
−	A	−
P	−	−

A third class of strictly sign-symmetric templates consists of the following template and its equivalent templates:

+	0	−
+	A	+
−	0	+

These three classes, together with the templates in Section 6, form all the 3×3 strictly sign-symmetric templates. Out of the 81 possible types of strictly sign-symmetric 3×3 templates, 53 have been proved to be stable almost everywhere. The rest form the three classes above.

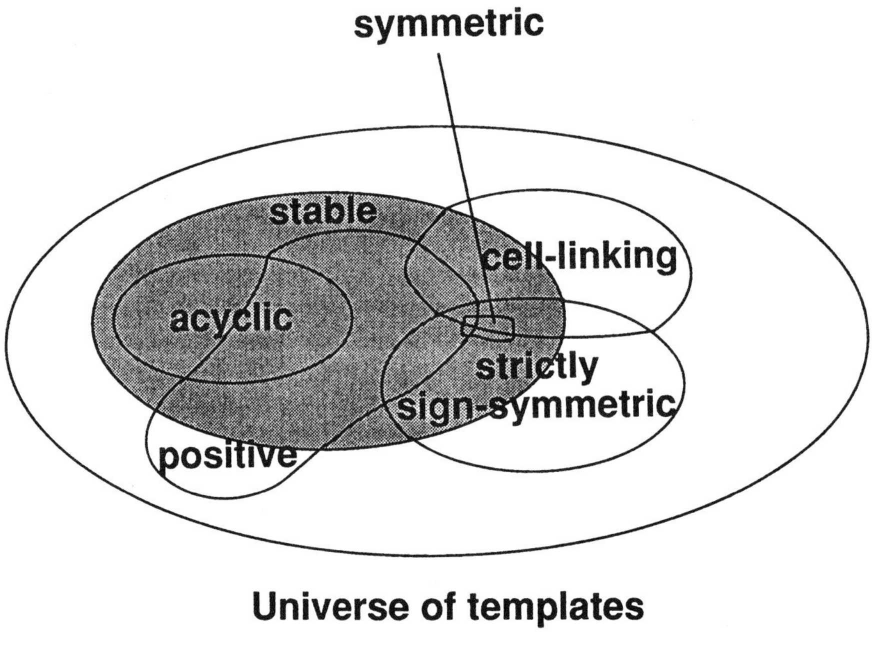

Figure 8. Venn diagram illustrating the relationship between the different classes of templates. The universe of templates stable almost everywhere is shown shaded

Similarly, we find that out of the 6561 possible types of 3×3 templates with $A(i, j; i, j) > R_x^{-1}$, 965 have been proved to be stable almost everywhere. This forms a relatively large subset of the universe of all 3×3 templates. Figure 8 illustrates the relationship between the various types of templates discussed. For the definitions of the different classes see Definitions 5, 7 and 11 and Section 7. The yet unsolved problem is to identify the *complete* universe of *all stable* CNNs.†

Remark 9 The above results can be extended to non-linear templates which are strictly increasing and C^1 in the non-local variable.[17]

ACKNOWLEDGEMENTS

The authors would like to thank Professor Morris Hirsch and Mr. Bertram Shi for helpful discussions.

This work is supported in part by the Office of Naval Research under grant N00014-89-J-1402, by the National Science Foundation under grant MIP 89-12639, by the Semiconductor Research Corporation under contract 91-DC-008 and by the George C. and Helen N. Pardee Scholarship.

REFERENCES

1. L. O. Chua and L. Yang, 'Cellular neural networks: theory', *IEEE Trans. Circuits and Systems*, **CAS-35**, 1257–1272 (1988).
2. L. O. Chua and L. Yang, 'Cellular neural networks: applications', *IEEE Trans. Circuits and Systems*, **CAS-35**, 1273–1290 (1988).
3. T. Matsumoto, L. O. Chua and T. Yokohama, 'Image thinning with a cellular neural network', *IEEE Trans. Circuits and Systems*, **CAS-37**, 638–640 (1990).
4. T. Roska, T. Boros, P. Thiran and L. O. Chua, 'Detecting simple motion using cellular neural networks', *Proc. IEEE Int. Workshop on Cellular Neural Networks and Their Applications*, 1990, IEEE, New York, 1991, pp. 127–138.
5. L. O. Chua and T. Roska, 'Stability of a class of nonreciprocal cellular neural networks', *IEEE Trans. Circuits and Systems*, **CAS-37**, 1520–1527 (1990).
6. M. W. Hirsch, 'Systems of differential equations that are competitive or cooperative. II: Convergence almost everywhere', *SIAM J. Math. Anal.*, **16**, 423–439 (1985).
7. L. Vandenberghe, S. Tan and J. Vandewalle, 'Cellular neural networks: dynamic properties and adaptive learning algorithm', in L. B. Almeida and C. J. Wellekens (eds), *Proc. EURASIP Workshop on Neural Networks, Lecture Notes in Computer Science*, Vol. 412, Springer, Berlin, New York, 1990, pp. 141–150.
8. M. W. Hirsch, 'Convergent activation dynamics in continuous time networks', *Neural Networks*, **2**, 331–349 (1989).
9. M. W. Hirsch, 'Convergent in neural nets', in M. Caudill and C. Butler (eds), *Proc. IEEE First Int. Conf. on Neural Networks*, IEEE, New York, 1987, pp. 115–125.
10. T. Matsumoto, L. O. Chua and H. Suzuki, 'CNN cloning template: connected component detector', *IEEE Trans. Circuits and Systems*, **CAS-37**, 633–635 (1990).
11. T. Matsumoto, L. O. Chua and R. Furukawa, 'CNN cloning template: hole-filler', *IEEE Trans. Circuits and Systems*, **CAS-37**, 635–638 (1990).
12. T. Matsumoto, L. O. Chua and H. Suzuki, 'CNN cloning template: shadow detector', *IEEE Trans. Circuits and Systems*, **37**, 1070–1073 (1990).
13. G. Seiler, 'Small object counting with cellular neural networks', *Proc. IEEE Int. Workshop on Cellular Neural Networks and Their Applications*, 1990, IEEE, New York, 1991, pp. 114–123.
14. L. O. Chua and C. W. Wu, 'On the universe of stable cellular neural networks', *ERL Memorandum UCB/ERL M91/31*, University of California, Berkeley, 1991.
15. H. L. Smith, 'Systems of ordinary differential equations which generate an order preserving flow. A survey of results', *SIAM Rev.*, **30**, 87–113 (1988).
16. L. O. Chua and B. E. Shi, 'Multiple layer cellular neural networks: a tutorial', in E. F. Deprettere (ed.), *Algorithms and Parallel VLSI Architectures*, Elsevier, Amsterdam, 1991, pp. 137–168.
17. T. Roska and L. O. Chua, 'Cellular neural networks with nonlinear and delay-type template elements', *Proc. IEEE Int. Workshop on Cellular Neural Networks and Their Applications*, IEEE, New York, 1991, pp. 12–25.

† In general, the stability of a CNN cannot be found solely by looking at the sign pattern of the A-templates. For example, the stability of | + | A | − | templates seems to depend on the actual values of the template.[5] Furthermore, the input values and the B-templates can also play a role in the stability of a CNN. What we have presented here are A-templates which are stable (almost everywhere) under any constant input and any B-template while specifying only the signs of the elements in the A-template.

A PATH-FOLLOWING METHOD FOR FINDING MULTIPLE EQUILIBRIUM POINTS IN CELLULAR NEURAL NETWORKS†

LIEVEN VANDENBERGHE AND JOOS VANDEWALLE

Department of Electrical Engineering, ESAT Laboratory, Katholieke Universiteit Leuven, Kardinaal Mercierlaan 94, B-3001 Heverlee, Belgium

SUMMARY

Owing to the piecewise linear character of the non-linearities, the computation of equilibrium points in a cellular neural network could be accomplished in a trivial way by solving 3^n linear systems (where n is the number of neurons). This is clearly infeasible even for moderate n and it is the purpose of this paper to present a more efficient path-following method for computing equilibrium points. The method is guaranteed to find at least one solution. The possibility of finding all solutions is also investigated and this leads to new theoretical insight to the relation between the properties of the interconnection matrix and the number of equilibrium points.

1. INTRODUCTION

Cellular neural networks depend crucially on non-linear dynamics, even if stability can be proved for many important examples. Proper operation often requires the existence of multiple equilibrium points and it is therefore important to have efficient analysis tools for obtaining a global picture of the dynamic behaviour, the equilibrium pattern and the basins of attraction in a given network. This paper is concerned with the problem of finding equilibrium points in nets described by the equation[1]

$$\mathrm{d}\boldsymbol{\xi}/\mathrm{d}t = -\boldsymbol{\xi} + \mathbf{AS}(\boldsymbol{\xi}) + \mathbf{Bu} + \boldsymbol{\chi} \tag{1}$$

where $\boldsymbol{\xi}$ is the n-dimensional state vector, $\mathbf{u}$ is an external input, $\boldsymbol{\chi}$ is a constant vector, $\mathbf{A}$ and $\mathbf{B}$ are the feedback and feedforward coefficient matrices respectively and $\mathbf{S}(\cdot)$ is the diagonal mapping defined by (see Figure 1(a))

$$S_i(\mathbf{x}) = \begin{cases} 1 & \text{if } x_i \geqslant 1 \\ x_i & \text{if } -1 \leqslant x_i \leqslant 1 \\ -1 & \text{if } x_i \leqslant -1 \end{cases}$$

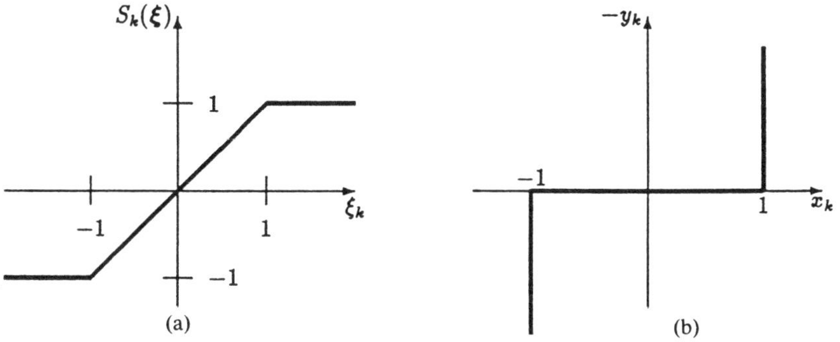

Figure 1. (a) Non-linear function $\mathbf{S}(\cdot)$. (b) Constitutive relation of the resistors connected to the n-port $\mathbf{y} = \mathbf{Dx} + \mathbf{c}$

† Part of this work has been presented at the CNNA-90 Workshop in Budapest[20] and ECCTD-91 in Copenhagen.[21]

In addition, the matrices **A** and **B** have a very sparse and structured form owing to the 'template' structure of the coefficients. This will not be taken into account here, although exploitation of these properties should increase the computational efficiency considerably.

2. AN EQUIVALENT PROBLEM AND THE INTERPRETATION IN THE STATE SPACE

Computing equilibrium points of (1) amounts to solving the set of piecewise linear equations

$$\boldsymbol{\xi} = \mathbf{A}\mathbf{S}(\boldsymbol{\xi}) + \mathbf{B}\mathbf{u} + \boldsymbol{\chi} \tag{2}$$

For ease of notation the $n \times n$ matrix **A** will be split as $\mathbf{I} - \mathbf{D}$, where **I** is the identity matrix. Since the input **u** is kept constant during the operation of the network, the vector $\mathbf{Bu} + \boldsymbol{\chi}$ will be abbreviated as one single constant, $-\mathbf{c}$. Also, the output of the non-linearity **S** will be written as $\mathbf{x} = \mathbf{S}(\boldsymbol{\xi})$. This vector **x** is constrained to lie in the cube $C = \{\mathbf{x} \in \mathbb{R}^n \mid -1 \leqslant x_k \leqslant 1\}$. With these conventions equation (2) can be written as

$$\mathbf{x} = \mathbf{S}(\mathbf{x} - \mathbf{D}\mathbf{x} - \mathbf{c}) \tag{3}$$

and this can easily be shown to be equivalent to the following: a point $\mathbf{x}^* \in C$ is an equilibrium point iff for $k = 1, \ldots, n$

$$\begin{aligned} x_k^* = 1 \quad &\Rightarrow \quad (\mathbf{D}\mathbf{x}^* + \mathbf{c})_k \leqslant 0 \\ -1 < x_k^* < 1 \quad &\Rightarrow \quad (\mathbf{D}\mathbf{x}^* + \mathbf{c})_k = 0 \\ x_k^* = -1 \quad &\Rightarrow \quad (\mathbf{D}\mathbf{x}^* + \mathbf{c})_k \geqslant 0 \end{aligned} \tag{4}$$

Equations (4) can be seen as the description of a resistive n-port $\mathbf{y} = \mathbf{D}\mathbf{x} + \mathbf{c}$, terminated by n resistors with characteristics as depicted in Figure 1(b).

Clearly, this problem has much in common with diode–resistor circuits and the method studied in this paper is closely related to the globally convergent methods developed for the analysis of diode–resistor circuits.[2,3] Problem (4) is also known as a *linear complementarity problem with upper and lower bounds* and the method presented here is based on recent work in Reference 4, with an extension for finding multiple solutions.

The equilibrium condition (3) has a simple geometrical meaning. First note that the mapping $\mathbf{S}(\cdot)$ performs a projection on the cube C as shown in Figure 2(a). With this in mind, equation (3) can be visualized as in Figure 2(b).

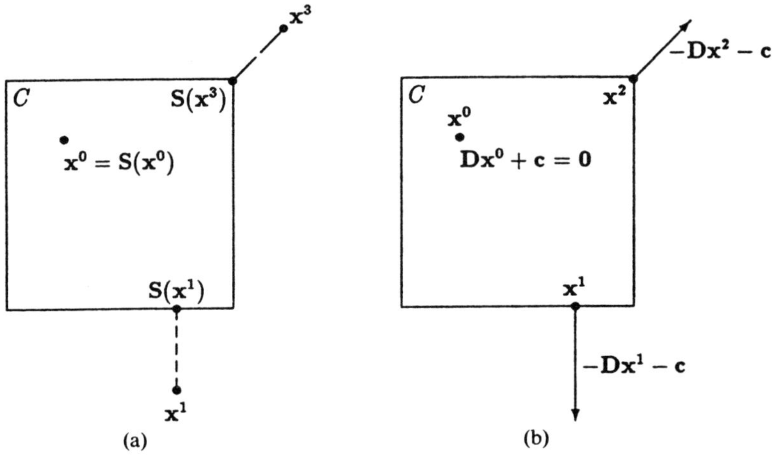

Figure 2. (a) Geometrical interpretation of the mapping $\mathbf{S}(\cdot)$ as the projection on the cube C. (b) Points $\mathbf{x}^0$, $\mathbf{x}^1$ and $\mathbf{x}^2$ are equilibrium points of (1)

3. BASIC PRINCIPLE OF THE ALGORITHM

3.1. Continuous deformation of the equations

The idea in Reference 4 is based on continuous deformation. Problem (4) is embedded in a one-parameter family of similar problems by replacing the characteristics of Figure 1 by those in Figure 3, in which $0 \leqslant t < \infty$. The algorithm starts with $t = 0$, $\mathbf{x} = 0$ and then follows a one-dimensional piecewise linear curve by varying t. Each crossing of the path with the boundary of the cube C (i.e. each point on the path with $t = 1$) corresponds to a solution. Eventually the path goes to infinity or gets stuck in a solution to $\mathbf{Dx} + \mathbf{c} = \mathbf{0}$. In the latter case, if the endpoint lies inside the cube C, it is a solution to (4) as well. In the former case it will be shown below that the algorithm can still be continued provided that the matrix $\mathbf{D}$ is non-singular.

Let us first introduce the concept of *sign vector* which will be needed for a formal description of the method: a sign vector $\mathbf{s}$ is a non-zero vector from $\{-1, 0, 1\}^n$ and indicates on which part of the three-segment characteristic of Figure 3 each pair (x_k, y_k) lies:

$$\begin{aligned} s_k = 1 &\quad \Rightarrow \quad x_k = t \text{ and } y_k \leqslant 0 \\ s_k = 0 &\quad \Rightarrow \quad -t \leqslant x_k \leqslant t \text{ and } y_k = 0 \\ s_k = -1 &\quad \Rightarrow \quad x_k = -t \text{ and } y_k \geqslant 0 \end{aligned} \tag{5}$$

The mechanism of the method will now be illustrated by two examples in $\mathbb{R}^2$. A more formal description follows.

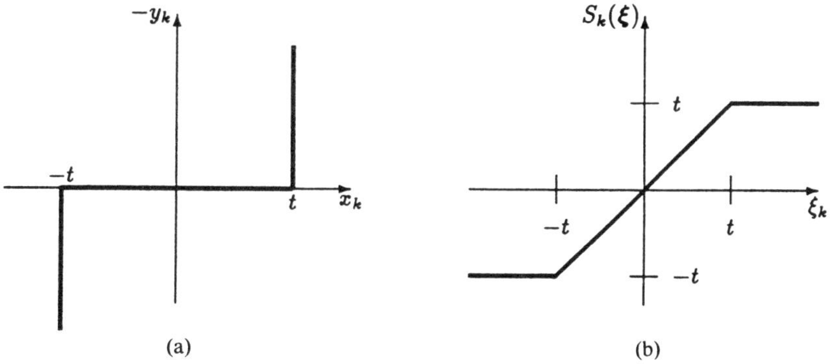

Figure 3. (a) Modified constitutive relation and (b) modified function $\mathbf{S}(\cdot)$

Example 1

We take

$$\mathbf{D} = \begin{bmatrix} 2 & 1 \\ -1/3 & 1/2 \end{bmatrix}, \qquad \mathbf{c} = \begin{bmatrix} 2 \\ 1 \end{bmatrix}$$

The path followed is shown in Figure 4(a) and consists of two segments.

(i) On the first segment $\boxed{0}-\boxed{1}$ the sign vector is $(-1, -1)$ because $y_1 > 0$ and $y_2 > 0$ at the origin. The path is described by

$$x_1 = x_2 = -t, \qquad y_1 = -3t + 2, \qquad y_2 = -\tfrac{1}{6}t + 1$$

This is only valid up to point $\boxed{1}$ because y_1 and y_2 must remain positive if the sign vector is $(-1, -1)$.

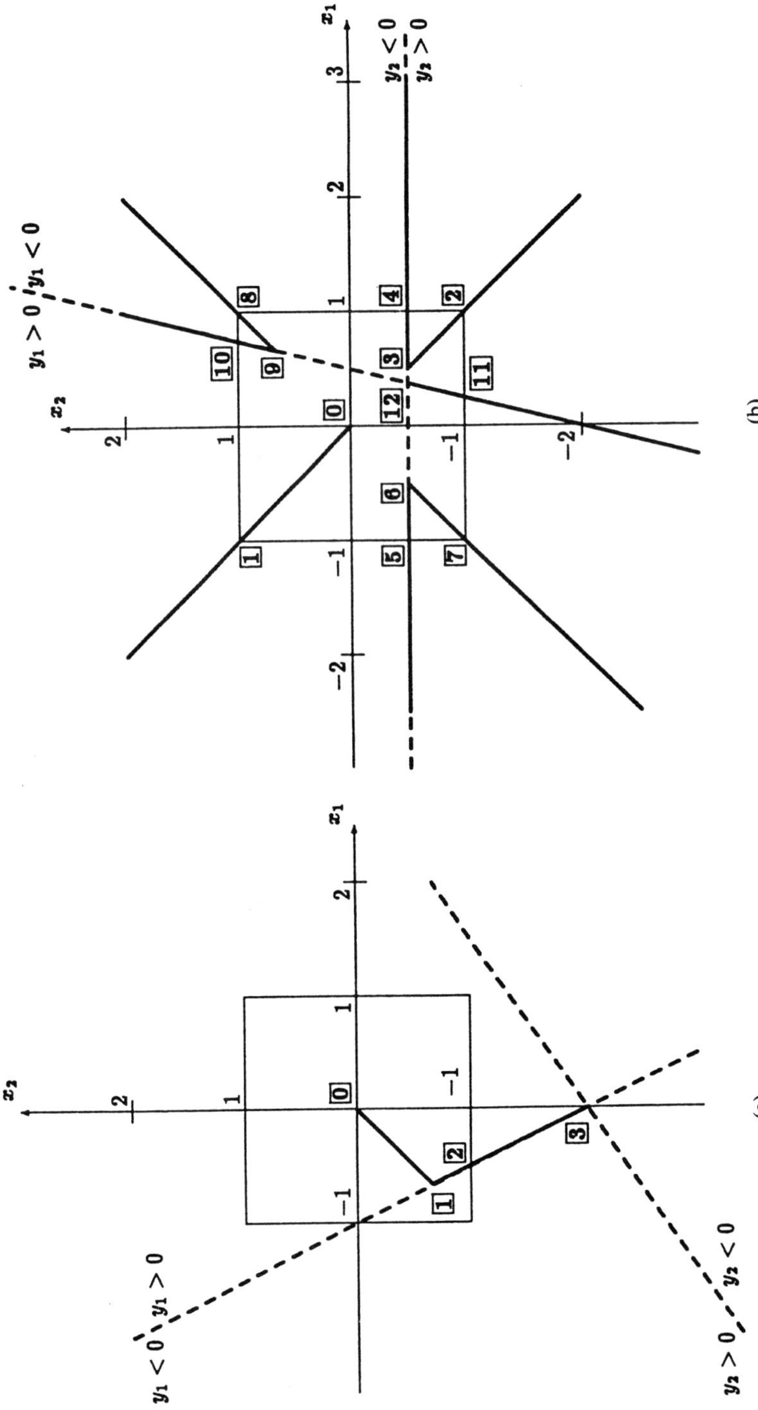

Figure 4. (a) Example 1 and (b) Example 2

(ii) For the second segment [1]–[3] the sign vector becomes (0, −1) and one has

$$x_2 = -t, \qquad y_1 = 0, \qquad y_2 = -\tfrac{1}{3}x_1 - \tfrac{1}{2}t + 1$$

which is valid as long as $-t \leqslant x_1 \leqslant t$ and $y_2 \geqslant 0$. At point [3] y_2 becomes zero and the algorithm ends at this point since increasing t no longer has an effect on $\mathbf{x}$.

(iii) The segment [1]–[3] intersects the boundary of the unit cube at $(x_1, x_2) = (-0{\cdot}5, -1)$ (point [2]). As expected, this is a solution to (4) since $(y_1, y_2) = (0, \tfrac{2}{3})$. It can also be verified that this is the only solution.

Example 2

A more complicated example is shown in Figure 4(b):

$$\mathbf{D} = \begin{bmatrix} -1 & 0{\cdot}25 \\ 0 & -1 \end{bmatrix}, \qquad \mathbf{c} = \begin{bmatrix} 0{\cdot}5 \\ -0{\cdot}5 \end{bmatrix}$$

The path now consists of eight segments.

(i) Since $y_1 > 0$ and $y_2 < 0$ at the origin, the algorithm is started with sign vector (−1, 1) and we have

$$x_1 = -t, \qquad x_2 = t, \qquad y_1 = 1{\cdot}25t + 0{\cdot}5, \qquad y_2 = -t - 0{\cdot}5$$

This segment is unbounded and crosses the boundary of C at the solution $(x_1, x_2) = (-1, 1)$ (point [1]).

(ii) The key observation needed to continue the algorithm is that if there is an unbounded segment with sign vector **s**, there can be expected to be one with sign vector −**s** as well. This needs to be proved in general, but it is clear in the two-dimensional case of Figure 4(b). Reversing **s** to **s** = (1, −1), one can proceed by decreasing t along the path

$$x_1 = t, \qquad x_2 = -t, \qquad y_1 = -1{\cdot}25t + 0{\cdot}5, \qquad y_2 = t - 0{\cdot}5$$

This satisfies the restrictions of (5) as long as $y_1 \leqslant 0$ and $y_2 \geqslant 0$, i.e. up to point $(x_1, -x_2) = (0{\cdot}5, 0{\cdot}5)$ (point [3]). In the meantime, a second solution has been found at $(x_1, x_2) = (1, -1)$ (point [2]).

(iii) At point [3] the sign vector switches to **s** = (1, 0) and the description of the path changes to

$$x_1 = t, \qquad y_2 = 0, \qquad y_1 = -t + 0{\cdot}25x_2 - 0{\cdot}5$$

with the restriction that $-t \leqslant x_2 \leqslant t$. This segment is again unbounded.

(iv) Continuing this procedure until a zero of $\mathbf{Dx} + \mathbf{c}$ (point [12]) is reached, a total of nine solutions are found (points [1], [2], [4], [5], [7], [8], [10], [11] and [12]). It can also be verified that the algorithm finds all solutions in this case.

In general the algorithm starts at the origin and generates a sequence of sign vectors **s**. Each sign vector defines a linear segment by (5). Two consecutive segments either have one endpoint in common or are unbounded and related by a reflection step. Eventually the algorithm must stop at the point $\mathbf{x} = -\mathbf{D}^{-1}\mathbf{c}$. Each time t takes the value unity, one has an equilibrium point of the neural network, and if the endpoint of the path lies in the interior of C, it is an equilibrium point as well. In general there is no guarantee that all equilibrium points can be found in this way, since besides the path generated by the algorithm there may be cyclic sequences of segments that cannot be detected by the algorithm. A sufficient condition to find all solutions will be given in Section 4.2 and it is hoped that this condition can be further generalized in the future.

A few remarks on the origins of the algorithm may be appropriate at this point. The algorithm is basically Wright's homotopy algorithm for solving sets of non-linear equations $\mathbf{f}(\mathbf{x}) = \mathbf{0}$,[5] which was studied in connection with DC analysis of electrical circuits in Reference 6. The algorithm studied in the present paper is derived from the algorithm in Reference 5 applied to the linear system $\mathbf{f}(\mathbf{x}) = \mathbf{Dx} + \mathbf{c}$, but it bears a number of distinct features as compared to the general procedure. Firstly, one is not interested in the endpoint of the path (i.e. in the zero of $\mathbf{Dx}+\mathbf{c}$) but in all crossings of the path with the boundary of the cube C. Secondly, the algorithm will be extended here by applying the reflection steps as in the examples. The idea of using this kind of reflection step in homotopy algorithms has been suggested in several places.[7,8]

3.2. *Description of the algorithm and the path traced by it*

3.2.1. Notation We now proceed to give a formal description of the algorithm. The following notational conventions will be used. If $\mathbf{A}$ is an $m \times n$ matrix, $\alpha \subseteq \{1, \ldots, m\}$ and $\beta \subseteq \{1, \ldots, n\}$, then the submatrix consisting of the rows of $\mathbf{A}$ indexed by α and the columns of $\mathbf{A}$ indexed by β will be written as $\mathbf{A}_{\alpha,\beta}$. Similarly, $\mathbf{A}_{\cdot,\beta}$ and $\mathbf{A}_{\alpha,\cdot}$ will refer to the matrix formed by the columns indexed by β and the matrix formed by the rows indexed by α respectively. For the sake of conciseness, $\alpha \cup \{k\}$ and $\alpha \setminus \{k\}$ will be written as $\alpha \cup k$ and $\alpha \setminus k$.

As can be seen from the examples in Section 3.1, the algorithm traces a piecewise linear path with each segment uniquely related to a particular sign vector. The defining equations for the segment corresponding to sign vector $\mathbf{s}$ can be expressed as follows. Define $\alpha = \{k \mid s_k = 0\}$ and $\beta = \{k \mid s_k \neq 0\}$. From Figure 3 and equations (5) it can be seen that $\mathbf{x}_\beta = \mathbf{s}_\beta t$, $\mathbf{y}_\alpha = \mathbf{0}$ and the independent variables are $\mathbf{x}_\alpha$, $\mathbf{y}_\beta$ and t. These will be gathered in one single $n+1$ vector $\mathbf{z}$:

$$\mathbf{z}_\alpha = \mathbf{x}_\alpha, \qquad \mathbf{z}_\beta = \mathbf{y}_\beta = (\mathbf{Dx}+\mathbf{c})_\beta, \qquad z_{n+1} = t$$

Also observe from Figure 3 that $t = z_{n+1}$ is in fact the infinity norm $\|\mathbf{x}\|_\infty$ of $\mathbf{x}$.

Define an $n \times (n+1)$ matrix $\mathbf{M}^{\mathbf{s}}$ as

$$\mathbf{M}^{\mathbf{s}}_{\cdot,\alpha} = \mathbf{D}_{\cdot,\alpha}, \qquad \mathbf{M}^{\mathbf{s}}_{\cdot,\beta} = -\mathbf{I}_{\cdot,\beta}, \qquad \mathbf{M}^{\mathbf{s}}_{\cdot,n+1} = \sum_{k \in \beta} \mathbf{D}_{\cdot,k} s_k = \mathbf{D}_{\cdot,\beta}\mathbf{s}_\beta \tag{6}$$

With this notation the segment of the path corresponding to sign vector $\mathbf{s}$ is determined by all solutions to

$$\mathbf{M}^{\mathbf{s}}\mathbf{z} + \mathbf{c} = \mathbf{0} \tag{7}$$

$$-z_{n+1} \leqslant z_k \leqslant z_{n+1} \quad \forall k \in \alpha \tag{8}$$

$$s_k z_k \leqslant 0 \quad \forall k \in \beta \tag{9}$$

$$z_{n+1} \geqslant 0 \tag{10}$$

If $\mathbf{M}^{\mathbf{s}}$ is of rank n, the solution set to equations (7)–(10) must be empty or a one-dimensional segment (closed or semiclosed and unbounded). The degenerate case where $\mathbf{M}^{\mathbf{s}}$ is not of full row rank should be precluded by the use of a suitable perturbation scheme (see below).

The basic step of the algorithm consists of moving from one breakpoint of the piecewise linear path to the next. This is called pivoting. It is convenient to express each segment as a set of points $\mathbf{z} = \hat{\mathbf{z}} + \gamma\tilde{\mathbf{z}}$ $(0 \leqslant \gamma \leqslant \bar{\gamma})$ satisfying equations (7)–(10). The vector $\hat{\mathbf{z}}$ is a breakpoint. It satisfies (7)–(10) and one of the inequalities in (8)–(10) with equality (exactly one of the inequalities except in the case of degeneracy). The vector $\tilde{\mathbf{z}}$ gives the direction of the segment. It must be computed from $\mathbf{M}^{\mathbf{s}}\tilde{\mathbf{z}} = \mathbf{0}$. In order to give $\tilde{\mathbf{z}}$ the right orientation, one considers the inequality in (8)–(10) that is satisfied with equality by $\hat{\mathbf{z}}$.

(i) If $-\hat{z}_{n+1} = \hat{z}_k$ for some $k \in \alpha$, then take $\tilde{z}_{n+1} + \tilde{z}_k > 0$.
(ii) If $\hat{z}_{n+1} = \hat{z}_k$ for some $k \in \alpha$, then take $\tilde{z}_{n+1} - \tilde{z}_k > 0$.
(iii) If $\hat{z}_k = 0$ for some $k \in \beta$, then take $s_k \tilde{z}_k < 0$.
(iv) If $\hat{z}_{n+1} = 0$, then take $\tilde{z}_{n+1} > 0$.

The initial segment is given by $s_k = 1$ if $c_k < 0$, $s_k = -1$ if $c_k > 0$ (assuming $c_k \neq 0$, $\forall k$; see below), $\hat{\mathbf{z}} = [\mathbf{c}^{\mathrm{T}}\ 0]^{\mathrm{T}}$ and $\tilde{\mathbf{z}} = [(\mathbf{Ds})^{\mathrm{T}}\ 1]^{\mathrm{T}}$.

3.2.2. Traversing a bounded segment Assume that one has given an endpoint $\hat{\mathbf{z}}$ and a direction $\tilde{\mathbf{z}}$ of the segment corresponding to sign vector $\mathbf{s}$. Then one takes the following steps to go to the next segment.

Let Γ be defined by

$$\Gamma = \{k \mid k \in \alpha \text{ and } \tilde{z}_{n+1} - \tilde{z}_k < 0\} \cup \{-k \mid k \in \alpha \text{ and } \tilde{z}_{n+1} + \tilde{z}_k < 0\} \cup \{k \mid k \in \beta \text{ and } s_k\tilde{z}_k > 0\} \quad (11)$$

If this set Γ is empty, the segment is unbounded. We first consider the case of a bounded segment; the unbounded case is treated in the next section.

For each element k of Γ a ratio r_k can be computed as

$$r_k = \begin{cases} -(\hat{z}_{n+1} - \hat{z}_k)/(\tilde{z}_{n+1} - \tilde{z}_k) & \text{if } k \in \alpha \\ -(\hat{z}_{n+1} + \hat{z}_k)/(\tilde{z}_{n+1} + \tilde{z}_k) & \text{if } -k \in \alpha \\ -\hat{z}_k/\tilde{z}_k & \text{if } k \in \beta \end{cases} \quad (12)$$

Assume that these ratios have a unique minimum $\bar{\gamma} = r_j = \min_{k \in \Gamma} r_k$. A treatment of ties is given below. Then the second breakpoint of the segment is given by $\hat{\mathbf{z}} + \bar{\gamma}\tilde{\mathbf{z}}$. From both breakpoints it can easily be checked whether the segment intersects the boundary of C (and hence contains an equilibrium point).

The index j where the minimum occurs determines the next sign vector $\mathbf{s}$: only one component changes according to

$$\begin{aligned} j \in \beta \quad &\Rightarrow \quad s_j := 0 \\ j \in \alpha \quad &\Rightarrow \quad s_j := 1 \\ -j \in \alpha \quad &\Rightarrow \quad s_{-j} := -1 \end{aligned}$$

3.2.3. The case of an unbounded segment If Γ defined in (11) is empty, the points $\hat{\mathbf{z}} + \gamma\tilde{\mathbf{z}}$ satisfy (7)–(10) for all $\gamma \geqslant 0$ and hence lie on an unbounded segment. From (11) one has that

$$\tilde{z}_{n+1} - \tilde{z}_k \geqslant 0 \quad \text{and} \quad \tilde{z}_{n+1} + \tilde{z}_k \geqslant 0 \quad \forall k \in \alpha \quad (13)$$

$$s_k\tilde{z}_k \leqslant 0 \quad \forall k \in \beta \quad (14)$$

Assume that none of these inequalities holds with equality. This is another non-degeneracy assumption that will be relaxed below. Also, we know that $\tilde{z}_{n+1} > 0$, since $z_{n+1} = \| \mathbf{x} \|_\infty$ and the path is unbounded. Under these assumptions it can be verified that, provided γ is sufficiently negative, the points $\mathbf{z}$ defined by

$$z_k = \hat{z}_k + \gamma\tilde{z}_k, \quad k = 1, \ldots, n \qquad z_{n+1} = -(\hat{z}_{n+1} + \gamma\tilde{z}_{n+1})$$

satisfy equations (7)–(10) if $\mathbf{s}$ is replaced by $-\mathbf{s}$. This is the reflection step from Example 2 above. The required value of γ is the minimum of the ratios

$$r_k = \begin{cases} -(\hat{z}_{n+1} - \hat{z}_k)/(\tilde{z}_{n+1} - \tilde{z}_k) & \text{if } k \in \alpha \\ -(\hat{z}_{n+1} + \hat{z}_k)/(\tilde{z}_{n+1} + \tilde{z}_k) & \text{if } k \in \alpha \\ -\hat{z}_k/\tilde{z}_k & \text{if } k \in \beta \end{cases}$$

For each unbounded segment one can easily check whether it contains an equilibrium point.

3.2.4. Degeneracy Above we have made two non-degeneracy assumptions. The first type, a tie in the minimum ratio test (12), could be resolved by subtracting an infinitesimal positive vector $[\varepsilon \ \varepsilon^2 \ \ldots \ \varepsilon^n]^{\mathrm{T}}$ from $\mathbf{c}$. This would lead to standard lexicographical pivoting rules.[9] The second type of degeneracy occurs when Γ in (11) is empty but one of the inequalities (13) or (14) holds with equality. This makes it unclear whether a reflection must be applied or not, since the situation might as well be interpreted as a 'breakpoint at infinity'. Perturbing $\mathbf{c}$ does not help since the problem concerns the homogeneous system $\mathbf{M}^{\mathbf{s}}\tilde{\mathbf{z}} = \mathbf{0}$. An odd perturbation of the mapping $\mathbf{Dx} + \mathbf{c}$ would be required instead. A practical scheme of this nature was proposed in Reference 7 in connection with a different algorithm and will be modified below to suit our purposes.

Two assumptions will be made: *all components of the vector* $\mathbf{c}$ *are non-zero* and *the matrix* $\mathbf{D}$ *is non-singular*. Neither of these assumptions is very restrictive. As for the first one, in most applications a small random number can be added to a zero component c_k without much harm. In the case of a singular $\mathbf{D}$ the algorithm can still be started, but it will end at a point which does not satisfy $\mathbf{Dx} + \mathbf{c} = \mathbf{0}$.

To avoid degeneracy, one should add the requirement that the solutions to (7)–(10) be robust against small perturbations of the equations. The idea used here is to perturb the equations

$$x_k = s_k t \quad \forall k \in \beta, \qquad -t \leqslant x_k \leqslant t \quad \forall k \in \alpha$$

which define Figure 3, as

$$x_k = s_k(t + \varepsilon^k) \quad \forall k \in \beta, \qquad -t - \varepsilon^k \leqslant x_k \leqslant t + \varepsilon^k \quad \forall k \in \alpha$$

for infinitesimal positve ε.

It can be shown that provided the matrix $\mathbf{D}$ is non-singular, this perturbation is sufficient to resolve the ambiguities caused by degeneracy.[10]

4. FURTHER PROPERTIES OF THE PATH AND THE EQUILIBRIA

From now on it will be assumed that the degenerate situations described above do not occur. We also assume that the path crosses the boundary of the cube C transversally, i.e. that there are no continua of equilibrium points.

Let $\mathcal{Z}(\mathbf{s})$ be the solution set to the equations (7)–(10) for a particular choice of $\mathbf{s}$ and let $\mathcal{Z} = \bigcup_{\mathbf{s} \neq \mathbf{0}} \mathcal{Z}(\mathbf{s})$. The purpose of this section is to describe the geometry of the set $\mathcal{Z}$.

As was done above, each segment $\mathcal{Z}(\mathbf{s})$ can be parametrized by its 'arc-length' γ, i.e. it can be expressed as $\mathbf{z} = \hat{\mathbf{z}} + \gamma\tilde{\mathbf{z}}$, $0 \leqslant \gamma \leqslant \bar{\gamma}$, and $\tilde{\mathbf{z}}$ satisfies

$$\mathbf{D}_{\alpha,\alpha}\tilde{\mathbf{z}}_\alpha + \mathbf{D}_{\alpha,\beta}\mathbf{s}_\beta\tilde{z}_{n+1} = \mathbf{0}, \qquad \mathbf{D}_{\beta,\alpha}\tilde{\mathbf{z}}_\alpha - \tilde{\mathbf{z}}_\beta + \mathbf{D}_{\beta,\beta}\mathbf{s}_\beta\tilde{z}_{n+1} = \mathbf{0}$$

These equations can be solved by Cramer's rule to give

$$\tilde{z}_k = -\delta \det\begin{pmatrix} \mathbf{D}_{\alpha\backslash k,\alpha\backslash k} & \mathbf{D}_{\alpha\backslash k,\beta}\mathbf{s}_\beta \\ \mathbf{D}_{k,\alpha\backslash k} & \mathbf{D}_{k,\beta}\mathbf{s}_\beta \end{pmatrix}, \quad k \in \alpha \tag{15}$$

$$\tilde{z}_k = \delta \det\begin{pmatrix} \mathbf{D}_{\alpha,\alpha} & \mathbf{D}_{\alpha,\beta}\mathbf{s}_\beta \\ \mathbf{D}_{k,\alpha} & \mathbf{D}_{k,\beta}\mathbf{s}_\beta \end{pmatrix}, \quad k \in \beta \tag{16}$$

$$\tilde{z}_{n+1} = \delta \det \mathbf{D}_{\alpha,\alpha} \tag{17}$$

(where the determinant of an empty matrix is assumed to be $+1$). In these equations $\delta = \pm 1$ depending on the orientation in which the segment is traversed. Initially $\delta = 1$, since $\tilde{z}_{n+1}$ is increasing on the first segment while α is empty. It can be shown that δ remains the same on two connected segments if they are traversed in the same sense and that δ is reversed with each reflection of the algorithm. This is stated by the following two theorems.

Theorem 1

Let $\mathcal{Z}(\mathbf{s}^1)$ and $\mathcal{Z}(\mathbf{s}^2)$ be two segments that share one endpoint $\hat{\mathbf{z}}$ and let the corresponding vectors $\tilde{\mathbf{z}}^1$ and $\tilde{\mathbf{z}}^2$ be given by equations (15)–(17) with δ^1, $\mathbf{s}^1$, α^1, β^1 and δ^2, $\mathbf{s}^2$, α^2, β^2 respectively. Then if both segments are traversed in the same sense, one has $\delta^1 = \delta^2$.

Proof The sign vectors $\mathbf{s}^1$ and $\mathbf{s}^2$ can differ in only one component, say component j. Assume $s_j^1 \neq 0$ and $s_j^2 = 0$. The other possibilities can be proved similarly. For ease of notation we write $\alpha = \alpha^1$, $\beta = \beta^1$, $\mathbf{s} = \mathbf{s}^1$, $\alpha^2 = \alpha \cup j$ and $\beta^2 = \beta \backslash j$.

By (16) one knows that

$$s_j\tilde{z}_j^1 = s_j\delta^1 \det\begin{pmatrix} \mathbf{D}_{\alpha,\alpha} & \mathbf{D}_{\alpha,\beta}\mathbf{s}_\beta \\ \mathbf{D}_{j,\alpha} & \mathbf{D}_{j,\beta}\mathbf{s}_\beta \end{pmatrix}$$

and hence by (15) and (17)

$$\begin{aligned} s_j\tilde{z}_j^1 &= s_j\delta^1 \det\begin{pmatrix} \mathbf{D}_{\alpha,\alpha} & \mathbf{D}_{\alpha,\beta\setminus j}\mathbf{s}_{\beta\setminus j} \\ \mathbf{D}_{j,\alpha} & \mathbf{D}_{j,\beta\setminus j}\mathbf{s}_{\beta\setminus j} \end{pmatrix} + \delta^1 \det(\mathbf{D}_{\alpha\cup j,\alpha\cup j}) \\ &= (\delta^1/\delta^2)(\tilde{z}_{n+1}^2 - s_j\tilde{z}_j^2) \end{aligned}$$

Furthermore, one knows that either $s_j\tilde{z}_j^1 > 0$ and $\tilde{z}_{n+1}^2 - s_j\tilde{z}_j^2 > 0$ or $s_j\tilde{z}_j^1 < 0$ and $\tilde{z}_{n+1}^2 - s_j\tilde{z}_j^2 < 0$ if both segments are traversed in the same sense. From this $\delta^1 = \delta^2$ follows. □

Theorem 2

Let $\mathcal{Z}(\mathbf{s}^1)$ and $\mathcal{Z}(\mathbf{s}^2)$ be two unbounded segments with opposite s-vectors $\mathbf{s}^1 = -\mathbf{s}^2$ and let the corresponding vectors $\tilde{\mathbf{z}}^1$ and $\tilde{\mathbf{z}}^2$ be given by equations (15)–(17) with δ^1, $\mathbf{s}^1$, α^1, β^1 and δ^2, $\mathbf{s}^2$, α^2, β^2 respectively. If one segment is traversed in the direction of increasing t and the other in the direction of decreasing t, then one has $\delta^1 = -\delta^2$.

Proof This follows immediately from (17) and from $\alpha^1 = \alpha^2$. □

4.1. The index and the stability of an equilibrium point

In analogy with the definition of the index of solutions of a set of non-linear equations,[11] one can introduce an index for each equilibrium point of the neural network.

Definition 1

If $\mathbf{x}$ satisfies (4) with corresponding sign vector $\mathbf{s}$, i.e.

$$\begin{aligned} x_k = s_k \quad &\text{and} \quad s_k(\mathbf{Dx}+\mathbf{c})_k \leqslant 0 \quad \forall s_k \neq 0 \\ -1 \leqslant x_k \leqslant 1 \quad &\text{and} \quad (\mathbf{Dx}+\mathbf{c})_k = 0 \quad \forall s_k = 0 \end{aligned}$$

then the index of $\mathbf{x}$ is defined as sign det $\mathbf{D}_{\alpha,\alpha}$, where $\alpha = \{k \mid s_k = 0\}$. If the point $\mathbf{x} = -\mathbf{D}^{-1}\mathbf{c}$ is an equilibrium point, its index is defined as sign det $\mathbf{D}$.

The following observations follow from Theorems 1 and 2.

1. The number of solutions found by the algorithm is odd and the algorithm finds at least one solution with index +1.
2. Assume that the algorithm finds one solution $\mathbf{z}^1$ and after r reflections finds a second solution $\mathbf{z}^2$. Then the indices of $\mathbf{z}^1$ and $\mathbf{z}^2$ are opposite if r is even and equal if r is odd.

It is also clear that the spectrum of the matrix $\mathbf{D}_{\alpha,\alpha}$ determines the local stability of the equilibrium point. Equilibrium points with all cells in saturation ($|\xi_k| > 1$, $k = 1, \dots, n$) are always stable. The global stability, however, and the basins of attractions are much harder to assess.

4.2. Uniqueness of equilibrium and the possibility of finding all solutions

From (17) it follows that the path behaves monotonically if det $\mathbf{D}_{\alpha,\alpha} > 0$ for all $\alpha \subset \{1, \dots, n\}$. In this simple case it can be proved that all solutions are found by the algorithm. It should be stressed, however, that in general the parameter t need not increase or decrease monotonically and therefore the condition below is expected to be much too strong.

Theorem 3

If det $\mathbf{D}_{\alpha,\alpha} > 0$ $\forall\alpha \subset \{1, \dots, n\}$ (det $\mathbf{D} > 0$ is not required), then the network has either one or three solutions and all of these are found by the algorithm. If, in addition, det $\mathbf{D} > 0$, the network has a unique equilibrium.

Proof Since det $\mathbf{D}_{\alpha,\alpha} > 0$ $\forall\alpha$, the sign of $\tilde{z}_{n+1}$ can only change because of a reflection (change of δ in (17)). Hence there can be no closed loops of connected segments and the path traced by the algorithm is the only component of $\mathscr{Z}$. In addition, there can be at most one reflection. A reflection will occur if det $\mathbf{D} < 0$. This can be seen from (16): on the final segment of the path, $\mathbf{s}$ has only one non-zero component, say j, and $s_j\tilde{z}_j > 0$. Then with (16) one has $s_j\tilde{z}_j = \delta$ det $\mathbf{D} > 0$. If the algorithm ends before a reflection occurs, one has $\delta = +1$ and hence det $\mathbf{D} > 0$. If it ends after one reflection, one must have det $\mathbf{D} < 0$. In this case of negative determinant of $\mathbf{D}$ there can be one or three equilibrium points depending on the value of $\mathbf{c}$. In the case of positive det $\mathbf{D}$ the number of solutions is always exactly one. □

Example 3

A simple four-dimensional example of a $\mathbf{D}$-matrix with positive principal minors is

$$\mathbf{D} = \begin{pmatrix} d_{11} & 0 & 0 & d_{14} \\ d_{21} & d_{22} & 0 & 0 \\ 0 & d_{32} & d_{33} & 0 \\ 0 & 0 & d_{43} & d_{44} \end{pmatrix}$$

with $d_i > 0$, $i = 1, \dots, 4$. Here det $\mathbf{D} = d_{11}d_{22}d_{33}d_{44} - d_{21}d_{32}d_{43}d_{14}$.

The theorem states that a network where $\mathbf{D}$ has positive principal minors (except det $\mathbf{D}$ itself) can have one or three equilibrium points depending on the value of $\mathbf{c}$. If in such a network one neuron is deleted (i.e. one row and one column of $\mathbf{D}$ are deleted), the equilibrium point becomes structurally unique. In this sense the situation corresponds to the minimal structures known to be needed in non-linear electrical circuits to have multiple equilibrium points.[12–14]

5. NUMERICAL ASPECTS

5.1. Computational cost

A look at the algorithm shows that the computational load per pivot is mainly determined by the calculation of a vector $\tilde{\mathbf{z}}$ satisfying $\mathbf{M}^{\mathbf{s}}\tilde{\mathbf{z}} = \mathbf{0}$ with $\mathbf{M}^{\mathbf{s}}$ given by (6). Because of the structure of $\mathbf{M}^{\mathbf{s}}$, this reduces to solving

$$[\mathbf{D}_{\alpha,\alpha} \;\; \mathbf{D}_{\alpha,\beta}\mathbf{s}_\beta]\begin{bmatrix} \tilde{\mathbf{z}}_\alpha \\ \tilde{z}_{n+1} \end{bmatrix} = \mathbf{0}$$

and then computing $\tilde{\mathbf{z}}_\beta$ from

$$\tilde{\mathbf{z}}_\beta = \mathbf{D}_{\beta,\alpha}\tilde{\mathbf{z}}_\alpha + \mathbf{D}_{\beta,\beta}\mathbf{s}_\beta\tilde{z}_{n+1}$$

In a typical implementation one would therefore store $\mathbf{D}_{\alpha,\alpha}$ in factored form (e.g. LU or QR) and update the factorization at each pivot. This updating consists of adding one row and one column to $\mathbf{D}_{\alpha,\alpha}$ (when a new element is added to α) or deleting one row and one column (when an element is deleted from α) and can be executed in $O(m^2)$ floating point operations if m is the cardinality of α.[15] Since $\mathbf{D}$ is usually very sparse, the efficiency can be further increased by using sparse matrix techniques. This subject is well documented in the literature on large-scale linear programming.

Not much can be said about the required number of pivots, except that it is 3^n in the worst case and probably much less on the average. The problem is reminiscent of the gap between worst-case and average-case complexity of the simplex method for linear programming and of a number of related algorithms.[16,17]

5.2. *Examples*

We first take the connected component detector of Reference 18 with $n = 5$. The matrix **A** is obtained by using the template $[1\ \ 2\ \ -1]$. The external input $\mathbf{u} = \mathbf{0}$. This CNN detects the number of connected black components on a white background (or vice versa). In order to make the network detect the number of black components, it is assumed that a zeroth and an $(n+1)$th cell are added to the left and to the

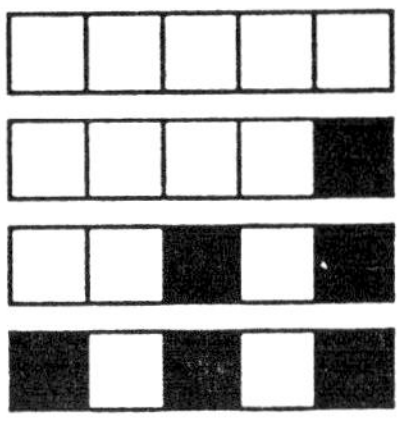

Figure 5. Four equilibrium patterns of the connected component detector

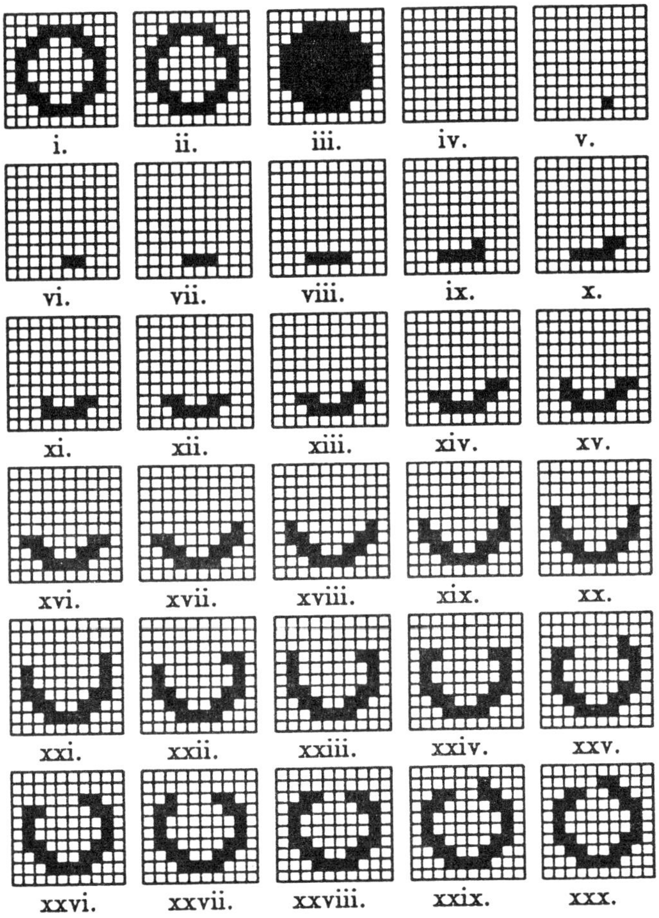

Figure 6. (i) Input to the hole-filling neural network. (ii)–(xxx) Equilibrium points found by the algorithm

right. The states of these cells are kept fixed at -1. The algorithm applied to this problem terminates in 84 pivots with 23 solutions. Only four of these lie on the corners of the hypercube C (Figure 5). These states are stable and correspond to the final states that can be generated under proper operation of the algorithm. The remaining equilibrium points have at least one component with absolute value smaller than unity and are therefore unstable.[1]

As a second example we apply the algorithm to the hole-filling network of Reference 19. The matrix **A** is defined by the template

$$\begin{bmatrix} 0 & 1 & 0 \\ 1 & 2 & 1 \\ 0 & 1 & 0 \end{bmatrix}$$

$B = 4$ (a scalar) and $\chi_k = -1 \quad \forall k$. We take a 10×10 net with the input shown in Figure 6(i). The algorithm finds the 29 equilibrium points shown in Figures 6(ii)–6(xxx). These were all found during the first 10 000 pivots. After 300 000 pivots the programme was interrupted. The equilibrium shown in Figure 6(iii) is the one the network is intended to reach. It is also clear that there must be more than 29 equilibrium points because of the symmetry in the image.

6. CONCLUSIONS

The algorithm presented here can be useful in the analysis of neural networks, e.g. during the design of templates for cellular neural networks. It can be applied to large networks provided that the extreme sparsity and the structure of the coefficients are exploited. The method can also be generalized for neurons with smooth non-linear output functions, but this requires more complicated path-following techniques. In general, there does not seem to be much hope for an efficient way to find all equilibrium points in a given neural network unless appropriate guidelines are followed during the synthesis process. Theorem 3 might be helpful in this respect, although the condition stated there is very strong. Further work on this matter is therefore desirable.

ACKNOWLEDGEMENTS

L. Vandenberghe is a research assistant of the Belgian National Fund for Scientific Research (NFWO). This research work was carried out at the ESAT Laboratory of the Katholieke Universiteit Leuven in the framework of a Concerted Action Project of the Flemish Community entitled Applicable Neural Networks. The scientific responsibility is assumed by its authors.

REFERENCES

1. L. O. Chua and L. Yang, 'Cellular neural networks: theory' and 'Cellular neural networks: applications', *IEEE Trans. Circuits and Systems*, **CAS-35**, 1257–1272 and 1273–1290 (1988).
2. S. N. Stevens and P. M. Lin, 'Analysis of piecewise-linear resistive networks using complementary pivot theory', *IEEE Trans. Circuits and Systems*, **CAS-28**, 429–441 (1981).
3. W. M. G. van Bokhoven, *Piecewise-linear Modelling and Analysis*, Kluwer, Deventer, 1980.
4. G. van der Laan and A. J. J. Talman, 'An algorithm for the linear complementarity problem with upper and lower bounds', *J. Optim. Theory Appl.*, **62**, 151–163 (1989).
5. A. H. Wright, 'The octahedral algorithm, a new simplicial fixed point algorithm', *Math. Programm.*, **21**, 47–69 (1981).
6. L. Vandenberghe and J. Vandewalle, 'Variable dimension algorithms for solving resistive circuits', *Int. J. Cir. Theor. Appl.*, **18**, 443–474 (1990).
7. M. J. Todd and A. H. Wright, 'A variable-dimension simplicial algorithm for the antipodal fixed-point problem', *Numer. Funct. Anal. Optim.*, **2**, 155–186 (1980).
8. G. van der Laan, 'On the existence and approximation of zeroes', *Math. Programm.*, **28**, 1–24 (1984).
9. M. J. Todd, *The Computation of Fixed Points and Applications*, Springer, New York, 1976.
10. L. Vandenberghe and J. Vandewalle, 'A path-following method for finding multiple equilibrium points in neural networks', *ESAT-SISTA Report 1990-16*, Katholieke Universiteit Leuven, 1990.
11. J. M. Ortega and W. C. Rheinboldt, *Iterative Solutions of Nonlinear Equations in Several Variables*, Academic, New York, 1970.

12. R. O. Nielsen and A. N. Willson, 'A fundamental result concerning the topology of transistor circuits with multiple equilibria', *Proc. IEEE*, **68**, 196–208 (1980).
13. T. Nishi and L. O. Chua, 'Topological criteria for nonlinear resistive circuits containing controlled sources to have a unique solution', *IEEE Trans. Circuits and Systems*, **CAS-31**, 722–741 (1984).
14. M. Fosséprez, *Topologie et Comportement des Circuits Non Linéaires Non Réciproques*, Presses Polytechniques Romandes, Lausanne, 1989.
15. P. E. Gill, G. H. Golub, W. Murray and M. A. Saunders, 'Methods for modifying matrix factorizations', *Math. Comput.*, **28**, 505–535 (1974).
16. K. H. Borgwardt, *The Simplex Method. A Probabilistic Analysis*, Springer, New York, 1987.
17. Y.-H. Wan, 'On the average speed of Lemke's algorithm for quadratic programming', *Math. Programm.*, **35**, 236–246 (1986).
18. T. Matsumoto, L. O. Chua and H. Suzuki, 'CNN cloning template: connected component detector', *IEEE Trans. Circuits and Systems*, **CAS-37**, 633–635 (1990).
19. T. Matsumoto, L. O. Chua and R. Furukawa, 'CNN cloning template: hole-filler', *IEEE Trans. Circuits and Systems*, **CAS-37**, 635–638 (1990).
20. L. Vandenberghe and J. Vandewalle, 'Finding multiple equilibrium points of cellular neural networks without enumeration', *Proc. 1990 IEEE Int. Workshop on Cellular Neural Networks and Their Applications*, Budapest, 1990, IEEE, New York, 1991, pp. 45–54.
21. L. Vandenberghe and J. Vandewalle, 'The computation of equilibrium points in cellular neural networks using complementary pivoting', *Proc. 1991 Eur. Conf. on Circuit Theory and Design*, Copenhagen, 1991, The Technical University of Denmark, Lyngby, 1991, pp. 30–38.

CELLULAR NEURAL NETWORKS: THEORY AND CIRCUIT DESIGN

JOSEF A. NOSSEK AND GERHARD SEILER‡

Institute for Network Theory and Circuit Design, Technical University of Munich, Arcisstr. 21, W-8000 München 2, Germany

TAMÁS ROSKA

Computer and Automation Institute of the Hungarian Academy of Sciences, PO Box 63, H-1502 Budapest, Hungary

AND

LEON O. CHUA

Department of Electrical Engineering and Computer Sciences, University of California, Berkeley, CA 94720, U.S.A.

SUMMARY

Cellular neural networks or CNNs are a novel neural network architecture introduced by Chua and Yang which is very general and flexible, has some important properties desirable for design applications and can be efficiently implemented on custom hardware based on analogue VLSI technology.

In this paper an abstract normalized definition of cellular neural networks with arbitrary interconnection topology is given. Instead of stability, the property of convergence is found to be of central importance: large classes of convergent CNNs in practice always asymptotically approach some stable equilibrium where each component of the corresponding output is binary-valued.

A highly efficient CMOS-compatible CNN circuit architecture is then presented where a basic cell consists of only two fully differential op amps, two capacitors and several MOSFETs, while a variable interconnection weight is realized with only four MOSFETs. Since all these elements are standard components in the current analogue IC technology and since all network functions are implemented directly on the device level, this architecture promises high cell and interconnection densities and extremely high operating speeds.

1. CELLULAR NEURAL NETWORKS

Cellular neural networks or *CNNs* as introduced by Chua and Yang[1] are a specialized neural architecture that lends itself to efficient implementation using VLSI technology. There has also been a strong emphasis on applications-oriented research continually since their introduction.

In contrast to most other neural architectures, CNNs have a special local interconnection pattern and are defined in terms of ordinary differential equations. This allows their interconnection topology to be much more general, since in addition to the usual feedforward interconnections a complete feedback structure can also be meaningfully defined where the output of any cell may again be connected to the input of itself and any other cell.

Definition 1. Cellular neural network

Let the *cell grid* $\mathscr{CG}$ be a discrete set of neurons or *cells*. Introduce the *input* $\mathbf{u}(\tau)$, the *state* $\mathbf{x}(\tau)$, the *output* $\mathbf{y}(\tau)$ and the *offset* $\mathbf{i}$ as vectors with real components such as $i^c \in \mathbb{R}$, and the *feedback operator* $\mathbf{a}$ and *control operator* $\mathbf{b}$ as objects with real components a_d^c, $b_d^c \in \mathbb{R}$. The cell grid serves as the domain

‡Author to whom correspondence should be addressed.

of all indices: $c, d \in \mathscr{CG}$. Then (using the Einstein summation convention explained in Appendix I) define a dynamical system with respect to *normalized time* τ by the *state equation*

$$\frac{\mathrm{d}x^c(\tau)}{\mathrm{d}\tau} = -x^c(\tau) + a^c_d y^d(\tau) + b^c_d u^d(\tau) + i^c \tag{1}$$

and the *output equation*

$$y^c(\tau) := f(x^c(\tau)) \tag{2}$$

where the *cell output function* $f: \mathbb{R} \to [-1, +1]$ is given by

$$f(x^c) := \begin{cases} -1 & \text{if } x^c \leqslant -1 \\ x^c & \text{if } |x^c| < 1 \\ +1 & \text{if } x^c \geqslant +1 \end{cases} \tag{3}$$

The system is called a *cellular neural network* or *CNN* if additionally the initial state is given as $\mathbf{x}(0)$ at $\tau = 0$ and its range and the range of the input $\mathbf{u}(\tau)$ are restricted by the *constraint conditions*

$$x^c(0) \in [-1, +1] \tag{4}$$

$$u^c(\tau) \in [-1, +1] \tag{5}$$

If $\mathbf{u}(\tau)$ is constant (which is the standard case), the cellular neural network is said to have *constant input*, otherwise it has *time-variant input*. □

The cell output function f imposes a restriction similar to the constraint conditions (4) and (5) on the range of the output $\mathbf{y}$:

$$y^c(\tau) \in [-1, +1] \tag{6}$$

This is obvious from the graph of f, which is shown in Figure 1 together with roman numeral labels for its three regions.

An often convenient abbreviation for the last two terms in (1) is the introduction of the (possibly time-variant) *bias* vector $\mathbf{k}(\mathbf{u})$. Its components, the *cell bias values* k^c, are defined as

$$k^c(\mathbf{u}) := b^c_d u^d + i^c \tag{7}$$

Using this, the state equation (1) becomes

$$\frac{\mathrm{d}x^c(\tau)}{\mathrm{d}\tau} = -x^c(\tau) + a^c_d y^d(\tau) + k^c(\mathbf{u}) \tag{8}$$

Finally, it should be noted that both the initial state $\mathbf{x}(0)$ and the input $\mathbf{u}(\tau)$ may represent information that is to be processed by the CNN.

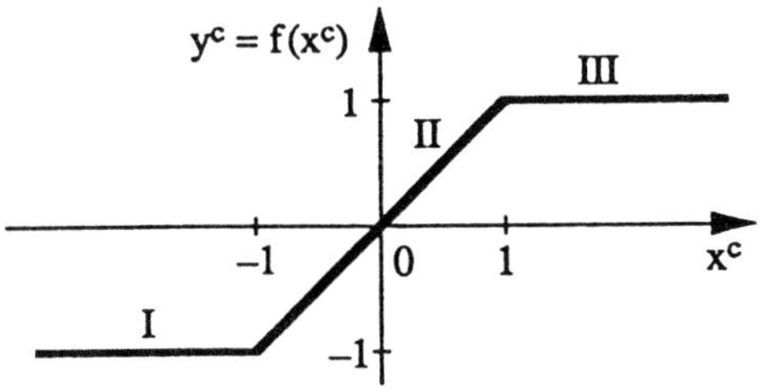

Figure 1. The graph of the cell output function

1.1. A small example CNN

As a small example let us consider a CNN with only two cells. If the numbers 1 and 2 are chosen as the unique symbols to identify the cells, the cell grid is the set

$$\mathscr{C}\mathscr{G} = \{1, 2\}$$

The expanded form of the state and output equations (1) and (2) then is still manageable:

$$\frac{dx^1}{d\tau} = -x^1 + a_1^1 y^1 + a_2^1 y^2 + b_1^1 u^1 + b_2^1 u^2 + i^1, \qquad y^1 = f(x^1)$$

$$\frac{dx^2}{d\tau} = -x^2 + a_1^2 y^1 + a_2^2 y^2 + b_1^2 u^1 + b_2^2 u^2 + i^2, \qquad y^2 = f(x^2)$$

The vectors all have $n = |\mathscr{C}\mathscr{G}| = 2$ elements as shown. The feedback and control operators **a** and **b** both have $n^2 = 4$ components, where the upper index always denotes the cell that is acted on, while the lower index gives the origin of this influence. For example, the quantity a_2^1 is the strength of signal flow *from* the output of cell 2 *to* the input of cell 1. Also, as will be usual practice from now on, the time dependences of the input, state and output have no longer been indicated explicitly.

Figure 2 illustrates the general structure of this two-cell CNN in a graph-like manner, with directed vertices representing interaction paths.

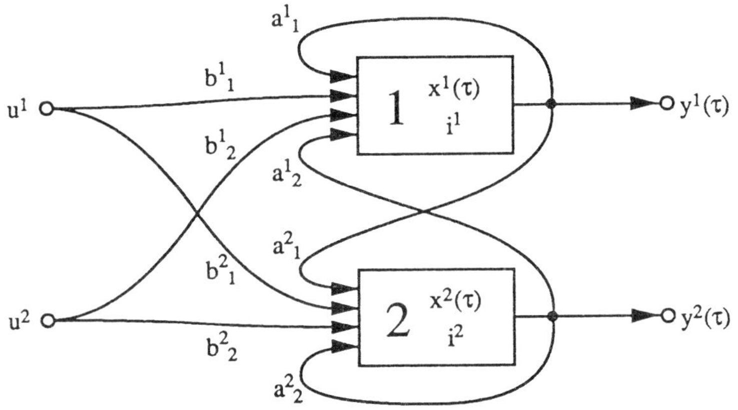

Figure 2. Signal flow in a fully interconnected cellular neural network with two cells

2. THE ORIGINAL CELL CIRCUIT

In the original publication[1] a cell c was defined by a circuit similar to the *original cell circuit* shown in Figure 3. Here the signals are defined as the voltages of the cell's three nodes with respect to ground:

(i) the *input voltage* v_u^c as provided by an independent voltage source
(ii) the *state voltage* v_x^c of the cell's only memory-possessing element, the capacitor C
(iii) the *output voltage* $v_y^c = F(v_x^c)$ of a non-linear voltage source controlled by v_x^c, whose transfer characteristic is shown in Figure 4.

Together with a linear resistor R_x, the circuitry described so far forms the *inner cell circuit*. Finally, a cell *offset current* I^c is injected into the inner node together with the output currents of a number of linear voltage-controlled current sources, which represent the incoming signals from other cells. The transfer conductance of such a source which is controlled by an input voltage v_u^d is called the *control conductance* B_d^c, while that of a source controlled by an output v_y^d is called the *feedback conductance* A_d^c.

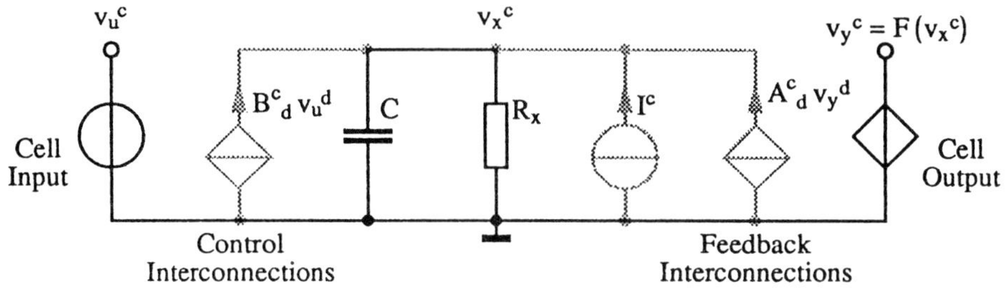

Figure 3. The original circuit of a cell c. The parts shown in black form the inner cell circuit, while the interconnection signal and offset sources are depicted in grey

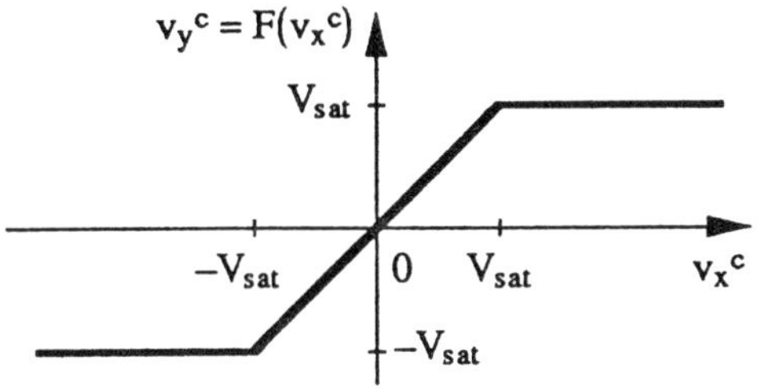

Figure 4. The transfer characteristic of the non-linear voltage-controlled voltage source for cell output

Analysis of the original cell circuit yields the equation

$$C\frac{dv_x^c}{dt} = -\frac{1}{R_x}v_x^c + A_d^c v_y^d + B_d^c v_u^d + I^c \tag{9}$$

The output voltage is defined by a non-linear read out map F:

$$v_y^c = F(v_x^c) \tag{10}$$

$$F(v) := \begin{cases} -V_{\text{sat}} & \text{if } v \leqslant -V_{\text{sat}} \\ v & \text{if } |v| < V_{\text{sat}} \\ +V_{\text{sat}} & \text{if } v \geqslant +V_{\text{sat}} \end{cases} \tag{11}$$

By convention the ranges of the initial state and input voltages $v_x^c(0)$ and v_u^c are restricted by

$$v_x^c(0) \in [-V_{\text{sat}}, +V_{\text{sat}}] \tag{12}$$

$$v_u^c \in [-V_{\text{sat}}, +V_{\text{sat}}] \tag{13}$$

Owing to the limited range of F, the same restriction also applies to the outputs v_y^c:

$$v_x^c(0) \in [-V_{\text{sat}}, +V_{\text{sat}}] \tag{14}$$

Equations (9)–(14) obviously parallel the equations (1)–(6) used in the abstract Definition 1 of cellular neural networks. This formal correspondence can be rigorously established by applying a normalization procedure which maps the voltage ranges in (12)–(14) into the real interval $[-1, +1]$ and renders all other quantities dimensionless in the most natural way.

Multiplying both sides of (9) by R_x/V_{sat} yields

$$\frac{R_x C}{V_{\text{sat}}}\frac{dv_x^c}{dt} = -\frac{v_x^c}{V_{\text{sat}}} + R_x A_d^c \frac{v_y^d}{V_{\text{sat}}} + R_x B_d^c \frac{v_u^d}{V_{\text{sat}}} + \frac{R_x I^c}{V_{\text{sat}}} \tag{15}$$

Table I. The full set of normalization equations for the original cell circuit

Object	Normalization (definition)	Denormalization
Time	$\tau := t/R_xC$	$t = R_x C\tau$
Input	$u^c := v_u^c/V_{\text{sat}}$	$v_u^c = V_{\text{sat}}u^c$
State	$x^c := v_c^x/V_{\text{sat}}$	$v_c^x = V_{\text{sat}}x^c$
Output	$y^c := v_y^c/V_{\text{sat}}$	$v_y^c = V_{\text{sat}}y^c$
Offset	$i^c := R_xI^c/V_{\text{sat}}$	$I^c = (V_{\text{sat}}/R_x)\, i^c$
Feedback	$a_d^c := R_xA_d^c$	$A_d^c = (1/R_x)\, a_d^c$
Control	$b_d^c := R_xB_d^c$	$B_d^c = (1/R_x)\, b_d^c$

Further, a *normalized transfer function* f is defined by requiring

$$\frac{v_y^c}{V_{\text{sat}}} = f\left(\frac{v_x^c}{V_{\text{sat}}}\right) \tag{16}$$

to hold for the same pairs (v_x^c, v_y^c) as (10) does. Obviously, this is identical to f as defined in (3).

Equations (15) and (16) are equivalent to (1) and (2) if the *normalized variables* are defined as in the middle column of Table I. The reverse or *denormalization* equations for determining the physical quantities from the normalized ones are given in the right column. The whole systems (1)–(6) and (9)–(14) are equivalent under the set of substitutions constituting the normalization procedure.

3. BASIC THEORY

This section presents the most important results on the dynamics of cellular neural networks. Although the material is largely based on previously published ideas and proofs,[1,2] it has been completely redeveloped to cover CNNs on arbitrary grids and to provide slightly stricter statements of results and proofs. Also, the logical structure of the whole topic of stability has been reworked and expanded upon in order to provide a more practical tool for theory.

3.1. Boundedness of states

If CNNs are to be faithfully implemented as physical systems, signal ranges in the abstract system must be bounded. This will be shown in the first theorem.

Theorem 1. Boundedness of states[1]

The state $\mathbf{x}(\tau)$ of a CNN (with possibly time-variant input) is bounded. In particular, for all times $\tau \geqslant 0$ each cell c satisfies

$$x^c(\tau) \in [-r^c, +r^c] \tag{17}$$

where its *(cell) range* r^c is defined as

$$r^c := 1 + \sum_{d \in \mathscr{C}\mathscr{G}} |a_d^c| + \sum_{d \in \mathscr{C}\mathscr{G}} |b_d^c| + |i^c| \tag{18}$$

Proof Applying a standard theorem on the integration of a linear first-order ordinary differential equation,[3] the solution of (1) for a single cell c can be cast in the form

$$x^c = x^c(0)e^{-\tau} + \int_0^\tau e^{-(\tau-\bar{\tau})}(a_d^c y^d(\bar{\tau}) + b_d^c u^d(\bar{\tau}) + i^c)\, d\bar{\tau} \tag{19}$$

By taking absolute values for $\tau \geqslant 0$ and considering (4)–(6),

$$|x^c(\tau)| \leqslant |x^c(0)|\,\mathrm{e}^{-\tau} + \int_0^\tau \mathrm{e}^{-(\tau-\bar{\tau})}(|a_d^c|\,|y^d(\bar{\tau})| + |b_d^c|\,|u^d(\bar{\tau})| + |i^c|)\,\mathrm{d}\bar{\tau}$$

$$\leqslant \mathrm{e}^{-\tau} + \left(\sum_{d\in\mathscr{C}\mathscr{G}} |a_d^c| + \sum_{d\in\mathscr{C}\mathscr{G}} |b_d^c| + |i^c|\right) \int_0^\tau \mathrm{e}^{-(\tau-\bar{\tau})}\,\mathrm{d}\bar{\tau} \leqslant r^c \qquad \square$$

A *network range* $\hat{r}$ is introduced by

$$\hat{r} := \max_{c\in\mathscr{C}\mathscr{G}} r^c \tag{20}$$

such that for all cells c

$$x^c(\tau) \in [-\hat{r}, +\hat{r}] \tag{21}$$

3.2. Binary output property

A dynamical system where the orbit of every state converges to an equilibrium is called convergent by Hirsch.[4] This is equivalent to the existence of a unique limit point of every trajectory for $\tau \to \infty$.

Definition 2. Convergence

A cellular neural network is *(almost) convergent* if it has constant input $\mathbf{u}$ and for all $\mathbf{u}$ and (almost) all initial states $\mathbf{x}(0)$ the *eventual state* $\mathbf{x}(\infty)$ exists:

$$\mathbf{x}(\infty) := \lim_{\tau\to\infty} \mathbf{x}(\tau) \tag{22}$$

The analogously defined *eventual output* $\mathbf{y}(\infty)$ then also exists. □

Although convergence does not allow stable limit cycles or chaos, neither does it imply asymptotic stability, since the eventual states are not required to be stable. For example, one can choose an unstable equilibrium position as initial state $\mathbf{x}(0)$. Fortunately, such initial states usually constitute a set of measure zero, so that most convergent systems are stable in practice because of the presence of physical noise.

Currently, two major families of convergent CNNs are known: the reciprocal ones, which were first investigated by Chua and Yang[1] and later in more detail by Seiler and Hasler,[5] and the co-operative or positive cell-linking CNNs described by Chua and Roska.[6]

As a kind of built-in A/D conversion, which is very desirable for applications design, convergent cellular neural networks can be shown to produce discrete output if a simple assumption is satisfied.

Theorem 2: Binary output property[1]

A convergent (or almost convergent) CNN which also satisfies the *binary output condition*

$$a_{\{c\}}^{\{c\}} > 1 \tag{23}$$

for all cells c is called a *binary output CNN*. The eventual output $\mathbf{y}(\infty)$ of a binary output CNN is *binary-valued* in the sense that for all cells c, all inputs $\mathbf{u}$ and almost every initial state $\mathbf{x}(0)$,

$$y^c(\infty) \in \{-1, +1\} \tag{24}$$

Proof Consider any single fixed cell c of the CNN and let

$$\tilde{a}_d^c := \begin{cases} a_d^c & \text{if } c \neq d \\ 0 & \text{if } c = d \end{cases}$$

$$e(\tau) := \tilde{a}_d^c y^d(\tau) + k^c(\mathbf{u})$$

Since the CNN is at least almost convergent, the limit $e(\infty) := \lim_{\tau\to\infty} e(\tau)$ exists for almost all initial states $\mathbf{x}(0)$, and $\varepsilon(\tau) := e(\tau) - e(\infty)$ satisfies $\lim_{\tau\to\infty} \varepsilon(\tau) = 0$. Then (1) can be rewritten as

$$\frac{\mathrm{d}x^c}{\mathrm{d}\tau} = a^{\{c\}}_{\{c\}} y^c - x^c + e(\infty) + \varepsilon(\tau)$$

Now suppose that $x^c \in (-1, +1)$. Then $y^c = x^c$ and

$$\frac{\mathrm{d}x^c}{\mathrm{d}\tau} = \left(a^{\{c\}}_{\{c\}} - 1\right) x^c + e(\infty) + \varepsilon(\tau)$$

which, since $a^{\{c\}}_{\{c\}} - 1 > 0$ in view of (23), must tend to $\pm\infty$ (for all initial $x^c(0)$ but one) even when $\varepsilon(\tau) \not\equiv 0$. Hence for $\tau \to \infty$ and almost every initial state and input, $x^c \notin (-1, +1)$. It follows from the theorems on existence and uniqueness of the solutions of ordinary differential equations and the boundedness of states that the limit $x^c(\infty) := \lim_{\tau\to\infty} x^c(\tau)$ exists, $x^c(\infty) \in [-r^c, -1] \cup [+1, +r^c]$ and (24) now follows from (2). □

3.3. *Asymptotic stability*

The following simple lemma provides some loosely quantitative information on the eventual state.

Lemma 1. Eventual state component values[1]

In a binary output CNN the components of the eventual state almost always satisfy

$$x^c(\infty) := k^c(\mathbf{u}) + \sum_{d \in \mathscr{C}\mathscr{G}} (\pm a^c_d) \tag{25}$$

Proof In a convergent CNN the $\mathbf{x}(\infty)$ are equilibria; hence from (8),

$$x^c(\infty) := a^c_d y^d(\infty) + k^c(\mathbf{u}) \tag{26}$$

Equation (25) now follows from the binary output property, since for almost all initial states the $y^d(\infty)$ exist and may equal either $+1$ or -1. □

Since the initial state of a physical system is never exactly known, practically useful systems are desired to be *asymptotically stable*: an open neighbourhood $\mathscr{N}(\bar{\mathbf{x}})$ of almost all $\bar{\mathbf{x}} \in [-1, 1]^n$ (in view of (4)) must exist such that the trajectories for all initial states $\mathbf{x}(0) \in \mathscr{N}(\bar{\mathbf{x}})$ asymptotically approach each other as $\tau \to \infty$. With convergent CNNs this means that for almost every $\mathbf{x}(0)$ the state $\mathbf{x}$ must converge towards a sink: an asymptotically stable equilibrium state.

Although stability is usually difficult to prove, in the case of convergent CNNs it can be easily derived from the binary output property; but first, some more terminology must be introduced.

An output $\mathbf{y}_q$ every component of which is ± 1 is called a *binary output*. Its corresponding *(closed) saturation region* $\mathscr{S}_\mathrm{c}(\mathbf{y}_q)$ is the set of all states whose corresponding output by (2) is $\mathbf{y}_q$:

$$\mathscr{S}_\mathrm{c}(\mathbf{y}_q) := \{\mathbf{x} \mid \forall c \in \mathscr{C}\mathscr{G}\colon f(x^c) = y^{\{c\}}_q\} \tag{27}$$

The *boundary* $\partial\mathscr{S}(\mathbf{y}_q)$ of $\mathscr{S}_\mathrm{c}(\mathbf{y}_q)$ consists of all of its vectors, one component of which is ± 1, and the *open saturation region* $\mathscr{S}_\mathrm{o}(\mathbf{y}_q)$ is the interior of $\mathscr{S}_\mathrm{c}(\mathbf{y}_q)$:

$$\partial\mathscr{S}(\mathbf{y}_q) := \{\mathbf{x} \in \mathscr{S}_\mathrm{c}(\mathbf{y}_q) \mid \exists c \in \mathscr{C}\mathscr{G}\colon |x^c| = 1\} \tag{28}$$

$$\mathscr{S}_\mathrm{o}(y_q) := \mathscr{S}_\mathrm{c}(\mathbf{y}_q) \backslash \partial\mathscr{S}(\mathbf{y}_q) \tag{29}$$

Lemma 2. Asymptotic stability of binary output CNNs

A binary output CNN is asymptotically stable if no equilibrium state belongs to the boundary of any saturation region. This is guaranteed if no eventual state component equals $+1$ or -1.

Proof There are three classes of equilibria in a CNN:

(1) equilibria in the inside of any saturation region, which (as can be verified easily) are stable
(2) equilibria on the boundary of a saturation region, which are usually degenerate
(3) equilibria outside the union of all closed saturation regions.

The binary output property guarantees that in a binary output CNN the trajectory of almost all initial states will asymptotically approach an equilibrium state that does not belong to the above class 3. Now if (as required) class 2 is empty, the only remaining possibility is class 1: almost all initial states belong to the basin of attraction of a stable equilibrium in an open saturation region. □

The above treatment of stability demonstrates that in the theory of CNN stability the crucial and desired property is in fact that of convergence: to prove stability of a family of CNNs that satisfies the binary output condition (23), it need only need be shown that the CNNs are almost convergent.

3.4. *Linear thresholding*

As a first step towards synthesis of cellular neural networks, which has been pioneered by Chua and Shi,[2] this section shows how linear decision making can be implemented by a specialized kind of CNN cell.

Definition 3. Linear thresholding cells

A cell c in a cellular neural network is a *linear thresholding cell* if

$$a_d^c = \begin{cases} 1+\mu^c & \text{if } d=c \\ 0 & \text{if } d \neq c \end{cases} \tag{30}$$

where $\mu^c > 0$ is called the *margin* of cell c and $\vartheta^c := -i^c$ its *threshold*. □

Now consider the differential equation of such a linear thresholding cell c,

$$\dot{x}^c = -x^c + a_{\{c\}}^{\{c\}} y^c + k^c(\mathbf{u}) \tag{31}$$

where the cell bias $k^c(\mathbf{u})$ can now also be expressed using the threshold ϑ^c:

$$k^c(\mathbf{u}) = b_d^c u^d - \vartheta^c \tag{32}$$

Since in this whole section only this single cell c will be discussed, the corresponding upper index c may as well be dropped everywhere. Then (31) and (32) become

$$\dot{x} = -x + a_c y + k(\mathbf{u}) \tag{33}$$

$$k(\mathbf{u}) = b_d u^d - \vartheta \tag{34}$$

The full indices will be retained in the statement of the results.

In preparation of the theorems, the different possible phase portraits of the differential equation (33) should be constructed. First we introduce the margin μ:

$$\dot{x} = -x + (1+\mu)y + k(\mathbf{u}) \tag{35}$$

This leads to the following set of equations for each of the three regions of f as labelled in Figure 1:

$$\begin{array}{lll} \text{(I)} & \dot{x} = -x-(1+\mu)+k & \text{if } x \leqslant -1 \\ \text{(II)} & \dot{x} = \mu x + k & \text{if } -1 \leqslant x \leqslant +1 \\ \text{(III)} & \dot{x} = -x+(1+\mu)+k & \text{if } +1 \leqslant x \end{array} \tag{36}$$

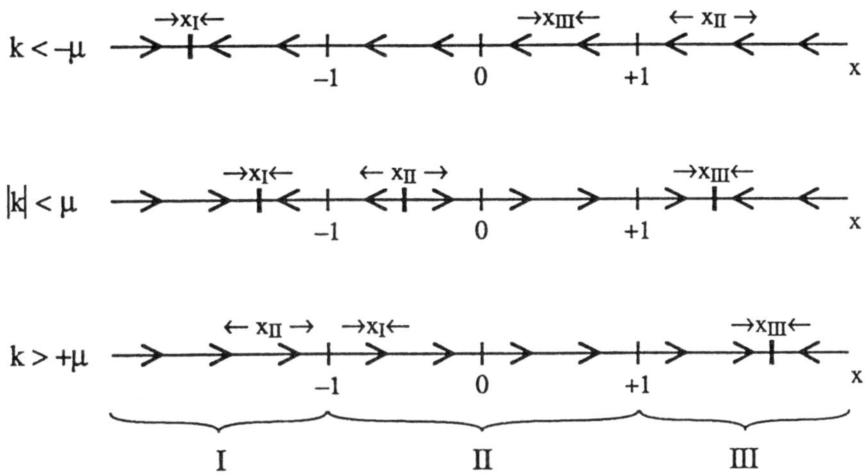

Figure 5. The different possible phase portraits of the state of a linear thresholding cell

The (actual or virtual) equilibrium positions x_{I}, x_{II} and x_{III} of the three regions are then

$$
\begin{array}{lll}
\text{(I)} & x_{\mathrm{I}} = -1-\mu+k & \begin{cases} <-1 & \text{if } k<+\mu \quad \text{actual} \\ >-1 & \text{if } k>+\mu \quad \text{virtual} \end{cases} \\
\text{(II)} & x_{\mathrm{II}} = -k/\mu & \begin{cases} >+1 & \text{if } k<-\mu \quad \text{virtual} \\ \in[-1,1] & \text{if } |k|<\mu \quad \text{actual} \\ <-1 & \text{if } k>+\mu \quad \text{virtual} \end{cases} \\
\text{(III)} & x_{\mathrm{III}} = +1+\mu+k & \begin{cases} <+1 & \text{if } k<-\mu \quad \text{virtual} \\ >+1 & \text{if } k>-\mu \quad \text{actual} \end{cases}
\end{array}
\tag{37}
$$

where obviously x_{I} and x_{III} are stable whereas x_{II} is unstable.

From this information, Figure 5 can be constructed, which shows the three different generic phase portraits of (35) which correspond to the cases $k < -\mu$, $|k| < \mu$ and $k > +\mu$.

Owing to the many cases to be distinguished, this method of construction of the phase portraits may seem a bit obscure. A more intuitive approach of doing it is to use the dynamic route method known from non-linear circuit theory: introducing an intermediate function $h(x)$, (33) can be rewritten as

$$\dot{x} = h(x) + k(\mathbf{u}) \tag{38}$$

$$h(x) := -x + (1+\mu)f(x) \tag{39}$$

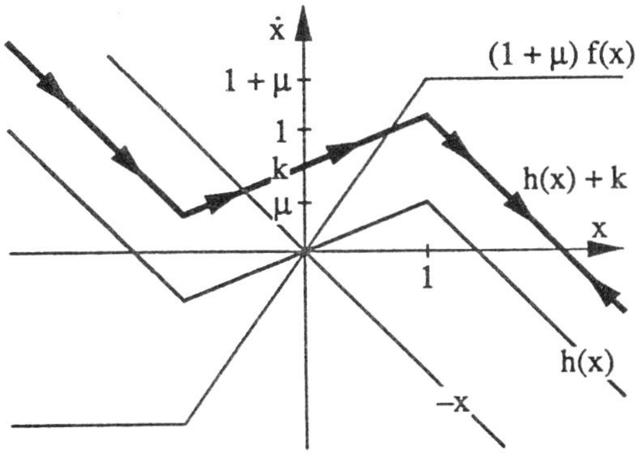

Figure 6. The construction of the dynamic route of a linear thresholding cell

In Figure 6 the derivative $\dot{x}$ is plotted over x. This graph can be marked by arrows indicating the direction of change, the *dynamic route*: above the x-axis it is to the right, below it to the left. The projections of the dynamic route on the x-axis for the different regions of k then also lead to the phase portraits shown in Figure 5. Specifically, the situation shown in Figure 6 corresponds to the bottom phase portrait of Figure 5, since here $k > +\mu$.

The next theorem shows that linear thresholding cells can sometimes operate meaningfully even when the input is time-variant. In this case the input function $\mathbf{u}(\tau)$ must only *eventually* satisfy some condition: there must exist a time $\tau_{\min}$ such that the condition is satisfied for all $\tau \geqslant \tau_{\min}$.

Theorem 3a. Robust linear thresholding[2]

Consider a linear thresholding cell c with threshold ϑ^c, margin μ^c and time-variant input $\mathbf{u}(\tau)$. Then

$$y^c(\infty) = \begin{cases} +1 & \text{if eventually } b^c_d u^d(\tau) > \vartheta^c + \mu^c \\ -1 & \text{if eventually } b^c_d u^d(\tau) < \vartheta^c - \mu^c \end{cases} \tag{40}$$

If the input is constant, there exists a finite *settling time* τ_{set} such that $y^c(\tau) = y^c(\infty)$ for all $\tau \geqslant \tau_{\text{set}}$.

Proof Consider only the case where eventually $b_d u^d(\tau) > \vartheta + \mu$ (the other case is similar).

Then $k > +\lambda$ for all $\tau > 0$ and at any time τ the only actual stable equilibrium of (35) is $x_{\text{III}}(\tau)$, which is larger than $+1$. This can also be seen from the corresponding phase portrait in Figure 5.

Consequently, the function $E(\tau)$ defined by

$$E(\tau) := \begin{cases} -x(\tau) & \text{if } x(\tau) < +1 \\ -1 & \text{if } x(\tau) \geqslant +1 \end{cases}$$

is strictly decreasing if $x(\tau)$ is in the open interval $(-\infty, 1)$. Hence $E(\infty) = -1$ and correspondingly $y(\infty) = +1$. If the input $\mathbf{u}$ is constant, then as long as $x < +1$ the rate $\dot{E} = \dot{x}$ is strictly less than zero, and owing to the bounded range of x, the limit will then be reached after some finite time τ_{set}. □

Unfortunately, the cell's output is not easily predictable if none of the above eventual conditions holds, and in particular if eventually $|b_d u^d(\tau) - \vartheta| \leqslant \mu$, which corresponds to a set of inputs with a width of 2μ around the separating plane of the linear inequality. Although in practical applications where this is a problem one might choose a small margin, the following supplementary theorem may be useful.

Theorem 3b. Exact linear thresholding

Consider a linear thresholding cell with initial state $x(0) = 0$, threshold ϑ^c and constant input $\mathbf{u}$. Then

$$y^c(\infty) = \begin{cases} +1 & \text{if } b^c_d u^d > \vartheta^c \\ -1 & \text{if } b^c_d u^d < \vartheta^c \end{cases} \tag{41}$$

Proof Again consider only the upper case where $b_d u^d(\tau) > \vartheta$. Then $x(0) > x_{\text{II}}$, and owing to the instability of region II of (35), $x(\tau)$ leaves it in the direction leading to positive values, away from x_{II}. Hence $x(\infty) \geqslant +1$ and $y(\infty) = +1$. □

4. EFFICIENT CIRCUIT ARCHITECTURE

In this section the original cell circuit (Figure 3) of the cellular neural network, which contains a large number of idealized circuit elements, will be systematically transformed first into an efficient op-amp-based circuit with linear resistors and finally into a CMOS-compatible one.

All op amps are assumed to have the same saturation voltage; the output voltage v_{out} of any op amp then satisfies

$$v_{\text{out}} \in [-V_{\text{sat}}, V_{\text{sat}}] \tag{42}$$

The abstract cell input and output u^c, $y^c \in [-1, +1]$ can be mapped into this range in the same way as with the original cell circuit:

$$v_u^c = V_{\text{sat}} u^c \tag{43}$$

$$v_y^c = V_{\text{sat}} y^c \tag{44}$$

In order to also map the abstract state $x^c \in [-\hat{r}, +\hat{r}]$ into the op amp output range, a suitable *reduction factor* $\bar{r}$ must first be chosen which satisfies

$$\bar{r} \geqslant \hat{r} \tag{45}$$

Then the *reduced state voltage* $\overline{v_x}^c$ defined by

$$\overline{v_x}^c := \frac{V_{\text{sat}}}{\bar{r}} x^c \tag{46}$$

lies well inside the linear operating region of the op amps:

$$\overline{v_x}^c \in \left[-\frac{\hat{r}}{\bar{r}} V_{\text{sat}}, +\frac{\hat{r}}{\bar{r}} V_{\text{sat}}\right] \subseteq [-V_{\text{sat}}, +V_{\text{sat}}] \tag{47}$$

Analogously to (9), a CNN cell circuit which operates with $\overline{v_x}^c$ instead of v_x^c as the voltage of the state node can be described by

$$\bar{C} \frac{\mathrm{d}\overline{v_x}^c}{\mathrm{d}t} = -\overline{G_x}\,\overline{v_x}^c + \bar{A}_d^c v_y^d + \bar{B}_d^c v_u^d + \bar{I}^c \tag{48}$$

where all parameters have been substituted by *modified* ones marked with an overbar. It is assumed that the capacitance $\bar{C}$ will be chosen by the designer of the circuit, as well as the *time unit* T which establishes the rapport between physical and normalized time:

$$t = T\tau \tag{49}$$

The substitutions (43), (44), (46) and (49) transform (48) into first

$$\frac{\bar{C} V_{\text{sat}}}{\bar{r} T} \frac{\mathrm{d}x^c}{\mathrm{d}\tau} = -\overline{G_x} \frac{V_{\text{sat}}}{\bar{r}} x^c + \bar{A}_d^c V_{\text{sat}} y^d + \bar{B}_d^c V_{\text{sat}} u^d + \bar{I}^c \tag{50}$$

and finally

$$\frac{\mathrm{d}x^c}{\mathrm{d}\tau} = \frac{\overline{G_x} T}{\bar{C}} x^c + \frac{\bar{r} T}{\bar{C}} \bar{A}_d^c y^d + \frac{\bar{r} T}{\bar{C}} \bar{B}_d^c u^d + \frac{\bar{r} T}{\bar{C} V_{\text{sat}}} \bar{I}^c \tag{51}$$

Table II. The design equations of the op-amp-based circuit architectures

Physical quantity	Design equation
Time	$t = T\tau$
Input	$v_u^c = V_{\text{sat}} u^c$
State	$\overline{v_x}^c = (V_{\text{sat}}/\bar{r})\, x^c$
Output	$v_y^c = V_{\text{sat}} y^c$
	$\overline{G_x} = \bar{C}/T$
Offset	$\bar{I}^c = (\overline{G_x}\, V_{\text{sat}}/\bar{r})\, i^c$
Feedback	$\bar{A}_d^c = (\overline{G_x}/\bar{r})\, a_d^c$
Control	$\bar{B}_d^c = (\overline{G_x}/\bar{r})\, b_d^c$

Since the normalized coefficient of x^c must equal -1, obviously

$$\overline{G_x} = \frac{\bar{C}}{T} \tag{52}$$

By comparing (51) with (1), explicit expressions for the modified feedback and control conductances $\bar{A}_d^c$ and $\bar{B}_d^c$ and the modified offset currents $\bar{I}^c$ can be obtained. This leads to the complete set of *design equations* assembled in Table II. Given the values fixed arbitrarily for $\bar{C}$ and T and a suitable choice of $\bar{r}$ consistent with (18), (20) and (45), the design equations completely specify the parameters of the circuits which will now be presented and discussed.

4.1. The practical cell circuit

Since the reduced node voltage $\overline{v_x}^c$ is guaranteed to be within the linear op amp operating range, the cell ground node of the original cell circuit (Figure 3) can be converted into a virtual one, at the inverting input of an op amp whose other input is connected to ground, as illustrated in Figure 7.

This *op amp inner cell circuit* offers two significant advantages

(i) Instead of injecting a current ΣI into the inner node, whose potential $\overline{v_x}^c$ varies with time, equivalently this same current can be drawn from the virtual ground node!

(ii) The inner node now maintains its potential $\overline{v_x}^c$ regardless of any load current I_{load}.

The output can now be realized by first using the full non-linear transfer characteristic of an inverting amplifier to map $\overline{v_x}^c$ non-linearly to $-v_y^c$, which will later be needed in the realization of some feedback and control signal paths, and then a second inverting amplifier to mirror this to the actual cell output v_y^c. The result of these transformations is the *practical inner cell circuit* shown in Figure 8.

Since In^c is a virtual ground node, a current can be drawn from it simply by connecting it to the controlling voltage via an Ohmian resistor! Its conductance will be the absolute value of the required transconductance, while the sign can still be chosen: in Figure 8 the reference direction of the controlled

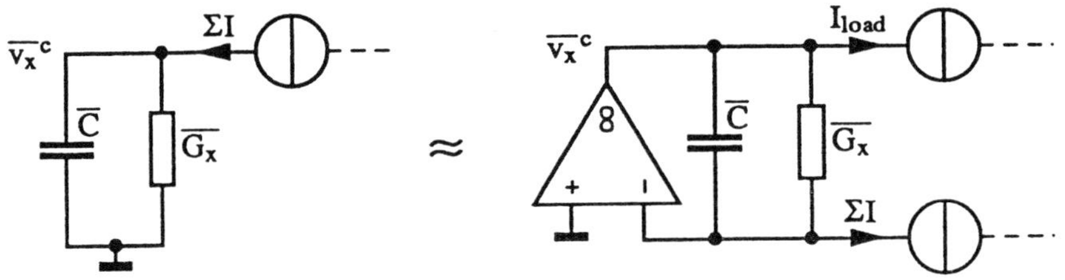

Figure 7. An equivalent realization of the inner cell circuit

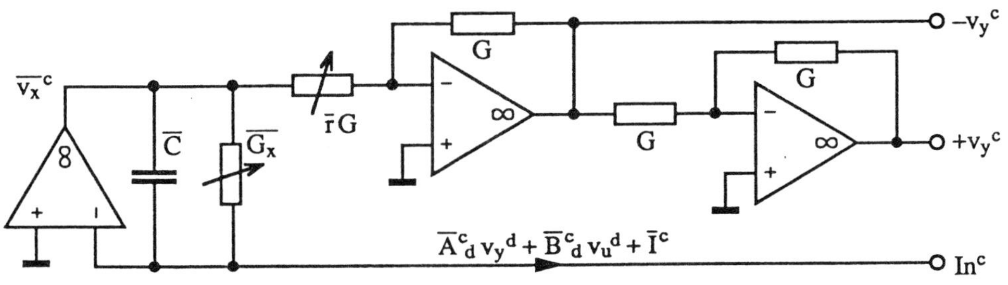

Figure 8. The practical inner cell circuit

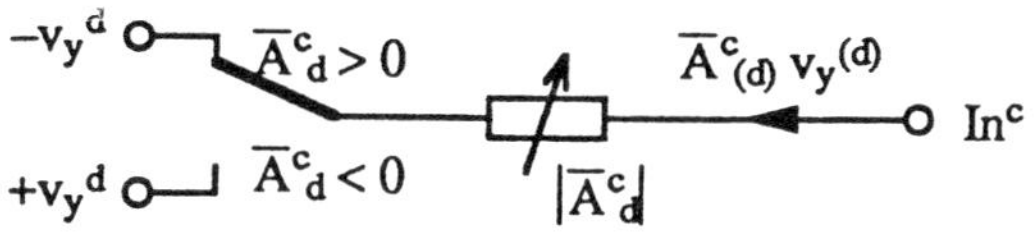

Figure 9. The realization of a variable feedback interconnection $\bar{A}^c_d$

current is directed away from In^c; to realize a positive transconductance, the resistor then has to be connected to the inverted controlling voltage, and to the uninverted signal otherwise.

Figure 9 shows the full circuit of a feedback interconnection with transconductance $\bar{A}^c_d$: the conductance of the Ohmian resistor is the absolute value $|\bar{A}^c_d|$ and the switch connects it either to $-v_y^d$ if $\bar{A}^c_d$ is positive or to $+v_y^d$ if $\bar{A}^c_d$ is negative.

The same circuit can be used to realize the cell offset current source $\bar{I}^c$, by connecting the resistor to the positive or negative saturation voltage $\underline{+} V_{sat}$ or $-V_{sat}$, as shown in Figure 10.

The value of the required linear resistor $G_i{}^c$ is obviously

$$\overline{G_i{}^c} := \frac{\bar{I}^c}{V_{sat}} \tag{53}$$

and finally, using the formula for $\bar{I}^c$ given in Table II,

$$\overline{G_i{}^c} := \frac{\overline{G_x}}{\bar{r}}\, i^c \tag{54}$$

Although in practice this circuit would naturally be connected to the supply, the calculations based on the saturation voltages are accurate for purposes of integrated circuit design: in an IC technology the switches would be implemented as MOSFETs. If $\pm V_B$ are the circuit's main supply voltages, V_B is also the highest possible gate voltage. Then the maximum output voltage of an FET switch one side of which is connected to V_B is about one transistor threshold voltage less than that, and with many-op amp architectures is equal to V_{sat}.

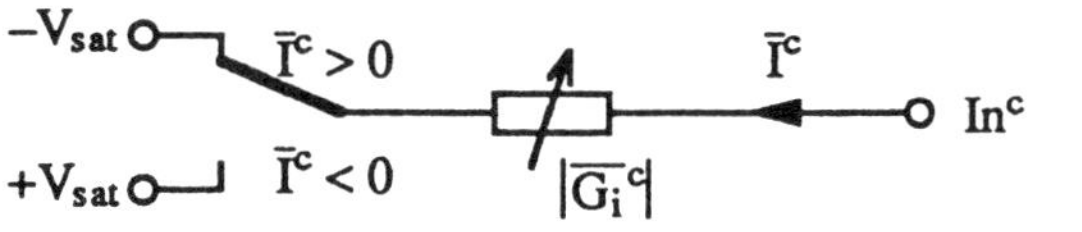

Figure 10. The realization of a variable cell offset current $\bar{I}^c$

4.2. *The balanced cell circuit*

Using the standard method (known from continuous-time MOSFET-C continuous filter design) of *balancing* the circuit, which means that every signal and voltage is provided symmetrically at both sides of ground, once positive and once negative, the practical cell circuit of Figure 8 is transformed into the *balanced inner cell circuit* shown in Figure 11. This circuit uses two *fully differential op amps*, which are available as standard blocks for analogue integrated circuit design. They have two outputs, labelled '+' and '−' where the '+'-output works just like the output of a standard op amp and the '−'-output provides exactly the negative of that voltage.

When compared with the practical circuit, the inner capacitor and resistor now have to be duplicated, since one element each must operate at the positive and negative voltage levels. On the other hand, one op amp has been economized, since the limiting inverting amplifier (the right half of Figure 11) already has output voltages of both signs as required.

The blocks labelled $\overline{G_x}$ and $\bar{r}G$ are *variable conductance blocks*. They can be visualized just as shown: as blocks containing two parallel variable resistors with the given conductance value, which is allowed

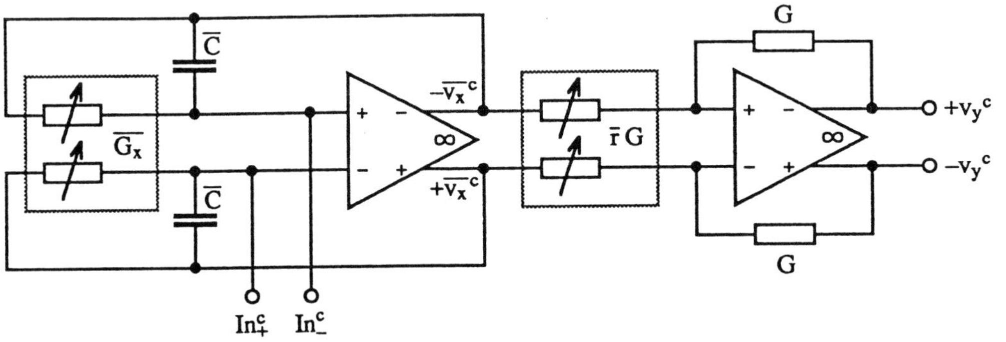

Figure 11. The balanced inner cell circuit

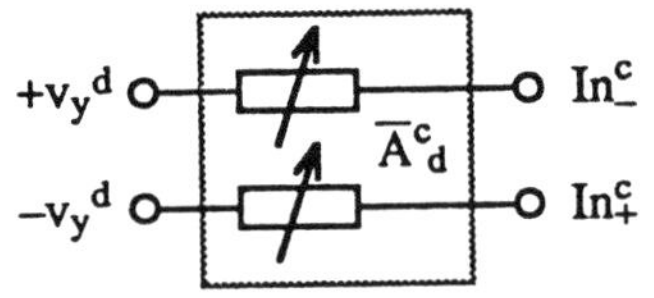

Figure 12. The balanced realization of a feedback interconnection $\bar{A}^c_d$

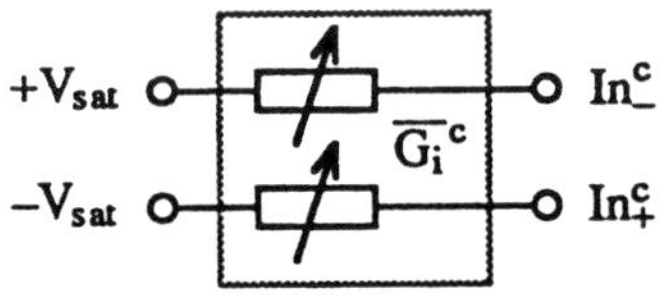

Figure 13. The balanced realization of a cell offset current $\bar{I}^c$

to be negative. All cell interconnections can now be realized by such variable conductance blocks. In Figure 12 this is illustrated with a single feedback interconnection. As shown in Figure 13, the conductance block is also used to realize cell offset currents. For the same reason as with the practical cell circuit, the fixed supply voltage for the current injection has been chosen to be V_{sat} instead of V_B.

The variable conductance block can be realized for both positive and negative conductance values G by the circuit shown in Figure 14. In essence, this circuit consists of two pass transistors with gate voltage

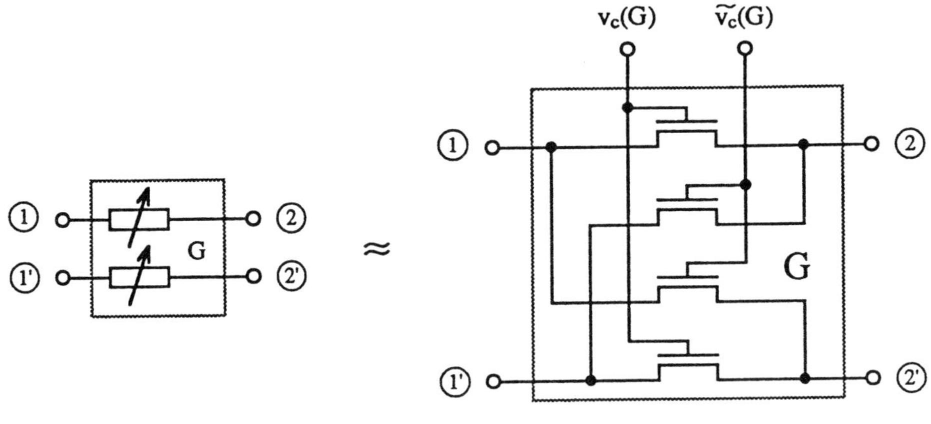

Figure 14. A high-precision CMOS circuit for a balanced variable conductance block

$v_c(G)$ which connect +In to +Out and −In to −Out, and two further ones with gate voltage $\tilde{v}_c(G)$ which implement a cross-over connection: from +In to −Out and from −In to +Out. If both $v_c(G)$ and $v_c(G)$ are larger than 3 V and the signal voltages are less than 2 V (or at most 3 V) absolute, this conductance block is highly linear: Tsividis *et al.*[7] claim that with bulk CMOS processes a relative deviation of less than 1% from linearity can be achieved.

The conductance G is approximately proportional to the difference $v_d(G)$ of the control voltages:

$$G \approx k_G v_d(G) \tag{55}$$

where the value of the proportionality constant k_G depends on the width-to-length ratio of the four MOSFETs (all of which should be identical) and $v_d(G)$ is defined by

$$v_d(G) := v_c(G) - \tilde{v}_c(G) \tag{56}$$

A further saving is possible by assuming one of the two control voltages to have a constant value $\tilde{v}_c(G)$ throughout the CNN circuit. The value of the conductance G is then chosen from over its whole range of positive and negative values by the single voltage $v_c(G)$ according to

$$v_c(G) = \tilde{v}_c(G) + v_d(G) \tag{57}$$

or, using (55) and (56)

$$v_c(G) = \tilde{v}_c(G) + \frac{G}{k_G} \tag{58}$$

Obviously, this simple variable conductance circuit offers many advantages over more conventional realizations of interconnection weights, in particular the following.

(i) The circuit contains only four devices.
(ii) Since only n-MOSFETs are used, no p-well is required. No part of this circuit needs to be wired to a p-well and no latch-up analysis is required.
(iii) Analogue selection of an interconnection weight value of both signs is possible using only a single control voltage.
(iv) Linearity is extremely good.

Unfortunately, this circuit also has a drawback, which is most easily seen by considering it as a Wheatstone bridge where the connections on the left side are the supply and the ones on the right are the transverse path. It is then obvious that the available transverse (output) current is only a fraction of the drain current of the FETs. Therefore the following trade-off exists.

(i) The current in the four FETs and consequently their power consumption are possibly large.
(ii) The drain currents can be reduced by increasing transistor length and consequently size.

Nevertheless, since this conductance block has already been applied in many linear filters, including some high-volume commercial ones, it constitutes a proven building block which should be well suited for CNN design.

5. CONTROL AND INITIALIZATION CIRCUITRY

The circuits described in the previous section implement the dynamics of CNN. This is not sufficient for the realization of a practical system, because many other issues must be considered and solved by suitable circuits. Examples of some problems are as follows.

(i) Input and output: circuitry must be provided to allow the transfer of information between the chip and the surrounding system; input and initial state voltages must be loaded into the circuit and final output voltages must be transferred to the world outside.
(ii) Interconnection weight values must either be stored on chip or be supplied by dedicated lines from the outside.

(iii) Control logic must be provided to co-ordinate all these diverse activities. Most importantly, one must distinguish between information transfer to the chip and initialization and the actual CNN transient.

Most of these problems will have to be solved while actually building a system. Here we will only consider some isolated points of general interest and applicability. An example circuit of a full CNN cell which uses the ideas presented here is given in Appendix II.

5.1. Analogue storage

The storage of analogue values is a key problem which arises in several contexts.

(i) The analogue input values (or perhaps the whole precalculated bias as values defined in (7)) must be stored.
(ii) If simultaneous initialization of all cells is not possible, separate storage for the initial state must possibly be provided.
(iii) Ideally, the control voltages for the interconnection weights should be stored inside the CNN. Owing to the higher accuracy required, it may for the time being, nevertheless, be more practical to have them generated by a separate external circuit.

The basic strategy for storing an analogue voltage v_c is to load it onto a dedicated capacitor C which is connected to a voltage follower. Owing to parasitic effects, C will not hold its charge indefinitely. In current IC technologies the dominant effect here is not the parasitic conductance of the capacitor but rather a fixed leakage current i_L which is generally proportional to the total drain area a_D of all FETs connected to the capacitor. Typically,

$$\frac{i_L}{a_D} = \frac{1 \text{ to } 10 \text{ pA}}{625\ \mu\text{m}^2} \tag{59}$$

If now over a total settling time $t_{set} = T\tau_{set}$ a voltage deviation of at most Δu_{tol} can be tolerated, then the minimum value of C must be

$$C_{min} = \frac{i_L t_{set}}{\Delta u_{tol}} \tag{60}$$

Ideally, a large C should be chosen, while a_D (and consequently i_L) should be kept as small as possible. A typical value of capacitance C per area a_c is

$$\frac{C}{a_c} = \frac{0{\cdot}25 \text{ to } 0{\cdot}5 \text{ fF}}{\mu\text{m}^2} \tag{61}$$

Figure 15 shows a realization for the practical cell circuit, where additionally the output is inverted. In order to allow it to be used in conjunction with control conductances of both signs, as for the realization of a positive conductance value, an inverted control voltage must be available.

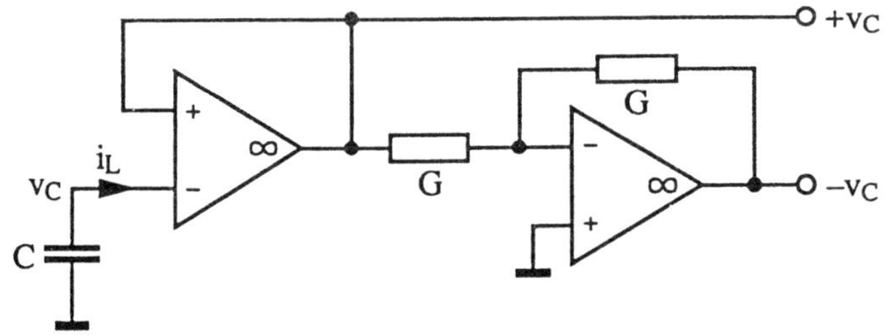

Figure 15. Analogue input storage for the practical cell circuit

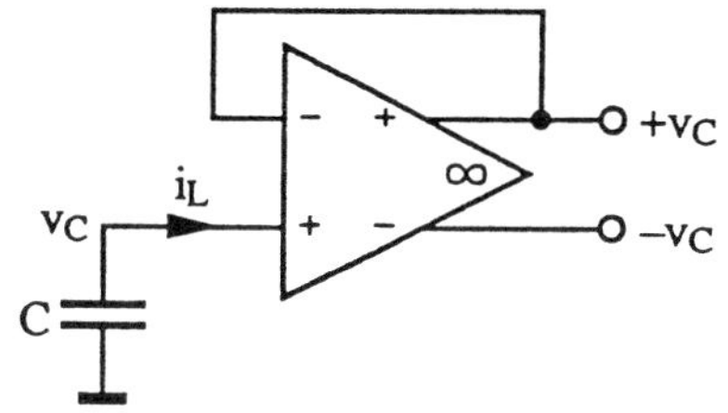

Figure 16. Analogue input storage for the balanced cell circuit

Figure 16 shows a storage circuit for the balanced structure, which uses only one storage capacitor, while the inverted output voltage is provided by the inverted output of the balanced op amp.

5.2. *Applying the initial state*

Before the CNN performs its dynamical operation, the state voltages of its cells must be initialized to proper values. Therefore one must distinguish between at least two different global operating modes of the CNN, which may be defined as follows.

(i) In the *load phase* the cell capacitors are initialized to the desired initial voltages.
(ii) In the *run phase* the actual CNN transient takes place.

To simplify matters, we will here assume the load and run phases to be complementary.

The best way to set the cell capacitor voltage in the practical cell circuit is to inject a suitable current into the virtual ground node using, as usual, a fixed voltage supply and an Ohmian resistor. This leads to the circuit shown in Figure 17.

The fixed initialization voltage v_{init}^c can conveniently be chosen as

$$v_{\text{init}}^c = v_x^c(0) \tag{62}$$

where v_x^c is defined as in Table I; hence

$$v_{\text{init}}^c = V_{\text{sat}} x^c(0) \tag{63}$$

With (12), v_{init}^c then has the largest voltage range possible on the circuit, namely

$$v_{\text{init}}^c \in [-V_{\text{sat}}, +V_{\text{sat}}] \tag{64}$$

This is particularly convenient if intermediate analogue storage of $-v_{\text{init}}^c$ should be required.

If in the above circuit the cell capacitor is fully charged, it is in equilibrium and $i_c = 0$. Consequently $i_x = i_{\text{init}}$ and the following equation holds:

$$\overline{v_x}^c(0)\overline{G_x} = v_{\text{init}}^c \overline{G_{\text{init}}} \tag{65}$$

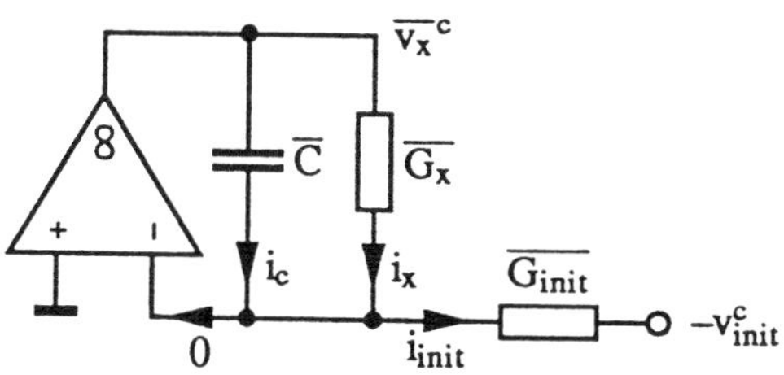

Figure 17. Initialization of the cell capacitor in the practical cell circuit

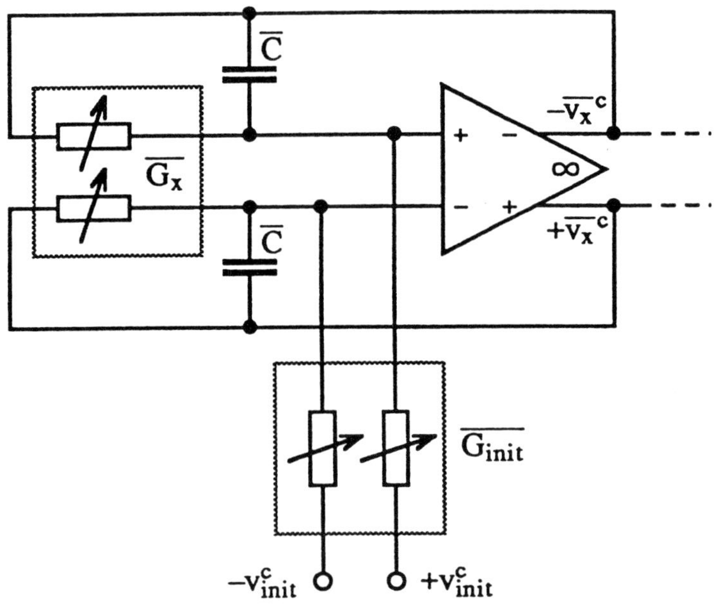

Figure 18. Initialization of the cell capacitors in the balanced cell circuit

Together with (46) and (63), this allows the required value of $\overline{G_{\text{init}}}$ to be calculated:

$$\overline{G_{\text{init}}} = \frac{\overline{G_c}}{\bar{r}} \tag{66}$$

As could be expected, the same value applies to all cells c in the network.

Figure 18 shows the corresponding initialization circuit for the balanced architecture. The relevant design equations of both Figures 17 and 18 are (63) and (66).

6. CONCLUSIONS

In this paper a theory of cellular neural networks with arbitrary interconnection topology has been presented. Interestingly enough, all the theorems presented in the original paper by Chua and Yang[1] and modifications thereof in this paper are valid for arbitrary topologies. Although current VLSI technology does not easily support the implementation of fully connected networks, it is important to keep the basic theorems as general as possible. The design of CNNs will then be less restricted by the underlying theoretical assumptions than by the VLSI manufacturing technology.

Novel, highly efficient circuits for implementing CNNs, both for discrete realizations and for fully monolithic CMOS VLSI realizations, have been described. As with the basic theorems, the circuit architectures are not restricted to specific cell grids and interconnection topologies. The gap between the basic theory described in normalized, abstract quantities and the practical circuit design is bridged with a set of design equations based on a rigorous (de)normalization procedure.

The architecture of these novel circuits, which has been designed for minimizing the expense of the interconnection weights (since there will be many more weights than cell cores in a general CNN), relies on the use of operational amplifiers in the core of the cells. These op amps have been in use in continuous-time MOSFET-C filters and in discrete-time switched capacitor circuits for many years. They have proven their robustness and insensitivity to manufacturing tolerances in many high-volume application-specific circuits.

Especially by using fully differential op amps, the implementation of the interconnection weights is possible with a very simple and also well-proven structure comprising only four n-MOS transistors, which provides high linearity and continuous programmability. This approach, which makes use of the vast experience obtained with analogue VLSI circuits in many successful applications, leads to less complicated, modular and well-structured circuits.

Initialization of CNNs and interfacing with a corresponding system architecture (input/output) are only touched on lightly and needs further detailed work.

Nevertheless, progress achieved both on the theory and circuit design fronts must be accompanied by research on application architectures as well as on learning algorithms and synthesis methods to bring CNNs from their impressive and promising start to large-scale real-world applications.

APPENDIX I: THE EINSTEIN SUMMATION CONVENTION

A short introduction to tensor calculus and the index notation used in this paper is given in the last chapter of Reference 8. Still, for purposes of a quick start, the following even shorter summary may serve.

Let $\mathbf{V}$ be an n-dimensional real vector space with base vectors $\mathbf{e}_1, \mathbf{e}_2, \ldots, \mathbf{e}_n$ and consider the decomposition of a vector $\mathbf{x} \in \mathbf{V}$ into components, where the indices i of the coefficients x^i are written as *upper* or *contravariant* indices, in contrast to the *lower* or *covariant* indices of the base vectors $\mathbf{e}_i$:

$$\mathbf{x} = x^1\mathbf{e}_1 + x^2\mathbf{e}_2 + \ldots + x^n\mathbf{e}_n = \sum_{i=1}^{n} x^i\mathbf{e}_i$$

In this decomposition it is even simpler to take the summation for granted and write only

$$\mathbf{x} = x^i\mathbf{e}_i$$

assuming that the domain of the index i (in this example the set $\{1, 2, \ldots, n\}$) is generally known, previously agreed upon or evident from the context. In a more formalized way this abbreviation (which was originally introduced by Albert Einstein, hence the name) is defined as follows.

Einstein summation convention

Whenever an (unparenthesized) index occurs both in the upper and the lower position in a product term in an expression, then this notation automatically implies a summation of this product term over the domain of the index. □

By requiring the summation indices to be unparenthesized, a loophole is provided for the few cases where automatic summation is not desired: for example, the ith component of the above vector $\mathbf{x}$ can now still be generally expressed as $x^{(i)}\mathbf{e}_{(i)}$ by putting the index i in parentheses.

Specifically with cellular neural networks, the summation convention offers several advantages.

(i) Summation symbols are almost completely eliminated.
(ii) Single elements or equations from whole sets or systems can be easily stated explicitly.
(iii) Objects with more than two indices can still be handled in a convenient way. Suppose for example that $\mathscr{CG}$ is a square grid, where each cell has two co-ordinates i and j, and define the cell index c as the whole ordered co-ordinate pair, $c := (i, j)$, which without punctuation is abbreviated to $c := ij$. The state equation for the square grid then can be written as

$$\dot{x}^{ij} = -x^{ij} + a^{ij}_{kl}y^{kl} + b^{ij}_{kl}u^{kl} + i^{ij}$$

This tensor notation even allows symmetry conditions on objects such as the components a^{ij}_{kl} of the feedback operator still to be represented in terms of the indices, a capability which would be lost during translation to matrix and vector notation.

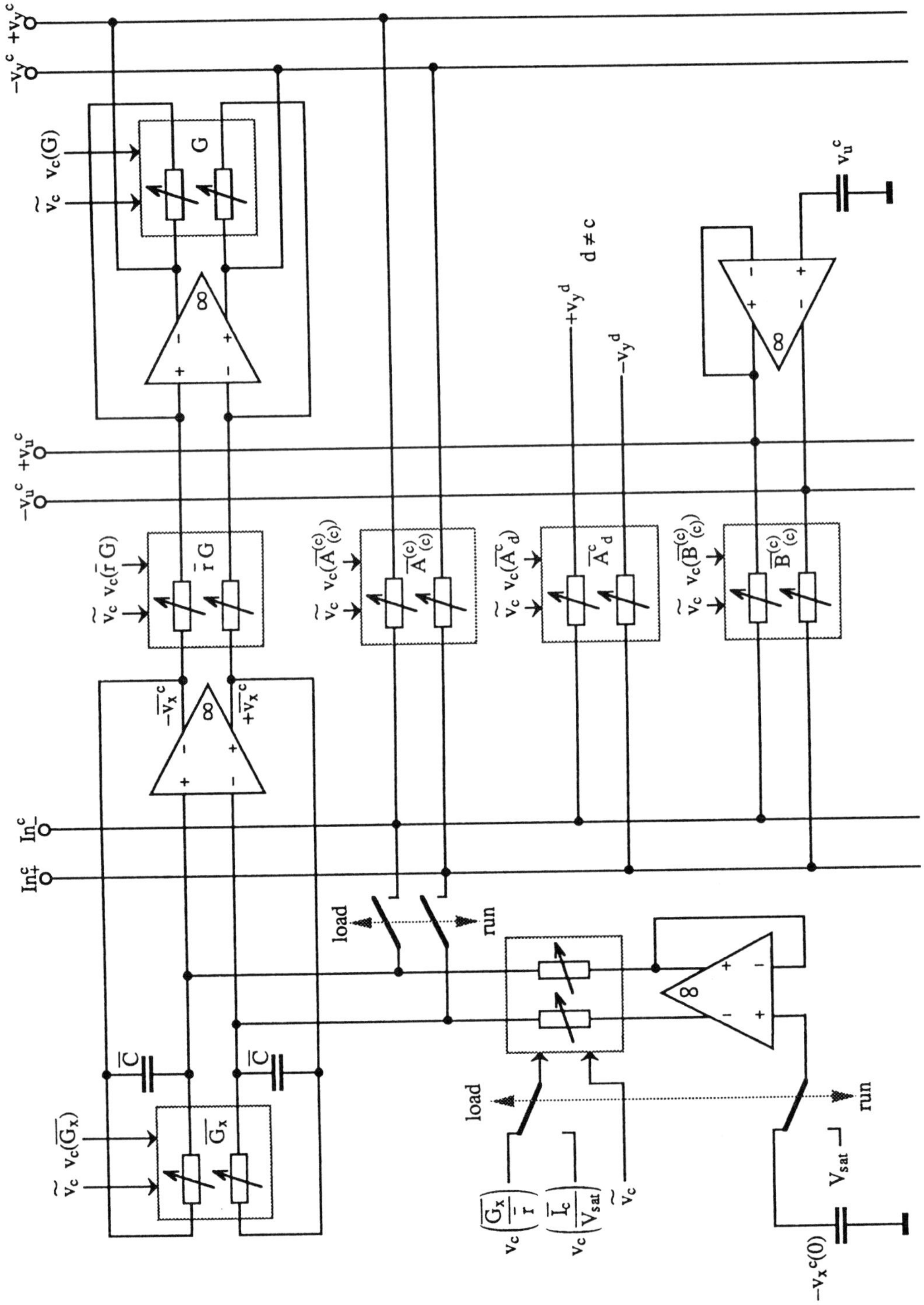

Figure 19. A full CNN cell circuit in the balanced architecture

APPENDIX II: A COMPLETE BALANCED CELL CIRCUIT

Figure 19 presents the schematic of a full CNN cell c which has been designed as described in this paper and which contains circuits for the following functions:

(i) the inner cell circuit
(ii) separate analogue storage for (the negative of) the initial condition, $-v_x^c(0)$
(iii) analogue storage for the input v_u^c
(iv) initialization circuitry for loading the initial state, which uses ideal switches, whose respective positions in the load and run phases are indicated by labelled arrows
(v) the self-feedback interconnection $\bar{A}_{\{c\}}^{\{c\}}$
(vi) a feedback interconnection $\bar{A}_d^c$ with another cell d
(vii) the control interconnection $\bar{B}_{\{c\}}^{\{c\}}$
(viii) the cell offset current source.

Special attention should be paid to the following architectural details:

- The balanced architecture is used for CMOS compatibility.
- The virtual ground, input and output lines, which are involved in cell interconnections, are shown as buses.
- The efficient transconductance circuit of Figure 14 is used wherever possible, including the inner cell circuit and the inverting amplifier.
- All variable transconductances are controlled by only one voltage each. The other control voltage of all transconductance circuits is fixed at v_c
- The ideal switches can be realized with FETs.
- The state initialization circuitry and the cell offset current source (shown in the lower left corner of Figure 19) share one fully differential op amp and one transconductance block.

REFERENCES

1. L. O. Chua and L. Yang, 'Cellular neural networks: theory', *IEEE Trans. Circuits and Systems,* **CAS-35**, 1257–1272 (1988).
2. L. O. Chua and B. E. Shi, 'Exploiting cellular automata in the design of cellular neural networks for binary image processing', *ERL Memo UCB/ERL M89/130,* University of California, Berkeley, November 1989.
3. V. I. Arnold, *Ordinary Differential Equations*, MIT Press, Cambridge, MA, 1973.
4. M. W. Hirsch, 'Convergent activation dynamics in continuous time networks', *Neural Networks*, **2**, 331–349 (1989).
5. G. Seiler and M. Hasler, 'Convergence of reciprocal cellular neural networks', *Technical Report TUM-LNS-TR-91-12*, Technical University of Munich, June 1991.
6. L. O. Chua and T. Roska, 'Stability of a class of nonreciprocal cellular neural networks', *IEEE Trans. Circuits and Systems,* **CAS-37**, 1520–1527 (1990).
7. Y. Tsividis, M. Banu and J. Khoury, 'Continuous-time MOSFET-C filters in VLSI', *IEEE Trans. Circuits and Systems*, **CAS-33**, 125–138 (1986).
8. M. R. Spiegel, *Vector Analysis*, Schaum's Outline, McGraw-Hill, New York, 1982.
9. L. O. Chua, C. A. Desoer and E. S. Kuh, *Linear and Nonlinear Circuits*, McGraw-Hill, New York, 1987.

DESIGN OF HIGH-SPEED, HIGH-DENSITY CNNS IN CMOS TECHNOLOGY

JOSÉ M. CRUZ AND LEON O. CHUA

Department of Electrical Engineering and Computer Sciences, University of California, Berkeley, CA 94720, U.S.A.

SUMMARY

An analogue VLSI circuit architecture for the CMOS implementation of cellular neural networks (CNNs) is presented. It is based exclusively on the use of small capacitors and operational transconductance amplifiers operating in continuous time. Integrated circuit implementations of this architecture are very well suited for processing applications requiring large array size and high speed. We describe a systematic design approach for those circuits and present the design, fabrication and testing of two chips. These chips are used for connected component detection applications and are the first working integrated circuit implementation of a CNN. They contain 2000 transistors and have been fabricated using 2 μm CMOS technology. The density is 32 cells per square millimetre of silicon and the time constant of the processing is of the order of 10^{-7} s. Experimental results of static and dynamic tests are given, including a complete image-processing example.

1. INTRODUCTION

A cellular neural network[1] is a novel analogue network architecture with many potential applications, especially in image processing.[2–4] It consists of a two-dimensional array of locally interconnected *analogue* processors. For image-processing applications, each pixel of the image to be processed is usually associated with one cell. The processing is therefore fully parallel and in principle should be adequate for high-speed real-time computation. Furthermore, as opposed to other neural network topologies, the interconnections between the processing elements are restricted to be only local. This represents a clear advantage over proposed parallel neural network architectures from the point of view of VLSI implementation.[5–8]

We present here a circuit architecture for the implementation of these networks in a CMOS technology. This circuit architecture is suitable for realizing high-density CNN integrated circuit implementations which are reliable and have a high speed/power ratio. It is based on the exclusive use of two elements: operational transconductance amplifiers (OTAs) and capacitors operating in continuous time. They can be considered as non-linear OTA-C circuits. We take advantage of the non-linear behaviour of the CMOS OTA block to implement the sigmoid non-linearity in each cell, making extremely compact designs possible. Using this circuit architecture, we have designed, fabricated and successfully tested two integrated prototypes for connected component detection.[4] This is an application useful in handwritten character recognition and in many other important feature extraction processes. It is just one example of a *global* image-processing task that can be realized by a CNN with a simple *local* interconnection topology.

The structure of this paper is as follows. First, Section 2 gives ideal circuit models of single-layer CNNs. In Section 3 we present a CNN circuit architecture suitable for CMOS implementation. Section 4 presents a systematic procedure for the design of CMOS CNNs and describes the complete design and fabrication of two prototypes for connected component detection in 2 μm technology. Section 5 shows the experimental results of static and dynamic testing of these chips. Finally, in Section 6 we give some conclusions.

2. IDEAL MODELS OF A CNN

A cellular neural network consists of a two-dimensional array of locally interconnected cells. For each cell C_{ij} define *input*, *state* and *output* variables denoted respectively by u_{ij}, x_{ij} and y_{ij}. The nature of the processing of the network is determined by a set of parameters. If for simplicity we consider the case of space-invariant templates, these parameters are:

(i) the neighbourhood size r, which determines the maximum distance between two cells which are interconnected; for convenience we define $\mathcal{N}_r = \{k, l \mid |k|, |l| \leqslant r\}$
(ii) the normalized feedback template, composed of a matrix of parameters denoted by $a_{k,l}$, $(k,l) \in \mathcal{N}_r$
(iii) the normalized control template, composed of a matrix of parameters denoted by $b_{k,l}$, $(k,l) \in \mathcal{N}_r$
(iv) the offset, denoted by x_o.

The dynamics of each cell is given by

$$R_x C_x \frac{\mathrm{d}x_{ij}}{\mathrm{d}t} = -(x_{ij} - x_o) + \sum_{k,l \in \mathcal{N}_r} a_{k,l} y_{i+k,j+l} + \sum_{k,l \in \mathcal{N}_r} b_{k,l} u_{i+k,j+l}, \qquad y_{i,j} = f_1(x_{ij}) \tag{1}$$

where $f_1(\cdot)$ is piecewise linear function defined by

$$f_1(x) = \tfrac{1}{2}(|x+1| - |x-1|) \tag{2}$$

Figure 1(a) shows an ideal circuit model of a cellular neural network cell. The *input*, *state* and *output* variables are represented by voltages at cell nodes. The offset x_o is implemented by a voltage source; the feedback and control operators are implemented by linear VCCSs with transconductance values given respectively by $a_{k,l}G_x$ and $b_{k,l}G_x$, where the parameters $a_{k,l}$ and $b_{k,l}$ are the dimensionless template coefficients and G_x is the conductance of the linear state resistor. This circuit was presented in Reference

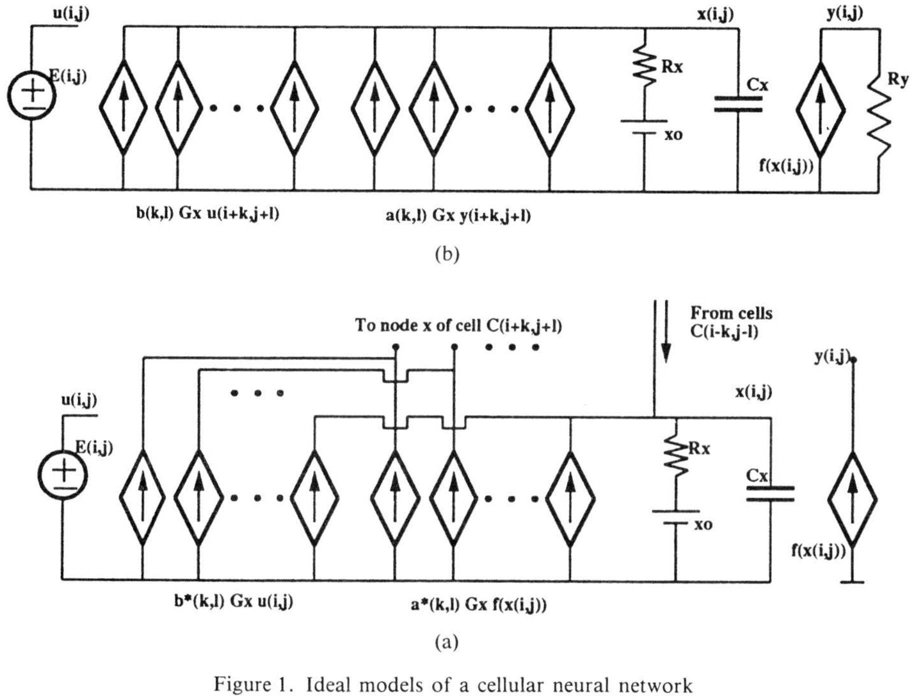

Figure 1. Ideal models of a cellular neural network

1 using the Norton equivalent of the DC source and non-normalized template parameters given by $A_{k,l} = a_{k,l}G_x$ and $B_{k,l} = b_{k,l}G_x$.

2.1 *An ideal circuit model for CMOS implementation*

Let us consider an equivalent circuit with the same topology but in which the PWL non-linearity is incorporated into the *feedback* VCCSs. Each of these current sources will now be controlled by the *state* variable, their current being given by $a_{k,l}G_xf_1(x_{i+k,j+l})$. The *control* VCCSs remains unchanged. Only two variables of each cell, i.e. those associated with the input node and the state node, are now significant for the processing.

The complete network composed of these cell circuits can be redrawn by associating with each cell the current sources controlled by its *own* input and state variables. Figure 1(b) shows a schematic of one of these new cells. The cell $C_{i,j}$ drives a current $i_{o_{k,l}}$ into the state node of cell $C_{i+k,j+l}$. This current is the sum of the contribution of the corresponding feedback and control currents:

$$i_{o_{k,l}} = i_{a_{k,l}} + i_{b_{k,l}} = a^*_{k,l}G_xf_1(x_{i,j}) + b^*_{k,l}G_xu_{i,j} \tag{3}$$

The new weight factors $a^*_{k,l}$ and $b^*_{k,l}$ are given by

$$a^*_{k,l} = a_{-k,-l}, \qquad b^*_{k,l} = b_{-k,-l} \tag{4}$$

Observe that we now require only one connection originating from cell $C_{i,j}$ to each other cell in its r-neighbourhood.

Figure 1(b) also includes a VCCS for generating the output variable. However, since the output variables do not have to be generated in real time during the processing, in a VLSI implementation this VCCS could be realized outside the cell, e.g. in the read-out circuitry.

In the implementation of the entire network it is necessary to compensate for the effects of the limited size of the array. This can be done by modifying the offset values of the cells near the edges (distance smaller than r). Another possibility is to use some extra cells outside the main array and set their state and input nodes to permanent voltages. Although this option requires more area in a VLSI implementation, it actually may be preferable to the first owing to symmetry and testability purposes.

The voltage ranges are limited in any analogue integrated circuit implementation. Therefore it is necessary to scale the two voltage variables of a cell: the *input* variable and the *state* variable. We define v_x and v_u as the scaled state and input voltages respectively. They are given by

$$v_x = n_x x, \qquad v_u = n_u u \tag{5}$$

where n_x and n_u are the normalizing factors and x and u are the normalized variables used until now. The same ideal circuit of Figure 1(b) can be used, but with a voltage source of value v_{x_o} and VCCSs whose transfer function is given by $i = a^*_{k,l}f_a(v_x)$ and $i = b^*_{k,l}f_b(v_u)$, when

$$v_{x_o} = n_x x_o, \qquad f_a(v_x) = G_x n_x f_1\left(\frac{v_x}{n_x}\right), \qquad f_b(v_u) = G_x \frac{n_x}{n_u} v_u \tag{6}$$

3. NON-LINEAR OTA-C CIRCUIT ARCHITECTURE

In this section we present a CMOS-compatible circuit architecture for the efficient implementation of CNNs. The circuit is based entirely on the use of capacitors and operational transconductance amplifiers (OTAs). Both elements can be integrated efficiently in CMOS technology.

In a first-order model an OTA block acts as a current source controlled by a differential voltage:

$$i = g(v) \tag{7}$$

If we are able to design OTAs with intrinsic transfer characteristics g similar to the linear and sigmoid functions given in the previous section, then the extremely compact circuit architecture shown in Figure

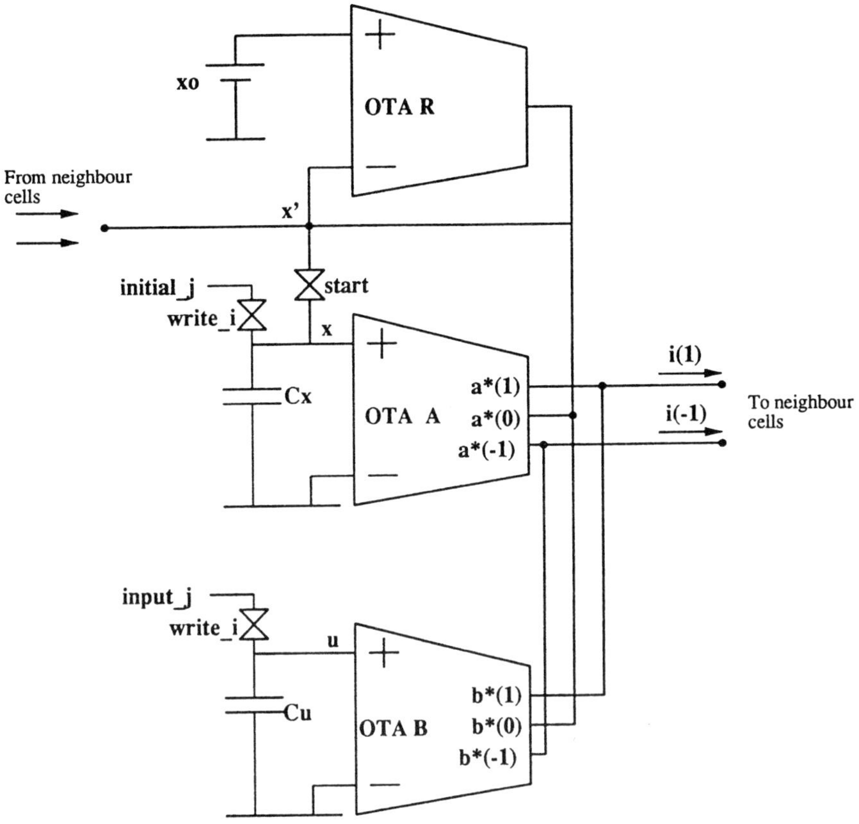

(a)

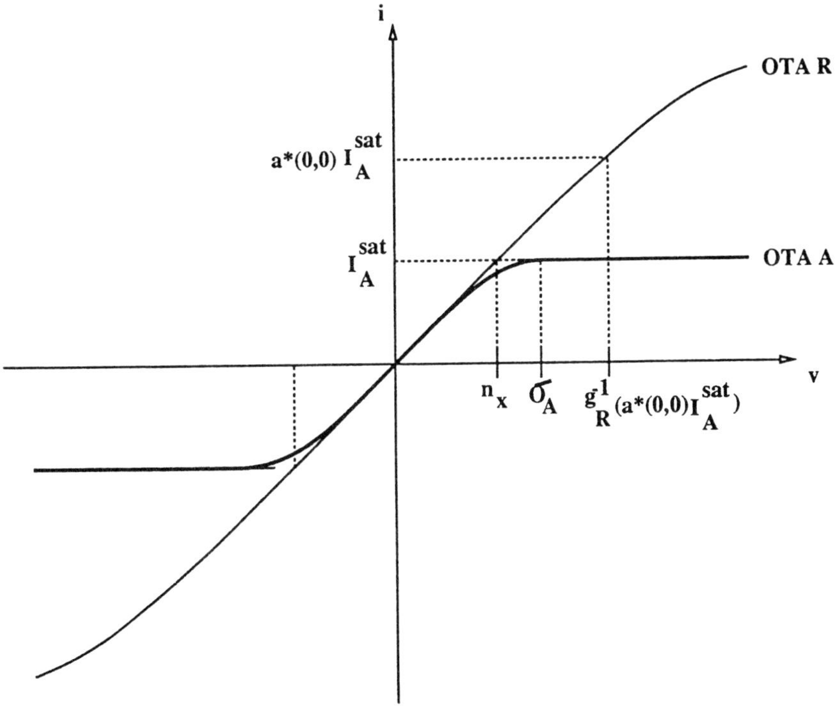

(b)

Figure 2. (a) OTA-C cell architecture. (b) Generic DC characteristics of OTAs A and R

2(a) can be used to implement each CNN cell. Only three OTAs and two capacitors are required. The values of corresponding pixels of the input and initial images are stored as voltages across C_u and C_x respectively. C_x is also one of the loads of the OTAs. For simplicity, the number of connections shown in the figure corresponds to a neighbourhood of size $r = 1$ in only one dimension.

OTA A is used to implement all the current sources controlled by the state voltage. It has multiple outputs, one for each element of the feedback template. The currents driven by them are given by

$$i_{k,l} = a^*_{k,l} g_A(v) \tag{8}$$

The transfer function g_A should ideally be identical to the piecewise linear function f_a given in (6).

OTA B is used to implement all the current sources controlled by the input voltage. It has also multiple outputs, one for each element of the control template. The currents driven by them are given by

$$i_{k,l} = b^*_{k,l} g_B(v) \tag{9}$$

This amplifier should have a linear transfer characteristic over the whole dynamic range of the input voltage v_u.

OTA R is connected to emulate the state resistor. It has a transconductance characteristic given by

$$i = g_R(v) \tag{10}$$

This OTA should be designed to have a linear transfer characteristic over the whole dynamic range of the state voltage v_x. The slope value—the amplifier transconductance G_x—should coincide with that of g_A in the central region. OTAs R and A can be designed such that both slopes exhibit the same dependence on the MOS transistor transconductance parameter $\varkappa_p$, allowing one to obtain values of $a_{k,l} = A_{k,l}/G_x$ insensitive in first order to $\varkappa_p$ variations.

Three switches are added to each cell for loading the input and initial image, starting the processing asynchronous operation and reading the results. They are denoted SW_{write}, SW_{start} and SW_{read} respectively.

This OTA-based circuit model is equivalent to the ideal CNN circuit model when considering an OTA model with characteristics identical to the functions f_a and f_b. Of course, in any real CMOS implementation of these amplifiers many second-order effects and non-idealities are present and should be taken into account in the CNN design.

In a CMOS OTA the relationship between the output current and the input differential voltage is determined mainly by the input transconductance stage. For most transconductance stages, including the differential pair, this function has a sigmoid characteristic with output saturation for $v > |\sigma|$, as shown in Figure 2(b). By selecting the value of σ adequately, the transconductance amplifier function can implement approximately both the piecewise linear and the linear transfer characteristics. These characteristics determine the value of the state normalizing factor

$$n_x = g_R^{-1}(I_A^{sat}) \approx R_x I_A^{sat} \tag{11}$$

The condition $a_{0,0} > 1$, given in Reference 1 to ensure that the network settles to a stable equilibrium point at which the outputs saturate, translates in this case to the condition $\sigma_A < g_R^{-1}(a_{0,0} I_A^{sat})$, where σ_A is the voltage value at which OTA A saturates. The possible state equilibrium points are not affected by the form of the sigmoid function g_A between the saturation limits if it is assumed that the derivative of this function is positive. However, the curvature of g_R introduces a deviation in the value of these stable points. In the case of a simple differential pair the stable points are given by

$$xn_x - \left(\frac{xn_x}{2\sigma_R}\right)^2 \tag{12}$$

The deviation term $(xn_x/2\sigma_R)^2$ should be smaller than n_x. We have studied by simulation how the use of the different sigmoid functions affects the behaviour of the entire network. Extensive examples of CNNs with the connected component detector template show that when we use the sigmoid function of the simple differential pair, the proper final stable points are always reached and the dynamical evolution of the state voltages is almost identical to that obtained using piecewise linear sigmoid functions.

The input voltages of the three OTAs must be maintained inside the bounds $v_{\mathrm{i}_{\max}}$ and $v_{\mathrm{i}_{\min}}$ or v_{iCMmin} limits. Exceeding these limits will drastically diminish the linearity of OTAs A and B and will introduce a non-zero slope in the saturation levels of OTA A. This may cause instability in some cases.[9] The design must be also done to maintain the output voltage of the OTAs inside the range $(v_{\mathrm{O}_{\min}}, v_{\mathrm{O}_{\max}})$. Not doing so would limit the state voltage to an artificially smaller range. This can cause the array to settle to different stable points. Therefore the following inequalities must be imposed:

$$v_x < \min(v_{\mathrm{O_{maxA}}}, v_{\mathrm{O_{maxB}}}, v_{\mathrm{O_{maxR}}}, v_{\mathrm{i_{maxA}}}, v_{\mathrm{i_{maxR}}}), \qquad v_x > \max(v_{\mathrm{O_{minA}}}, v_{\mathrm{O_{minB}}}, v_{\mathrm{i_{minA}}}, 2v_{\mathrm{i_{CMminR}}} - v_{x_{\mathrm{o}}}) \tag{13}$$

The speed of the processing is given by $\tau_x = C_x^{\mathrm{eq}} R_x$. R_x is the equivalent resistance of the OTA R configuration. Its values can be externally controlled through the OTA control bias. The capacitance C_x^{eq} is given by a double-poly capacitor C_x^{poly} and the parasitic capacitances associated with the nodes x and x':

$$C_x^{\mathrm{eq}} = C_x^{\mathrm{poly}} + C_x^{\mathrm{par}} + C_{x'}^{\mathrm{par}} \tag{14}$$

The parasitic capacitances associated with nodes of x and x' contribute to the stability of the network and can reduce the amount of poly-made capacitance C_x^{poly} required and the chip area associated with it. However, two effects impose a lower limit to C_x^{poly}. The first one is that the parasitic capacitances include non-linear junction capacitances whose values change with the state voltage and can therefore affect the dynamics of the network. The second effect is the charge sharing between nodes x and x' at the beginning of the processing. From Figure 2(a) it can be observed that before the start switch is enabled the nodes x and x' are unconnected. The voltage at node x' will depend on the voltage at nodes x and u of all the cells of its neighbour. If the initial value stored in x is denoted by $v_x(0)$, then the initial value in x' is given by

$$v_{x'_{i,j}}(0) = v_{x_{\mathrm{o}}} + \sum_{k,l \in \mathcal{N}_r} a_{k,l} f_{\mathrm{A}}(v_{x_{i+k,j+l}}(0)) + \sum_{k,l \in \mathcal{N}_r} b_{k,l} f_{\mathrm{B}}(v_{u_{i+k,j+l}}) \tag{15}$$

When the start switch is turned on at the beginning of the processing, the voltages at both nodes x and x' evolve to the value

$$v_{x_{i,j}}^{\mathrm{eff}} = v_{x_{i,j}}(0) + \frac{C_{x'}^{\mathrm{par}}}{C_x^{\mathrm{eq}}} \left(v_{x_{\mathrm{o}}} + \sum_{k,l \in \mathcal{N}_r} a_{k,l} f_{\mathrm{A}}(v_{x_{i+k,j+l}}) + \sum_{k,l \in \mathcal{N}_r} b_{k,l} f_{\mathrm{B}}(v_{u_{i+k,j+l}}) - v_{x_{i,j}}(0) \right) \tag{16}$$

This happens with a time constant given by $R_{\mathrm{start-sw}}^{\mathrm{on}} C_{x'}$ which is much smaller than the time constant of the processing, τ_x. $R_{\mathrm{start-sw}}^{\mathrm{on}}$ is the equivalent resistance of the *start* switch in the ON position. The maximum tolerated difference between $v_{x_{i,j}}(0)$ and v_x^{eff} sets a minimum value for the ratio of C_x over the parasitic capacitance $C_{x'}$.

Another effect is the frequency response of the VCCSs. The frequency dependence of the transconductance can be approximated by a high-frequency pole p_1 which is mainly caused by the current mirrors. It causes a phase lag which affects the dynamics of the network. This can be considered as a non-ideality[10] which sets another minimum in the value of C_x i.e. $1/C_x^{\mathrm{eq}} R_x \ll p_1$. However, this same effect—mainly a delay in the controlled current sources—can be beneficially used in CNNs with delay-type templates.

4. DESIGN OF CNNS FOR CONNECTED COMPONENT DETECTION

In this section we present the complete design of a CNN and its fabrication in CMOS technology.

4.1. Connected component detection

The CNN presented here is designed for detecting the number of connected components in each row of the image. This is just one example of global processing that can be performed by a CNN with a very simple local interconnection pattern. This processing application is realized using the connected component detector cloning template reported in Reference 4 and is given by

$$a = \begin{bmatrix} 0 & 0 & 0 \\ 1 & 2 & -1 \\ 0 & 0 & 0 \end{bmatrix}, \qquad b = 0, \qquad x_o = 0 \tag{17}$$

The image to be processed is input as the initial state voltage values in the cells of the network. It is assumed that the input image is binary-valued. An image pixel is represented as $x(0) = 1$ and a background pixel as $x(0) = -1$ in the corresponding cell. After processing, the number of cells with positive outputs in each row will be the number of original connected components in that row. These positive outputs alternate with negative outputs starting from the rightmost cell.

This simple cloning template illustrates very well one of the outstanding features of cellular neural networks; namely the propagation of information over time through the network. This property allows global computation with a very simple interconnection topology. We will demonstrate that this propagation can be so fast in some physical implementations of CNNs that the total processing time of the entire array is significantly shorter than that of other approaches such as conventional digital signal processing or fully connected neural networks.

4.2. Circuit structure of a cell

Since no control template is needed in this application, only the state capacitor C_x and two OTAs, one for the feedback template and the other for the resistor, are required. The resulting circuit architecture is shown in Figure 3(a). Each cell requires only one capacitor, two differential pairs, two current mirrors and some current sources, as indicated in Figure 3(b). The *start* control line is global to all the cells of the array. The control lines $write_i$ and $read_i$ are common for all the cells of row i and the lines in_j and out_j are common to all the cells of column j.

For the above application we have used CMOS OTA structures with simple input differential pairs and cascade output branches. The device-level schematic of the entire cell is shown in Figure 4(a). The nodes labelled *ol* and *or* are connected to the state nodes of the left-side cell and right-side cell respectively. To reduce the area, only an upper branch is used for each of the outputs. Two of these stages (one from the left-side cell and one from the right-side cell) are connected to node x. The DC current through these branches flows through the current source connected to the state node. OTA R has an additional output branch shown at the right of Figure 4(a) which provides an output current for the non-destructive sensing of the state voltage value. The four biasing current sources of the cell are controllable by five global lines. In this way the power consumption, the speed and other parameters can be tuned.

Table I gives a summary of the main parameters of the ORBIT 2 μm CMOS technology in which the designs have been integrated.

4.3. Transistor sizing

For dimensioning all the transistors of the circuit shown above, we have developed the design procedure described below.

We start by considering the following DC transconductance functions for OTAs A and R:

$$g_A = \begin{cases} I_A/2 & v \geqslant \sigma_A \\ (\beta_A/2)v\surd(2I_A/\beta_A - v^2) & -\sigma_A < v < \sigma_A \\ -I_A/2 & v \leqslant -\sigma_A \end{cases} \tag{18}$$

$$g_R = \begin{cases} I_R & v \geqslant \sigma_R \\ \beta_R v\surd(2I_R/\beta_R - v^2) & -\sigma_R < v < \sigma_R \\ -I_R & v \leqslant -\sigma_R \end{cases} \tag{19}$$

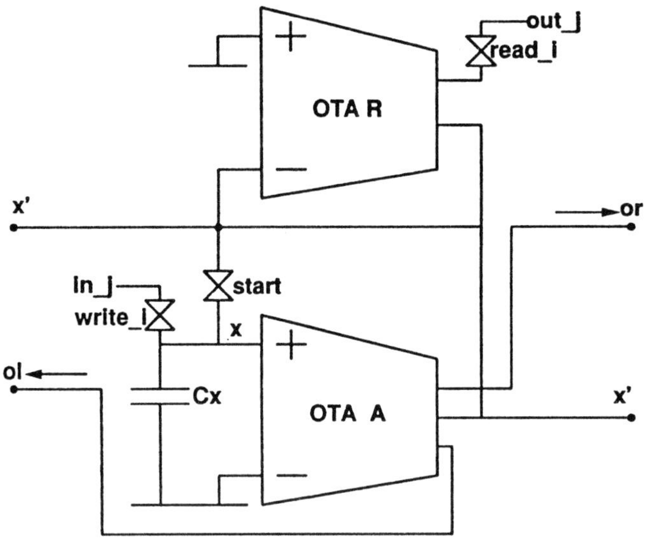

(a)

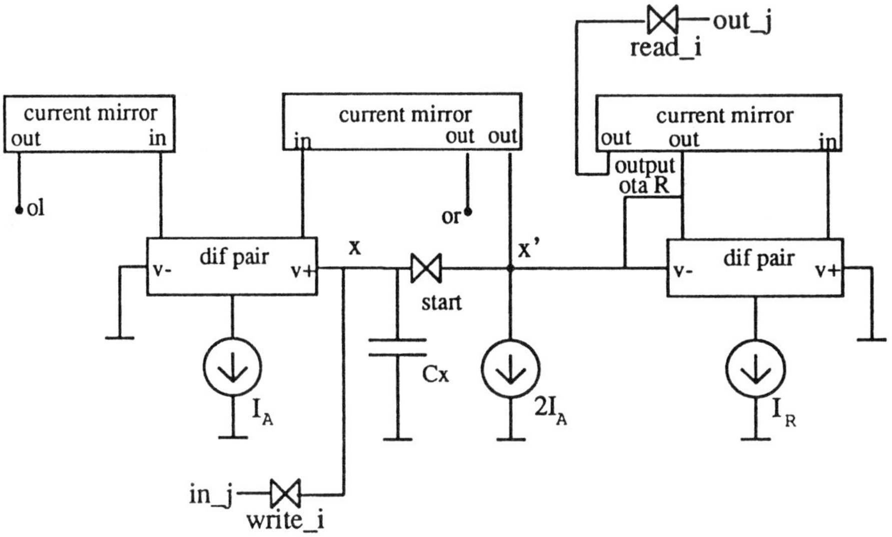

(b)

Figure 3. Cell circuit for connected component detection: (a) OTA-C diagram; (b) symbolic circuit diagram

Table I. Technological data

Parameter	N-channel	P-channel	Unit
V_{th}	1·0	0·8	V
$\varkappa$	23·6	11·6	$\mu A\ V^{-2}$
γ	1·06	0·45	$V^{1/2}$
ΔL	0·54	0·42	μm
ΔW	0·07	0·17	μm

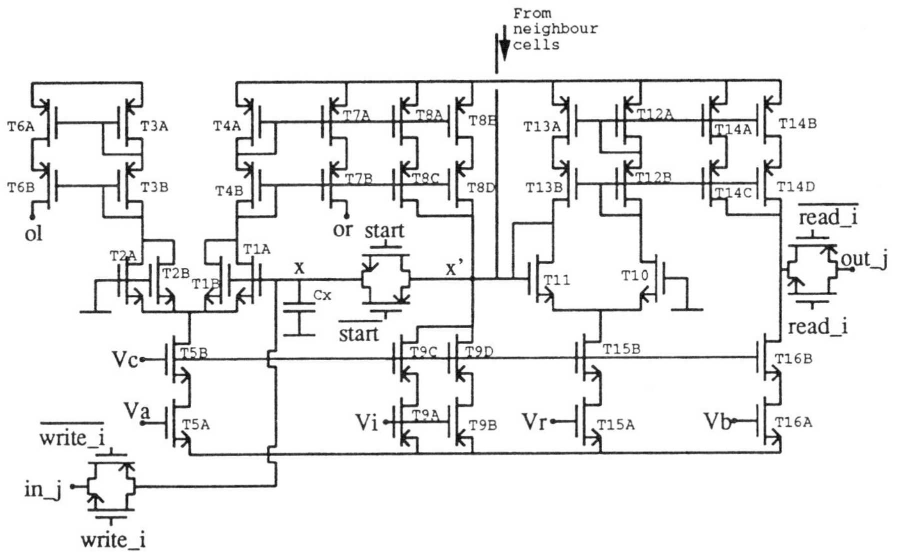

(a)

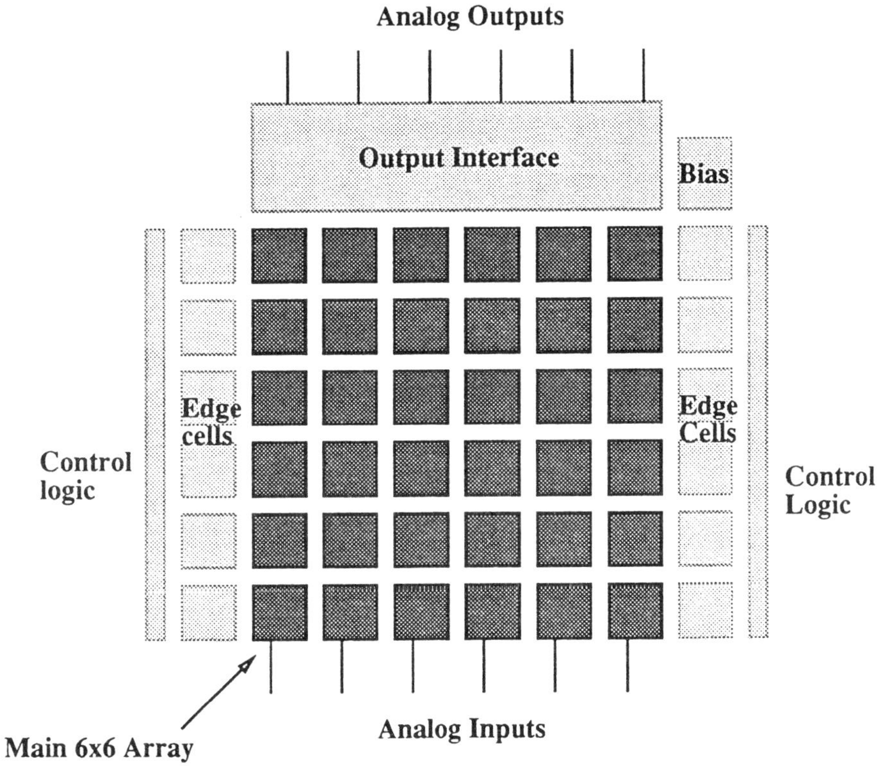

(b)

Figure 4. (a) Device schematic of a cell. (b) Chip structure

In the ideal model the slope of the function f_a in its linear region is equal to the conductance of the state resistor. In this OTA circuit we require that the derivatives of g_A and g_R at the origin, represented by g_{m_A} and g_{m_R} respectively, be equal:

$$g_{m_A} = g_{m_R} \tag{20}$$

The design procedure is summarized as follows.

1. An estimation of the maximum voltage range variation in the state node, $v_{x_{max}}$, is made based on the OTA structures to be used, the technology parameters and the supply voltage values.
2. The dynamic range of the normalized state variable, x_{max}, is calculated from

$$x_{max} = |x_o| + \sum_{k,l \in \mathcal{N}} (|a_{k,l}| + |b_{k,l}|) \tag{21}$$

   This is a slightly stricter limit than the one given in Reference 1.
3. The value of the state normalizing factor is calculated:

$$n_x < \frac{v_{x_{max}}}{x_{max}} \tag{22}$$

   Owing to noise considerations, this value should be chosen close to the maximum.
4. The value of σ_A is calculated from n_x:

$$\sigma_A = \sqrt{2 n_x} \tag{23}$$

   This equation is valid for simple differential pairs. For linearized transconductance structures the multiplication factor is closer to unity.
5. It must be verified that $a_{0,0} f_R^{-1}(I_A^{sat}) > \sigma_A$ to guarantee that the cell settles to a stable equilibrium point. In this case this inequality reduces to approximately $a_{0,0} > \sqrt{2}$.
6. The value of σ_R is calculated from the maximum relative variation ε_r of the DC characteristic of OTA R from a linear one over the state variable dynamic range:

$$\sigma_R = \frac{x_{max} n_x}{2\sqrt{\varepsilon_r}} \tag{24}$$

7. The geometry and current ratios can then be calculated as

$$\frac{(W/L)_A}{(W/L)_R} = 2\frac{\sigma_R}{\sigma_A}, \qquad \frac{I_A}{I_R} = 2\frac{\sigma_A}{\sigma_R} \tag{25}$$

8. The differential pair of OTA A has a smaller bias current but a larger W/L transistor ratio. The minimum allowable value of I_A is given by the condition that the OTA A transistors do not enter the weak inversion region. A value close to this should be selected to minimize the power.
9. The differential pair of OTA R has a larger bias current but a smaller W/L transistor ratio. The minimum allowable value of $(W/L)_R$ is given by the condition that the transistors of its bias current source do not enter their triode region.
10. The current source transistors are sized. They are designed to have a $V_{GS} - V_{th}$ of more than 0·4 V for robustness against variation of the threshold voltage of the MOS transistors. Also, dimensions of at least 4 μm have been considered for robustness against variation of ΔW and ΔL. They must also be designed so that the following inequality is satisfied:

$$-x_{max} n_x > \max(v_{i_{minA}}, 2v_{i_{CMminR}}) \tag{26}$$

11. The current mirrors are sized. They are also designed to have dimensions of at least 4 μm and a $V_{GS} - V_{th}$ of more than 0·4 V. A higher value of $V_{GS} - V_{th}$ will be used for increasing the value of the intrinsic pole p_1. The current mirror gain in OTA R is unity. In OTA A the gains are equal to the coefficients $a_{k,l}$ of the normalized feedback templates. These values are also typically of the

order of unity, which is adequate for maximizing the value of p_1. For robustness against uncertainty in the values of ΔW and ΔL we use transistor duplication rather than different W/L ratios for the branches associated with the different weights. The current mirrors must also be designed so that the following equation is satisfied:

$$x_{\max} n_x < \min(v_{\mathrm{i}_{\max\mathrm{A}}}, v_{\mathrm{i}_{\max\mathrm{R}}}, v_{\mathrm{O}_{\max}}) \tag{27}$$

12 Finally, the value of C_x^{poly} is determined according to the required accuracy in the effective initial state value and the maximum internal delays allowed. Simulations are required to set this limit. In our design, values of C_x^{poly} as small as 1 pF are enough.

Table II gives a summary of the values of all the design parameters that have been calculated manually following the above procedure. SPICE simulations are required for obtaining the final transistor dimensions. They are given in Table III. The circuit operates on a 6 V power supply voltage and at the nominal operating point the currents biasing the OTAs A and R are given by $I_{\mathrm{A}} = 7\ \mu\mathrm{A}$ and $I_{\mathrm{R}} = 17{\cdot}5\ \mu\mathrm{A}$. The power consumption of each cell is 0·37 mW.

Two values for the state capacitance have been considered. The prototype CNN_C1 contains a 1·1 pF capacitor in each cell, while the second prototype CNN_C6 contains a 5·6 pF capacitor in each cell.

Table II. Design values

Line	Value	Unit
$v_{\max}$	900	mV
$x_{\max}$	4	V
n_x	0·200	—
σ_{A}	320	mV
ε_{r}	0·08	—
σ_{R}	1·6	V
$(W/L)_{\mathrm{A}}/(W/L)_{\mathrm{R}}$	10	—
$I_{\mathrm{A}}/I_{\mathrm{R}}$	0·4	—

Table III. Mask device dimensions

Device	$W(\mu\mathrm{m})$	$L(\mu\mathrm{m})$
$T_{1A}, T_{1B}, T_{2A}, T_{2B}$	8	4
$T_{3A}, T_{4A}, T_{5A}, T_{6A}, T_{7A}, T_{8A}, T_{8B}, T_{9A}, T_{9B}, T_{12A}, T_{13A}, T_{14A}, T_{14B}$,	4	4
$T_{3B}, T_{4B}, T_{5B}, T_{6B}, T_{7B}, T_{8C}, T_{8D}, T_{9C}, T_{9D}, T_{12B}, T_{13B}, T_{14C}, T_{14D}$	4	2
T_{10}, T_{11}	4	10
T_{15A}, T_{16A}	9	4
T_{15B}, T_{16B}	9	2
$T_{\mathrm{startN}}, T_{\mathrm{startP}}$	8	2
$T_{\mathrm{writeN}}, T_{\mathrm{writeP}}, T_{\mathrm{readN}}, T_{\mathrm{readP}}$	4	2

4.4. Lay-out of a cell

To reduce the mismatch, we have placed all the transistors with the same orientation. The transistors of the input differential pairs are laid out in a common-centroid configuration. Each cell of both prototypes occupies an area of only 31 000 $\mu\mathrm{m}^2$ ($180 \times 177\ \mu\mathrm{m}^2$), corresponding to a density of 32 cells per square millimetre.

4.5. *Structure of the chip*

Figure 4(b) shows the structure of the complete designed chip. It contains a 6×6 array of cells, two 6×1 linear arrays for setting the edge conditions and for testing purposes, and the circuitry for initializing the network and to output the data. The analogue input data and output data are multiplexed row by row. In the read-out operation the output currents provided by the selected cells are converted to voltages using an OTA-based resistor structure similar to that used in the cells. The resulting characteristic of the voltage obtained in the sensing column j versus the voltage in node x' is almost linear, with a slope of -1, over the whole range of state voltage values. These column output voltages are buffered using Miller compensated op amps in the unity-gain configuration.

A die micrograph of one of the fabricated prototypes, CNN_C1, is shown in Figure 5. It contains about 2000 transistors. CNN_C6 has the same structure. The prototypes have been packaged in a dual-in-line 40-pin structure.

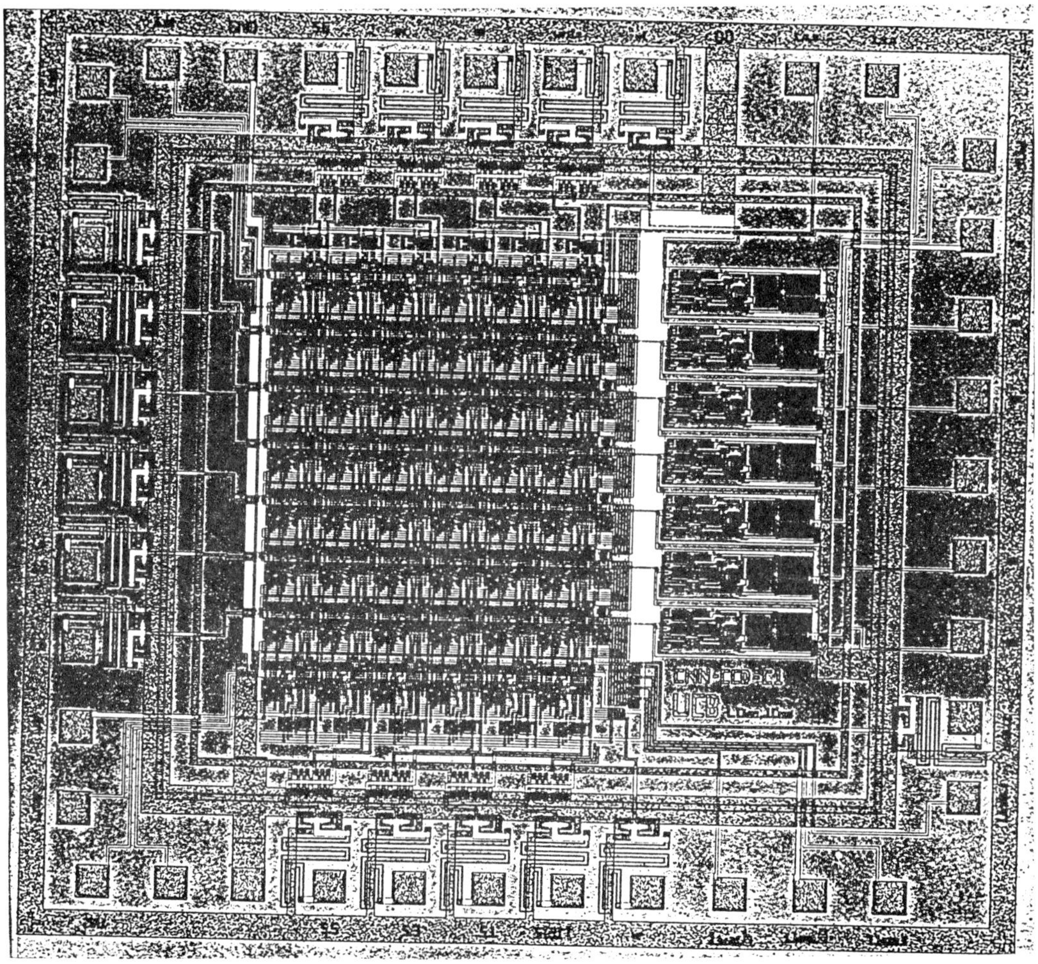

Figure 5. Chip micrograph

5. EXPERIMENTAL TESTING

The fabricated units of both prototypes, CNN_C1 and CNN_C6, worked correctly on the first fabrication run. Experimental tests have shown correct functional behaviour, with both static and dynamic characteristics confirming device-level simulations.

The chips can operate over a range of different biasing conditions. In addition to the power supply lines, the chip contains five additional lines for voltage and current biasing. These allow us to change the operating point and to adjust the power consumption, the speed of processing, the scale factor in the state node and the multiplication factor of the feedback template element values.

The results given here were measured at the chip's nominal operating point. Table IV shows the values of all its external bias voltages and currents.

Four types of tests have been conducted:

(1) static characterization of cell bias currents and the transfer characteristics of the amplifiers used in the cells and in the output interface
(2) testing to detect any catastrophic or parametric faults in the main cell array
(3) dynamical functional testing to determine if the network settles to the proper final stable point
(4) testing of the real-time transient evolution of cell state voltages.

Table IV. Nominal operating point

Line	Value	Unit
V_{DD}	2·5	V
V_{SS}	−3·5	V
V_{GND}	0·0	V
V_c	0·0	V
I_{bias_A}	14·4	μA
I_{bias_T}	14·4	μA
I_{bias_R}	14·4	μA
I_{bias_X}	14·4	μA

5.1. Static characterization of the chip

We make use of different test structures that have been added to the chip for extracting the DC characteristic of the chip. It is possible to measure the DC characteristic of OTAs A and R in any of the 12 edge cells. The cell to be tested is selected using the digital control lines. Figure 6 gives the DC transfer characteristic of both OTAs of one of these cells. Note that the characteristic of OTA A is a sigmoid and that the slope in the saturation region is zero. Table V gives the experimental values of the main parameters used to model the DC characteristics. The symbol δ stands for offset.

In the following three subsections we discuss the results of the last three types of tests. These tests require both loading and reading of the states of all cells in the array. We have built a test board for controlling the chip and realizing the data input and output. The chip contains six input and six output analogue channels, one for each column. They are internally multiplexed to access the cells of all rows. The reading of the state cell value ($v_{x'}$) is non-destructive, since in each cell the OTA R is used to sense the state voltage.

We will illustrate these tests with one image-processing example. The input image to be used is shown at the top of Figure 7. The black pixels of the input image correspond to an initial voltage v_x of $+1n_x$ V and the white pixels to $-1n_x$ V. The state values of all the external edge cells, not shown in the figure,

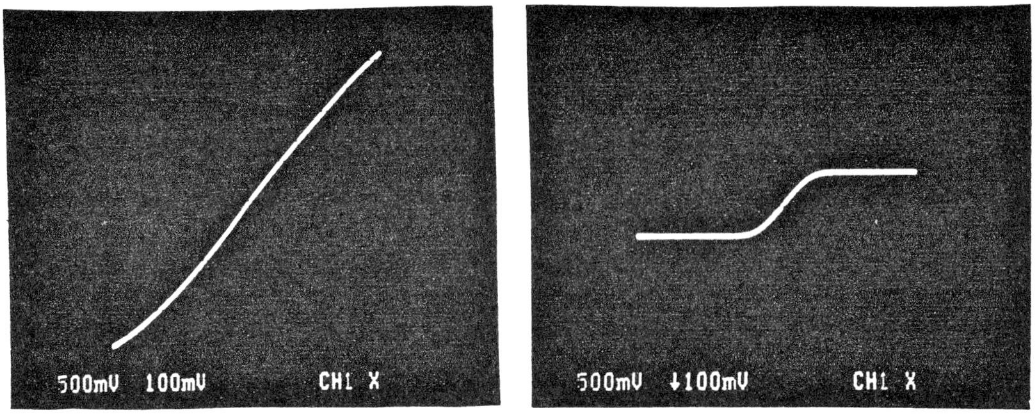

(a) (b)

Figure 6. DC measurements:. (a) transfer characteristic of OTA R; (b) transfer characteristic of OTA A

Table V. DC measurements

Parameter	Value	Unit
I_A	6·5	μA
I_R	17·5	μA
σ_A	0·4	V
δ_A	~0	V
I_A^{sat-}	0·0	μA
I_A^{sat-}	−6·3	μA
g_{m_A}	11·5	μA V^{-1}
σ_R	2·0	V
δ_R	−0·1	V
I_R^{sat+}	18·0	μA
I_R^{sat-}	15·0	μA
g_{m_R}	12	μA V^{-1}

have been set to $-1n_x$. The initial state normalized matrix is

$$X(0) = \begin{bmatrix} +1 & -1 & -1 & +1 & -1 & -1 \\ +1 & -1 & -1 & +1 & -1 & -1 \\ +1 & +1 & -1 & +1 & -1 & -1 \\ +1 & +1 & +1 & +1 & -1 & -1 \\ +1 & -1 & +1 & +1 & -1 & -1 \\ +1 & -1 & -1 & +1 & -1 & -1 \end{bmatrix} \tag{28}$$

This input image $X(0)$ is introduced through the six column input channels of the chip using a row-by-row multiplexing scheme. During the writing operation the chip control line *write* is enabled and the line *start* disabled.

5.2. Fault testing

While the chip control line *start* remain disabled, nodes x and x' in the chip cells are unconnected. The voltage values at the x'-nodes are given by equation (16), which in our case reduces to

$$v_{x'_{i,j}}(0) = \sum_{i=-1}^{1} a_{0,l} f_A(v_{x_{i,j+l}}(0)) \tag{29}$$

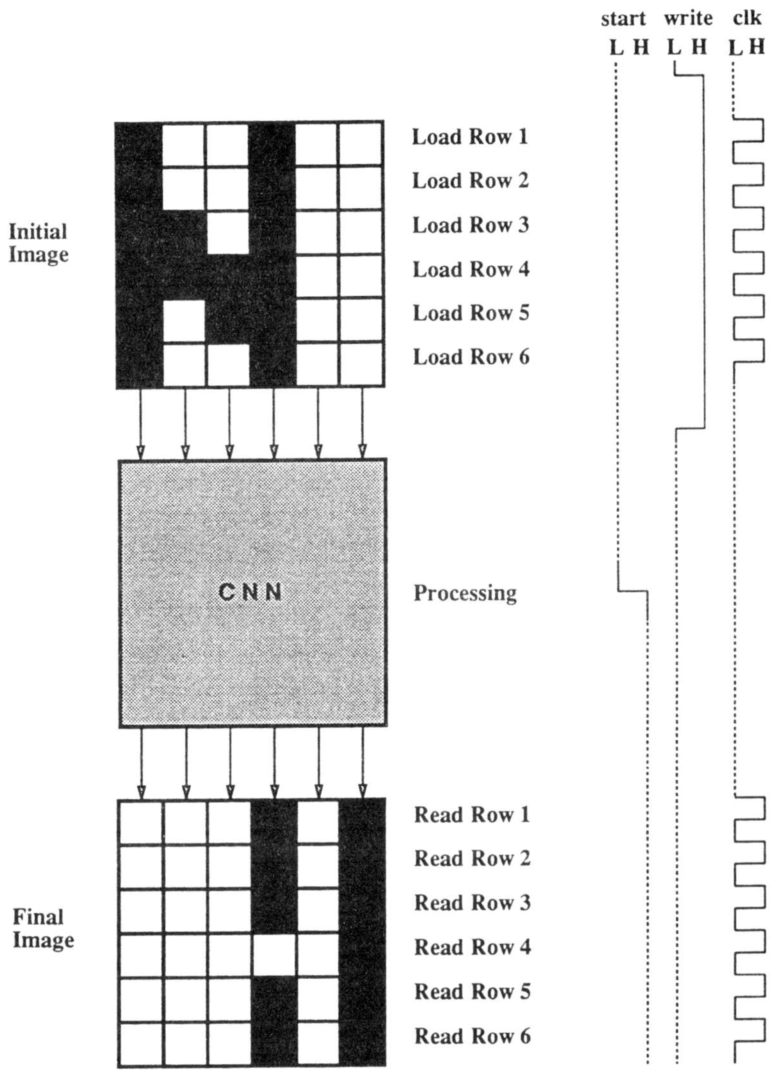

Figure 7. Image-processing example

The value of $v_{x'_{i,j}}(0)$ depends on all the cells in its neighbourhood. By using different X-inputs, we can test all the current sources in the entire CNN array. It is possible to locate precisely in the array any catastrophic or parametric fault in a current source or connection. Here we only present the results obtained by the input image example above. Using the above formula, the $v_{x'_{i,j}}(0)$ are given by

$$X'(0) = \begin{bmatrix} +2 & 0 & -4 & +2 & 0 & -2 \\ +2 & 0 & -4 & +2 & 0 & -2 \\ 0 & +4 & -2 & +2 & 0 & -2 \\ 0 & +2 & +2 & +4 & 0 & -2 \\ +2 & -2 & 0 & +4 & 0 & -2 \\ +2 & 0 & -4 & +2 & 0 & -2 \end{bmatrix} \tag{30}$$

Figure 8 shows the experimental oscilloscope trace of the chip output channel corresponding to the fourth column of the array. The trace is presented in the middle of the photograph (scale 500 mV/div).

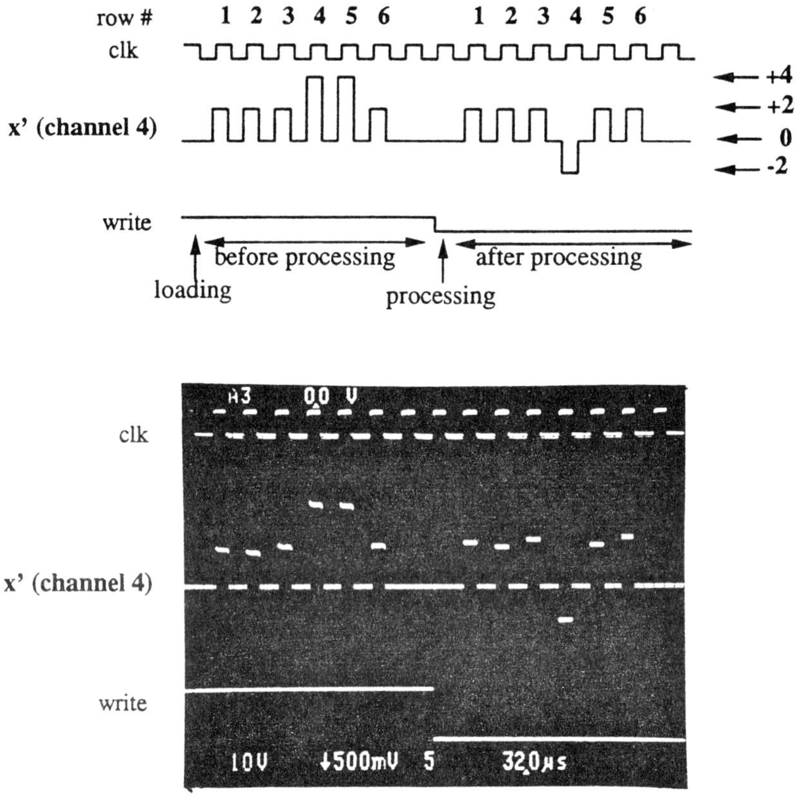

Figure 8. Output wave-form of channel 4, showing a sampling of the x'-voltages in every cell of column 4 before and after processing. The vertical scale for the analogue signal is 500 mV/div

In the photograph are also shown the clock signal at the top and the *write* signal at the bottom. In this example the *start* signal (not shown) is enabled just after *write* is disabled. The data input (not shown) and data output are made at each high period of the clock. It can be observed that the first six data samples, before *start* has been enabled, have the expected values $v_{x'_{i,4}}(0) = (2n_x, 2n_x, 2n_x, 4n_x, 4n_x, 2n_x)$, with scale factor $n_x = 200$ mV. The voltage standard deviation for cells with the same input conditions is 30 mV and the systematic offset is $+50$ mV. Both values are within the accuracy margins required for proper operation.

5.3. *Dynamic functional testing*

This test verifies that all the cells reach the proper stable points after processing. In our example the final normalized state values after the processing should be

$$X^{\text{steady}} = \begin{bmatrix} -2 & -2 & -4 & +2 & -2 & +2 \\ -2 & -2 & -4 & +2 & -2 & +2 \\ -2 & -2 & -4 & +2 & -2 & +2 \\ -2 & -2 & -2 & -2 & -4 & +2 \\ -2 & -2 & -4 & +2 & -2 & +2 \\ -2 & -2 & -4 & +2 & -2 & +2 \end{bmatrix} \tag{31}$$

Figure 8 shows the experimental oscilloscope trace of the chip output channel corresponding to the fourth column of the array. The six data samples on the right (after *start* has been enabled) contain the final stable states of the cells. They are, as expected, $v^{\text{steady}}_{x_{i,4}} = (2n_x, 2n_x, 2n_x, -2n_x, 2n_x, 2n_x)$. The scale factor and the deviation are the same as before.

Equivalent results are obtained for the other output channels of the chip. Our test board reads the six channels in parallel, stores the data in a digital memory and shows it in a graphical display. This allows us to perform extensive functional tests of our chips.

5.4. *Transient tests*

In this subsection we present the experimentally measured real-time evolution of the state voltages during processing. Figure 9 presents the dynamic evolution of the state voltage for all cells of the third row using the image example presented before.

Figure 9(a) shows the pixel values of this row of the input image. The pixel represented at the top of the figure is associated with the left-hand-side cell of the row.

Figure 9(b) shows the expected final binary output. The numerical values give, in normalized units, the expected final steady voltages of the corresponding state nodes.

Figure 9(c) shows the transient obtained by simulation using the ideal circuit model of a CNN given in Reference 1 and with a time constant of 0·43 μs, i.e. the R_xC_x-value for the prototype CNN_C6 at the nominal operating point. The vertical axis represents the state voltage in normalized units from −5 to +5. The horizontal axis is time, 1 μs/div.

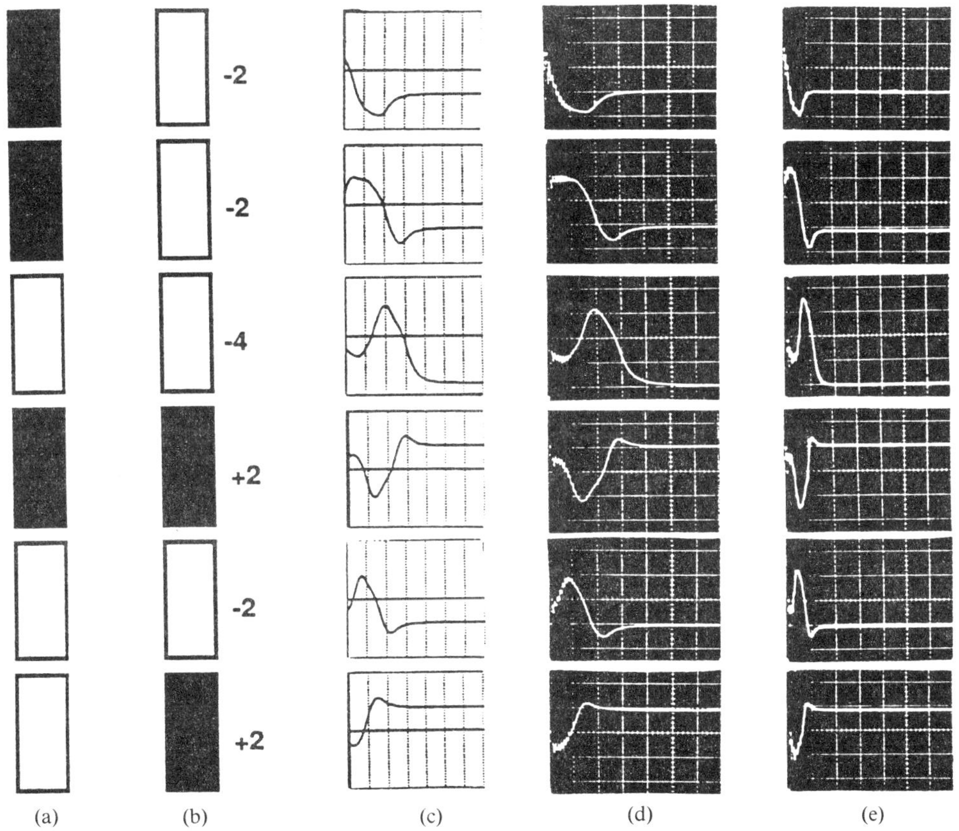

Figure 9. Test of the dynamic evolution of the state voltage for all cells of a row during CNN processing: (a) third row of the input image, where the top box itself corresponds to the leftmost circuit cell; (b) expected final output and steady state values; (c) simulated transient evolution of the state voltages using ideal CNN models; (d) experimental wave-forms of the transient evolution of the state voltages in prototype CNN_C6; (e) experimental wave-forms in the faster prototype CNN_C1 (the time scale is 1 μs/div in all cases)

Figure 9(d) shows the experimental transient wave-forms obtained in the prototype CNN_C6. The vertical axis again represents the state voltage. The scale is two normalized units per division, i.e. 0·4 V/div. The horizontal axis is time, 1 μs/div. The offset of 50 mV has been externally compensated. Observe that all the state nodes reach their proper final voltage values in less than 5 μs. Moreover, the transient responses are almost identical to the ideal case.

Figure 9(e) shows the equivalent experimental results in the high-speed prototype CNN_C1. The horizontal scale is also 1 μs/div. The response is qualitatively the same as for CNN_C6, except that the state voltages reach their steady value in less than 1·5 μs. The obtained increase in speed is in accordance with simulations in which we take into account the parasitic capacitances in parallel with the state node and with the column testing line.

6. CONCLUSIONS

Two CNN chips have been presented. They are the first working VLSI implementation of a CNN. They are based on a non-linear OTA-C architecture in which each cell is composed of only two simple CMOS OTAs and a small capacitor. They operate in continuous time at high speed, with a time constant of the order of 10^{-7}. The total settling times of the complete test arrays are of the order of microseconds. For the reported designs the area of each cell is $177 \times 180\ \mu m^2$, including the local and global interconnection wiring. With this architecture it is feasible to realize monolithic VLSI implementations of CNNs containing several thousand cells.

ACKNOWLEDGEMENTS

The authors would like to thank Dr Robert Brodersen and Dr Paul Gray for providing the resources (MOSIS) necessary in this work, Dr Kari Halonen for his contribution on the output buffers, and Bertram Shi for revising the manuscript.

This work is supported in part by a Fullbright/MEC Fellowship, the Semiconductor Research Corporation and the Office of Naval Research under Grant N00014-89-J-1402.

REFERENCES

1. L. O. Chua and L. Yang, 'Cellular neural networks: theory', *IEEE Trans. Circuits and Systems*, **CAS-35**, 1257–1272 (1988).
2. L. O. Chua and L. Yang, 'Cellular neural networks: applications', *IEEE Trans. Circuits and Systems*, **CAS-35**, 1273–1290 (1988).
3. L. O. Chua, L. Yang and K. R. Krieg, 'Signal processing using cellular neural networks', *J. VLSI Signal Process.*, **3**, 25–51 (1991).
4. T. Matsumoto, L. O. Chua and H. Suzuki, 'CNN cloning template: connected component detector', *IEEE Trans. Circuits and Systems*, **CAS-37**, 663–935 (1990).
5. L. Yang, L. O. Chua and K. Krieg, 'VLSI implementation of cellular neural networks', *Proc. IEEE ISCAS90*, 1990, IEEE, New York, 1990, pp. 958–961.
6. J. M. Cruz and L. O. Chua, 'A CNN chip for connected component detection', *IEEE Trans. Circuits and Systems*, **CAS-38**, 812–817 (1991).
7. K. Halonen, V. Porra, T. Roska and L. O. Chua, 'VLSI implementation of a reconfigurable cellular neural network containing local logic', *Proc. 1990 IEEE Int. Workshop on Cellular Neural Networks and Their Applications*, Budapest, December 1990, IEEE, New York, 1991, pp. 206–215.
8. A. Rodriguez-Vazquez, R. Dominguez-Castro and J. L. Huertas, 'Accurate design of analog CNN in CMOS digital technologies', *Proc. 1990 IEEE Int. Workshop on Cellular Neural Networks and Their Applications*, Budapest, December 1990, IEEE, New York, 1991, pp. 273–280.
9. K. Halonen and J. Vaananen, 'The non-idealities of the IC-realization and the stability of CNN-networks', *Proc. 1990 IEEE Int. Workshop on Cellular Neural Networks and Their Applications*, Budapest, December 1990, IEEE, New York, 1991, pp. 226–234.
10. E. Sanchez-Siencio, J. Ramirez-Angulo, B. Linares-Barranco and A. Rodriguez-Vazquez, 'Operational transconductance amplifier-based non-linear function syntheses', *IEEE J. Solid-state circuits*, **SSC-24**, 1576–1586 (1989).

PROGRAMMABLE ANALOGUE VLSI CNN CHIP WITH LOCAL DIGITAL LOGIC†

K. HALONEN AND V. PORRA
Electronic Circuit Design Laboratory, Helsinki University of Technology, Otakaari 5, SF-02150 Espoo, Finland

T. ROSKA
Computer and Automation Institute, Hungarian Academy of Sciences, H-1502 Budapest, Hungary

AND

L. CHUA
Department of Electrical Engineering and Computer Science, University of California, Berkeley, CA 94720, U.S.A.

SUMMARY

A new integrated circuit cellular neural network implementation with digitally or continuously selectable template coefficients is presented. Local logic and memory are added into each cell, providing a simple dual (analogue and digital) computing structure. Variable gain OTAs are used as the voltage-controlled current sources to programme the template element values. The cells have local switched feedback (both analogue and digital) to feed from the output to the input or state capacitor. Therefore this analogue array processor can be applied to solve problems with a sequence of different templates. A 4×4 CNN circuit is realized using the 2 μm analogue CMOS process.

1. INTRODUCTION

Analogue cellular neural networks[1,2] (CNNs) are extremely effective in some image-processing tasks. In practical situations, several different elementary functions defined by a series of templates are used consecutively. Hence programmability of analogue CNN VLSI arrays is a definite need. On the other hand, it is presupposed that in many intelligent tasks the analogue 'neural' type of processing and logical operations should be combined.[3] In what follows, a new VLSI architecture and its circuit realization are presented. The crucial new features of the design are:

(i) programming the A- and B-templates in prescribed discrete values and the I-template continuously between -3 and $+3$
(ii) local switched feedback within the cells, providing sequential operation (using several templates consecutively)
(iii) storing the present and last values of the cell outputs in logical memory and performing simple logical operations with them.

The designed circuit is a simple special case of a dual computing structure,[3] because some logic is built into each cell (CNNL). Hence the new CNN architecture contains not only a controllable CNN but also local logic and memory to perform digital operations on the results.

One of the major problems in VLSI implementations of cellular neural networks is the wiring needed for loading the template coefficients into the cells. The analogue type of programming has fewer control wires and provides a continuous spectrum of the template coefficients, e.g. providing learning capability.

† Part of this research has been reported in the Proceedings of the 1990 International Workshop on Cellular Neural Networks and Their Applications held in Budapest.

However, it is more sensitive to disturbances and process variations. Also, when the CNN network is controlled by a digital computer, D/A converters are needed to generate the analogue gain control signals.

In many applications, however, the signal processing requires only a few successively used standard template configurations.[1,2,4] Now the cell circuit may contain memory for all the templates and a digital control signal can be used to switch on each of the templates. In this case the size of the cells is increased as the number of possible templates increases, thus limiting the number of cells on a single chip.

The experimental circuit is designed for a 2 μm CMOS double-poly double-metal process. One complex cell requires 519 transistors, including both the analogue and digital parts, and an area of about 1 mm^2. The unit voltage value is 200 mV; the template element units are 0, $\pm 0\cdot 5$, ± 1, ± 2 and ± 3. The settling time of a typical noise removal circuit is about 3 μs according to HSPICE simulations. It is emphasized that this design contains many simpler CNN architectures as special cases. The circuit is under processing and has not been measured.

2. ARCHITECTURE OF THE CNN CHIP

In cellular neural networks the cells have connections with their neighbours.[1] In this study the simplest 1-neighbourhood (or 3×3 neighbourhood) is considered only with zero elements in the diagonal corners (e.g. Figure 4b in Reference 1b). As the number of neighbouring cells is increased, the more complex the cell structure and implementations will become.

A simplified schematic diagram of a cell C_{ij} with row and column indexes i and j is shown in Figure 1, where v_{uij} is the cell input voltage, v_{xij} is the state voltage and v_{yij} is the limited output voltage. The currents $I_{ya}(i, j; k, l), \ldots, I_{yd}(i, j; k, l)$ are the output currents fed out to the neighbouring cells, while $I'_{ya}(i, j; k, l), \ldots, I'_{yd}(i, j; k, l)$ are the output currents of the neighbouring cells fed in to the cell. Also, the cell has positive feedback connection with a gain of two, forcing the output v_{yij} to strive on either of the two saturation levels of the sigmoid. I_f is the constant current source. The gains of the voltage-controlled

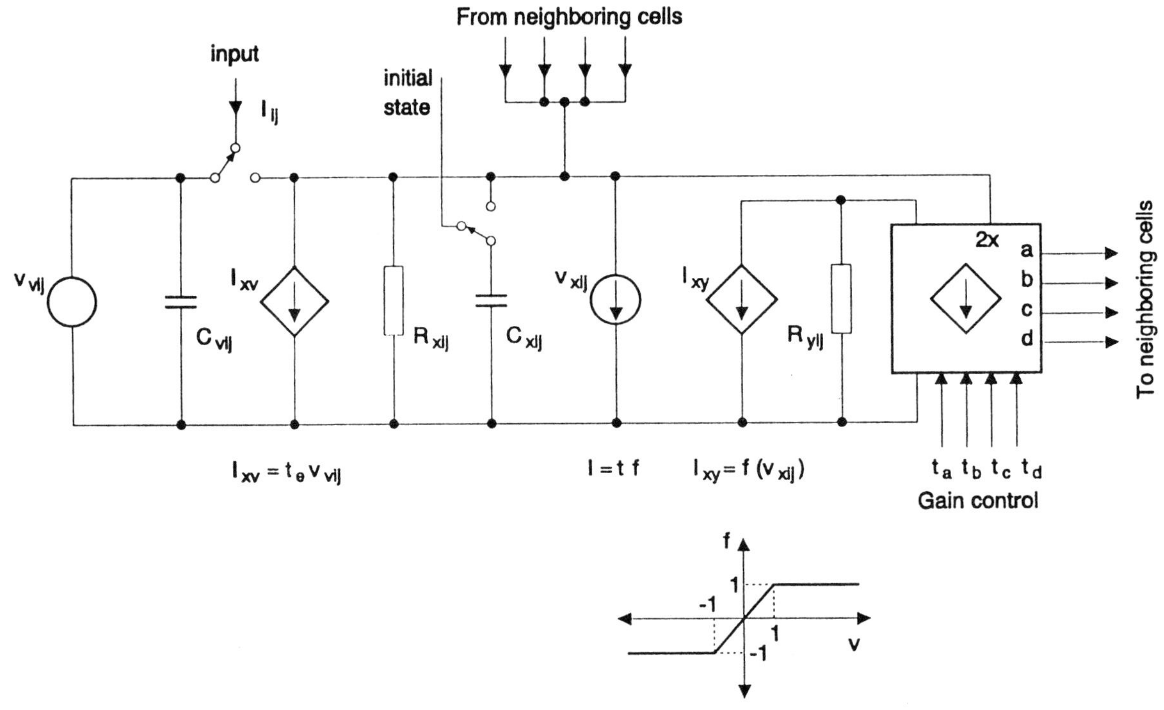

Figure 1. Principal CNN cell structure

current sources are given by the template matrices

$$\mathbf{A} = \begin{bmatrix} 0 & a & 0 \\ d & 2 & b \\ 0 & c & 0 \end{bmatrix}, \qquad \mathbf{B} = \begin{bmatrix} 0 & 0 & 0 \\ 0 & e & 0 \\ 0 & 0 & 0 \end{bmatrix}, \qquad \mathbf{I} = f \tag{1}$$

in which the scalars $a, \ldots, e$ may have the values 0, $\pm 0{\cdot}5$, ± 1, ± 2 and ± 3 and f can be varied continuously between -3 and $+3$. The values of the A- and B-template elements are selected using digital control signals $t_a, \ldots, t_e$. The value of the constant current source is selected with the analogue control signal t_f. Thus with the same hardware we are able to solve problems with different templates or problems which require a sequence of different templates.

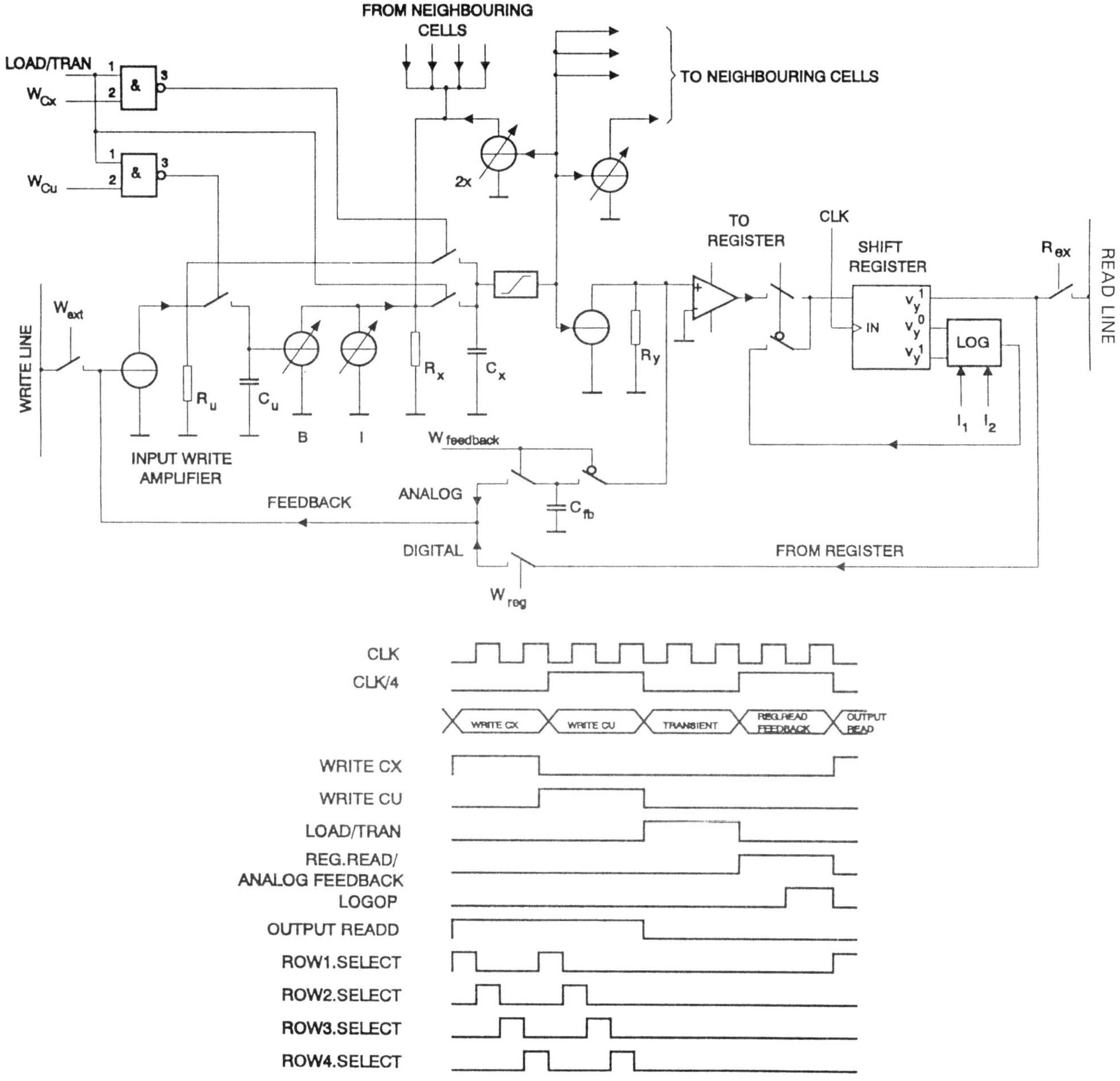

Figure 2. Principal circuit diagram of the realized programmable dual computing CNN cell and the control timing diagram

I_{yx} realizes the non-linear sigmoid-type characteristics as described in Reference 1. The input voltage is stored in the capacitor C_{uij}, which is switched to the input at the beginning of the transient. Also, an additional switch is used to load an initial state into the state capacitor C_{xij}.

3. INTEGRATED CIRCUIT REALIZATION OF THE PROGRAMMABLE CNN

The principal circuit diagram of the integrated circuit realization of one CNN cell is shown in Figure 2. It consists of the summation node, where all the input currents are summed, the input voltage-controlled variable gain current source, the sigmoid function and the variable gain voltage-controlled output current sources. In addition, we included analogue feedback of the settled output into the state capacitor and into the digital register, as well as allowing logic operations on the digital output values to be performed.

In CMOS technology the sigmoid is realized with the source coupled pair shown in Figure 3. The difference between the output currents I_1 and I_2 is described by the piecewise linear approximation

$$\Delta I = \begin{cases} g_m(V_1 - V_2), & |V_1 - V_2| < V_{asat} \\ I_{ss}, & |V_1 - V_2| > V_{asat} \end{cases} \tag{2}$$

where I_{ss} is the constant biasing current, g_m is the transconductance of the transistors and V_{asat} is the saturation voltage of the input transistors.

A voltage-controlled current source can be built with a source coupled pair. The most typical realization is the OTA structure (operational transconductance amplifier) of Figure 4, for which the output current is

$$I_{out} = B\Delta I \tag{3}$$

where B is the current mirror factor.

The variable gain OTA is obtained either by varying the biasing current of the differential input pair, I_{ss} in equation (2), or by varying the current mirror factor, B in equation (3). In the former the analogue

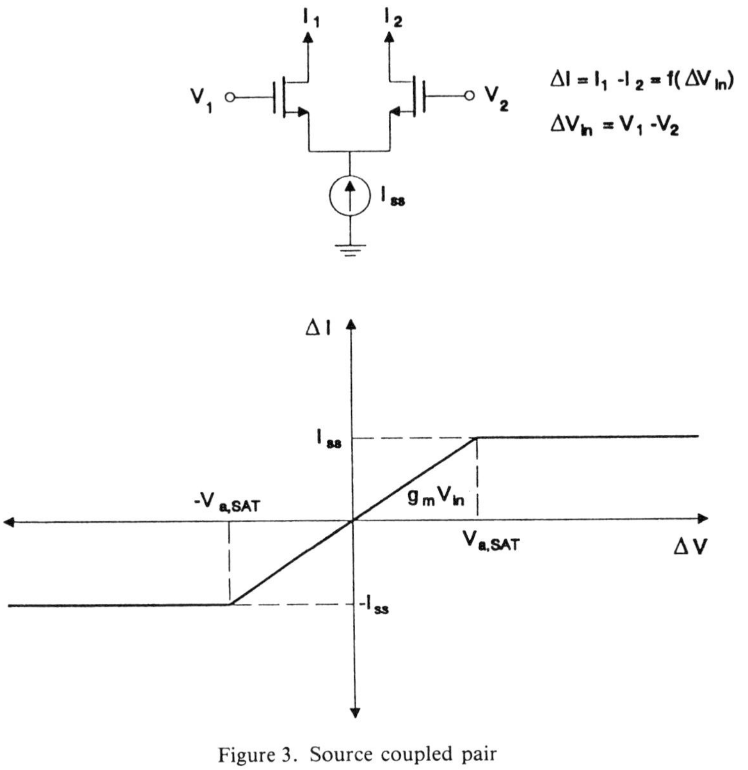

Figure 3. Source coupled pair

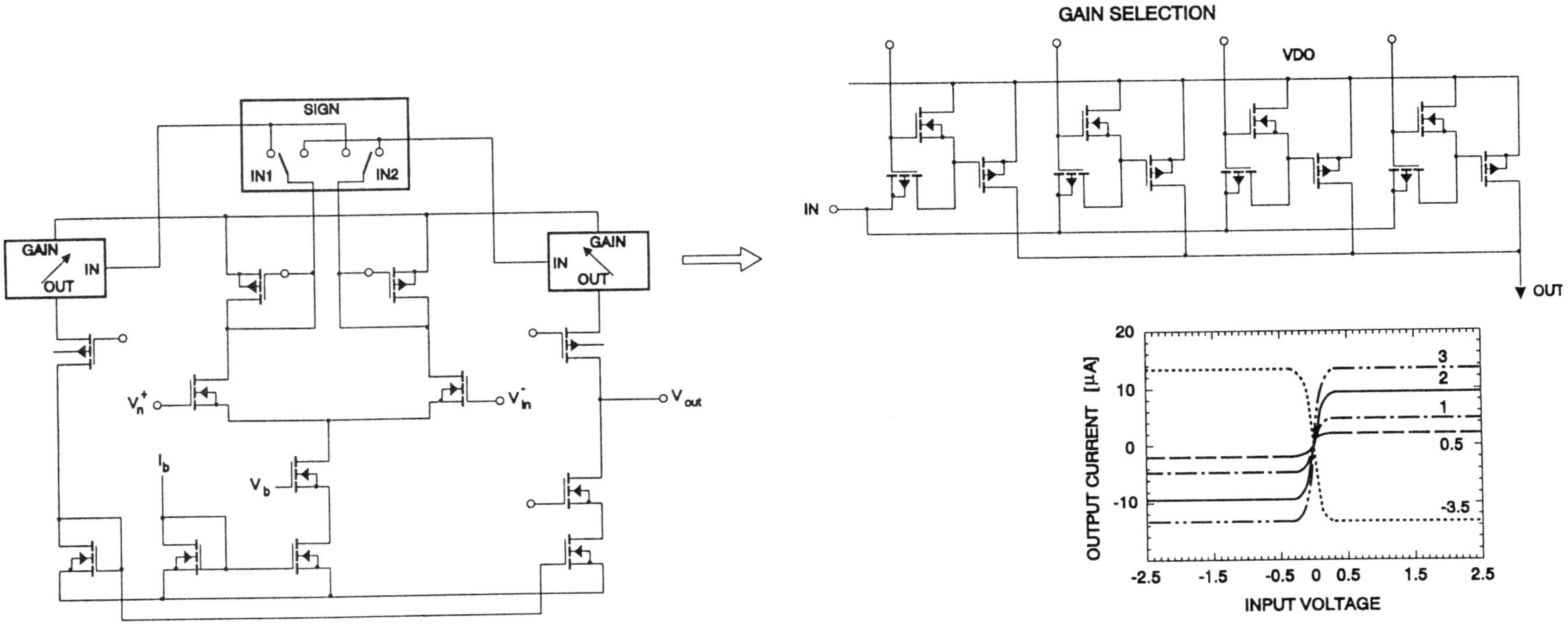

Figure 4. Variable gain OTA and SPICE-simulated output characteristics

gain select can be used. Because the transconductance of an MOS transistor depends on the square root of the biasing current, it leads to a non-linear relationship between the saturation current level and the current gain in the linear region and is therefore unacceptable.

To tune the current mirror factor B, we use a hardware programming method in which extra transistors are switched in parallel with the current mirror transistors (Figure 4). Digital control signals can be used directly to programme the gain factors. This method requires a large number of control lines and much hardware when the full 1-neighbourhood or a lot of gain levels, i.e. a large variety of template values, are required. In this simple case with only nine different gain levels we already have 20 control signals, even though the gain selection encoders are added into each cell. However, for problems of a reasonable range this method leads to the most reliable and both process- and noise-immune solution and therefore it was applied.

4. DUAL COMPUTING STRUCTURE

In addition to the analogue core processor each cell contains local memory and logic, i.e. a dual computing structure (Figure 2).[3] This digital memory consists of three one-bit words, called v_y^{-1}, v_y^0 and v_y^1, containing the output voltage information of the states of the last, present and calculated outputs respectively. The memory is scrolled down after each transient.

The v_y^1-value may also be calculated from the last, v_y^{-1}, and present, v_y^0, output values by using the logic operator $\boldsymbol{L}$ as follows:

$$v_y^1 = \boldsymbol{L}(v_y^{-1}, v_y^0) \tag{4}$$

After application of the operator $\boldsymbol{L}$ the memory is scrolled down again. In the current realization $\boldsymbol{L}$ can be selected with two digital control signals 11 and 12 to be one of the four simple Boolean operators NAND, NOR, NOT or XOR. However, depending on the application, more complex operations are also easily realized.

The output of the cell is read from the memory, thus providing the necessary buffering and conversion from the analogue to digital levels. Also, the feedback after logic operation can be performed from the output of the memory to the sate or input capacitor for the next analogue computation.

5. CONTROL FLOW OF THE PROGRAMMABLE CNN

The operation of our CNN network is divided into three phases (Figure 2):

(i) input and initial state loading
(ii) performing the transient
(iii) register read, analogue or digital feedback and output read.

The realized chip will consist of 4×4 cells with simple digital row select encoders included on the chip. The operation of the network is synchronized with the TRAN/LOAD control signal. During the LOAD phase the input voltage and the initial state are loaded into the input and state capacitors using multiplexing and writing on one row at a time.[5] The input write and output read lines could be multiplexed; however, in the current realization they are separate to ease testing of the chip. During the TRAN phase the neural network transient is performed. The output of the cell is automatically fed through the comparator into the digital register after each transient. Externally the output is read from the register output. During the next LOAD phase either the analogue or digital feedback can be performed.

The analogue feedback (see Figure 5(a)) is realized by sampling the analogue output voltage V_{yij} on the capacitor C_{fb} at the end of the neural network transient. This voltage is fed through a buffer amplifier to either the input capacitor C_{uij} or the state capacitor C_{xij}. The input write line is disabled from the cell input with W_{ext} low. The feedback operation is controlled by the W_{fb} digital control signal driving the analogue switches. Whether feedback is performed on the C_{uij} or C_{xij} can be selected with the $W_{cx/cu}$ control signal.

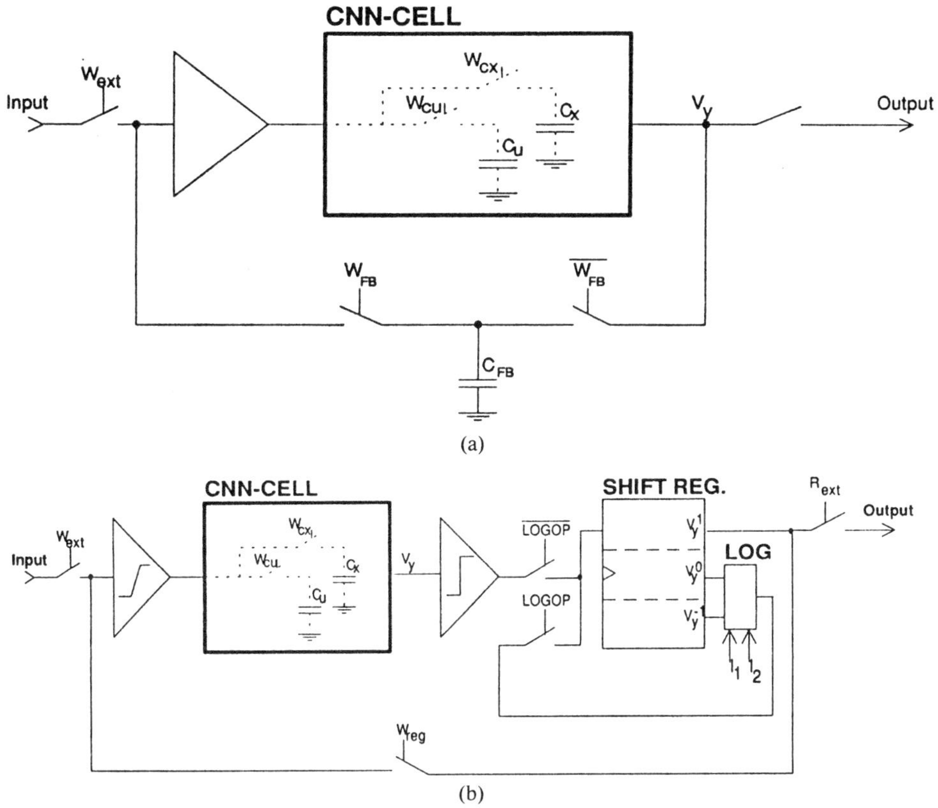

Figure 5. Principal flow graphs of (a) the analogue feedback and (b) the digital feedback

The digital equivalent of the analogue output V_{yij} is also formed at the end of each transient with a comparator so that neural network level 1 corresponds to logic level 1 and neural network level -1 corresponds to logic level 0. This value is stored in the digital register inherent in each cell (see Figure 5(b)). The output of the register can be fed out, but some logic operations can also be performed. The register is a shift register with three places in which the current and previous results have been stored. The logic operations can be any; howver, the current implementation includes INV, NOR, NAND and XOR. After the digital operation the result is feedback to the top of the shift register, from where it can be either read out or fed back to the input capacitor C_{uij} and/or the state capacitor C_{xij} as in the analogue feedback mode. However, from the digital levels the analogue levels corresponding to -1 and 1 are formed with a sigmoid.

An example of the analogue feedback and the template programming is shown in Figure 6. During the first cycle the hole filler template is applied on the input. As expected, the hole is filled. After the transient the limited output V_{yij} is fed back into the state capacitor and the original input is written on the input capacitor. The templates are modified to perform the XOR function on the content of the input and state capacitors. Thus cell 11 goes up while the rest will go down, indicating the position of the hole. This operation could also have been performed by the digital operators.

To test the circuit, a general-purpose control system has been designed (Figure 7) including both the programme and data memories. The programme memory consists of the sequence of operations to be performed, such as read and write operations, start transient, template selection, internal feedback selection and local logic operation selection. The data memory will have the initial values for the input and state capacitors of the CNN array as well as the template coefficient values. The photomicrograph of the realized CNN chip is shown in Figure 8.

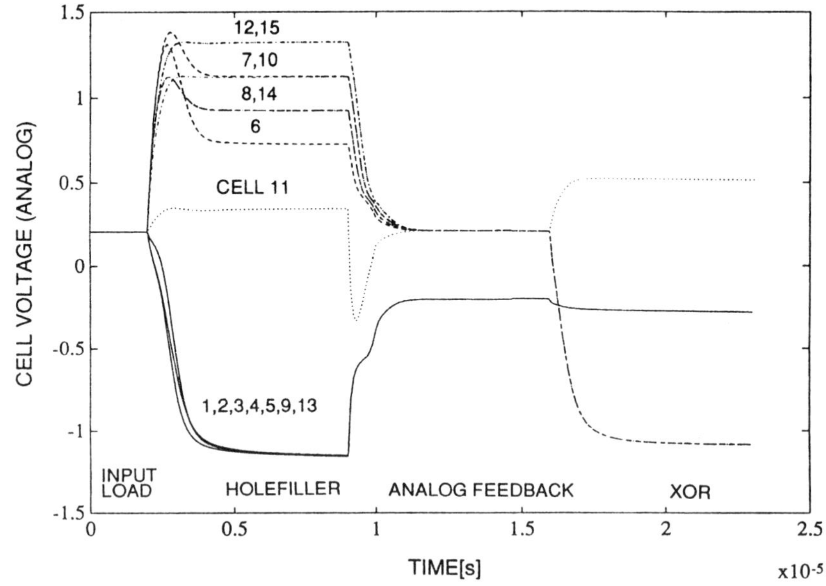

HOLEFILLER TEMPLATE:

$$A = \begin{matrix} 0 & 1 & 0 \\ 1 & 2 & 1 \\ 0 & 1 & 0 \end{matrix} \qquad B = \begin{matrix} 0 & 0 & 0 \\ 0 & -3 & 0 \\ 0 & 0 & 0 \end{matrix} \qquad I = 1.2$$

$$\text{INPUT} \quad \begin{matrix} -1 & -1 & -1 & -1 \\ -1 & 1 & 1 & 1 \\ -1 & 1 & -1 & 1 \\ -1 & 1 & 1 & 1 \end{matrix} \qquad \text{INIT.STATE} \quad \begin{matrix} 1 & 1 & 1 & 1 \\ 1 & 1 & 1 & 1 \\ 1 & 1 & 1 & 1 \\ 1 & 1 & 1 & 1 \end{matrix} \qquad \text{OUTPUT} \quad \begin{matrix} -1 & -1 & -1 & -1 \\ -1 & 1 & 1 & 1 \\ -1 & 1 & 1 & 1 \\ -1 & 1 & 1 & 1 \end{matrix}$$

XOR TEMPLATE:

$$A = \begin{matrix} 0 & 0 & 0 \\ 0 & 2 & 0 \\ 0 & 0 & 0 \end{matrix} \qquad B = \begin{matrix} 0 & 0 & 0 \\ 0 & -2 & 0 \\ 0 & 0 & 0 \end{matrix} \qquad I = -1$$

$$\text{INPUT} \quad \begin{matrix} -1 & -1 & -1 & -1 \\ -1 & 1 & 1 & 1 \\ -1 & 1 & -1 & 1 \\ -1 & 1 & 1 & 1 \end{matrix} \qquad \text{INIT.STATE} \quad \begin{matrix} -1 & -1 & -1 & -1 \\ -1 & 1 & 1 & 1 \\ -1 & 1 & 1 & 1 \\ -1 & 1 & 1 & 1 \end{matrix} \qquad \text{OUTPUT} \quad \begin{matrix} -1 & -1 & -1 & -1 \\ -1 & -1 & -1 & -1 \\ -1 & -1 & 1 & -1 \\ -1 & -1 & -1 & -1 \end{matrix}$$

Figure 6. HSPICE simulations of the 4×4 CNN circuit with the hole filler and XOR templates in sequence

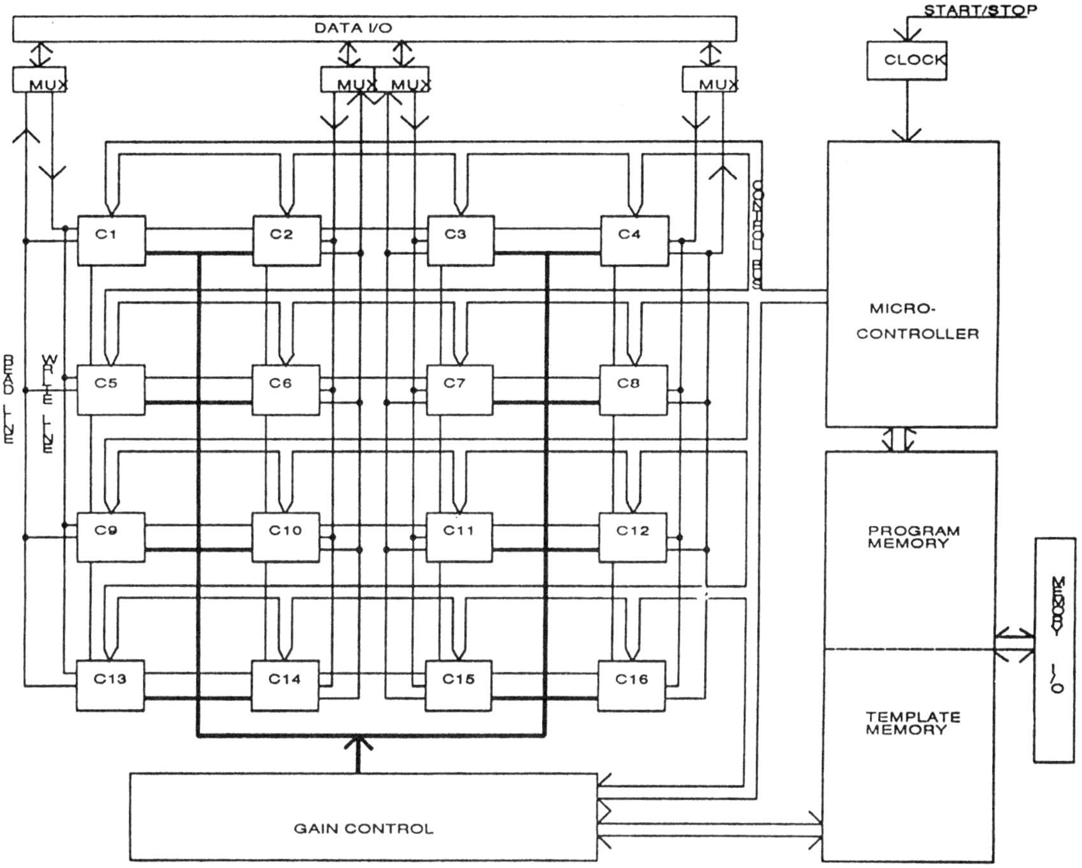

Figure 7. Control system for the CNN chip

6. APPLICATION POTENTIAL

The class of operators realizable by this CNNL chip is extremely wide. Many application possibilities are mentioned in References 1, 2, 4 and 6. Typical applications are character recognition and image processing. One application of interest is motion detection,[6] where a combination of the analogue dynamics and some simple local logic provides surprising effects. In realizing the template element values by OTAs, an inherent but small time delay is introduced. By increasing this delay deliberately, delay-type template elements[2] can also be realized.

7. CONCLUSIONS

A new VLSI implementation of a cellular neural network (CNN) has been presented in which the template coefficents are programmed digitally or continuously. The discrete coefficient values can be selected to be 0, $\pm 0{\cdot}5$, ± 1, ± 2 and ± 3. This enables the same hardware to be used in solving problems with a sequence of templates. As a penalty, the cell becomes more complex, the number of transistors and the cell size are increased and techniques to reduce them have to be looked for. Also, the connectivity problem may become apparent owing to the many control lines and switches needed to achieve programmability of the template. To test the circuit, a general-purpose control system has been designed whose interface is fully digital and can be connected e.g. to a personal computer.

Figure 8. Photomicrograph of the realized CNN chip

ACKNOWLEDGEMENTS

The authors would like to thank Professors Robert Brodersen and Paul Gray for providing the resources (MOSIS) necessary in this work.

REFERENCES

1. L. O. Chua and L. Yang, (a) 'Cellular neural networks: theory' and (b) 'Cellular neural networks: applications', *IEEE Trans. Circuits and Systems*, **CAS-35**, 1257–1272 and 1273–1290 (1988).
2. L. O. Chua and T. Roska, 'Cellular neural networks with nonlinear and delay-type template elements', *Proc. IEEE CNN'90 Conf.*, Budapest, December 1990, IEEE, New York, 1991, pp. 12–25.
3. T. Roska, 'Analog events and a dual computing structure using analog and digital circuits and operators', in P. Varaiya and A. B. Kurzhanski (eds), *Discrete Event Systems*, Springer, New York, 1988.
4. K. R. Krieg, L. O. Chua and L. Yang, 'Analog signal processing using cellular neural networks', *Proc. ISCAS'90 Conf.*, New Orleans, LA, May 1990, IEEE, New York, pp. 358–961.
5. L. Yang, L.O. Chua and K. R. Krieg; 'VLSI implementaton of cellular neural networks', *Proc. ISCAS'90 Conf.*, New Orleans, LA, May 1990, IEEE, New York, pp. 2425–2428.
6. T. Roska, T. Boros, P. Thiran and L. Chua, 'Detecting simple motion using cellular neural networks', *Proc. IEEE CNN'90 Conf.*, Budapest, December 1990, IEEE, New York, 1991, pp. 127–138.

TESTABILITY IN ANALOGUE CELLULAR NEURAL NETWORKS†

A. RUEDA AND J. L. HUERTAS

Dto. Diseño Analógico, Centro Nacional de Microelectrónica Univ. de Sevilla, Edificio CICA, Avda. Reina Mercedes, E-41012 Sevilla, Spain

SUMMARY

This paper addresses the problem of testing an ACNN by postulating the need for including some extra hardware to render feasible a postfabrication test. The work presented here deals with a test methodology based on adding some extra circuitry to every cell of a regular ACNN. This methodology is just an initial proposal for taking advantage of the network regularity to perform a global test which can be externally interpreted and hence has potential application for reconfiguring the network.

1. INTRODUCTION

VLSI implementations of analogue cellular neural networks (ACNNs) have recently been reported.[1-3] The difficulty of testing them as well as the significant influence of the yield on the final performance of the overall system have been recognized.[1] In this paper we propose testing procedures which take advantage of the very nature of the cellular architecture. We have tailored our approach to the presently reported implementations,[1-3] though it can be adapted to other implementation schemes. This contribution is aimed to raise a test awareness for ACNNs in order to motivate designers to consider simultaneously the synergy between design and test in the case of large-scale analogue networks.

2. ASSUMPTIONS

For the sake of simplicity we will restrict ourselves henceforth to ACNNs with the following constraints.‡

(a) *Complexity*. We consider ACNNs with $r = 1$ and whose cells have only four 'active' neighbours (i.e. only four coefficients in $A(i, j; k, l)$ can be non-null).

(b) *Inputs*. Cells fulfil $B(i, j; k, l) = 0 \quad \forall i, j, k, l$, i.e. only the feedback operator exists.

(c) *Structure*. Networks have a symmetrical template (i.e. the non-null coefficients in $A(i, j; k, l)$ must be equal and distributed in a cross-shaped way). For example,

$$A = \begin{bmatrix} 0 & 1 & 0 \\ 1 & 2 & 1 \\ 0 & 1 & 0 \end{bmatrix}$$

In addition, we will assume that no independent sources exist inside a cell, with the exception of the bias sources.

In general, the above constraints are not very restrictive and are applicable in most practical problems. Nevertheless, they can be released easily, our motivation for them being simplicity in dealing with implementations.

---

† Part of this research has been reported in the proceedings of the 1990 International Workshop on Cellular Neural Networks and Their Applications held in Budapest.

‡ In what follows we will use symbols and conventions from Reference 4.

Also, we will consider in this paper several assumptions concerning faults. These assumptions are intended to settle a meaningful framework for our purposes.

(a) All faults will be considered as permanent.
(b) Around a faulty cell, every neighbour is operational, i.e. if cell $C(i,j)$ is faulty, cell $C(k,l)$ with k,l either $i-1,\ j-1$ or $i+1,\ j+1$ is fully operational.
(c) The probability of a fault depends on the complexity. In other words, the probability of a fault for an op amp is much higher than the probability corresponding to a connecting wire.
(d) Beyond a given tolerance threshold, faults originating a slight degradation in voltages or currents will be treated as equivalent to stuck-at faults.

3. TESTING ALTERNATIVES

We can consider two ways for testing an ACNN, depending on whether the whole network has to be tested at once or not. We will refer to them respectively as *global* and *local* methods.

A *global* test relies on the successive application of input patterns to the whole network, followed by the interpretation of the global CNN state and/or output. Hence in a *global* testing approach a set of test patterns has to be applied, this set depending on the specific network. The derivation of an efficient test set is one of the main concerns. Both the test derivation time and the test complexity will increase with increasing network size and the time required for the application of the patterns will depend on the number of different patterns as well as on the number of different initial conditions.

In contrast, in a *local* test each cell is separated from the others and tested. Thus the problem is broken into testing either simultaneously or sequentially all cells in a network after isolating every one. The same input patterns are applied to all the cells to be tested and their states and/or outputs are interpreted. Then a *local* testing approach may be the best alternative since it is general (does not depend on the particular network), the number of patterns is much lower and their dimension is reduced. The penalty to be paid is the inclusion of some extra hardware to make every cell accessible during the test phase. Additionally, a method for testing an isolated cell must be developed.

4. CELL TEST

The next point is to derive a method to test every cell in any CNN. A simple procedure for testing can be proposed from the basic properties of these networks. In practice it is enough to explore the local equilibrium points associated with every cell.

We can formalize this procedure as follows. Let us consider the equivalent circuit of a cell in an ACNN as shown in Figure 1.[4] There are five circuit elements: a linear capacitor with a positive capacitance C, a linear resistor of value R_x, two piecewise linear voltage-controlled resistors and a time-varying current source whose output is given in our case by

$$g(t) = A(i,j;k,l)v_{ykl}(t)$$

because of our constraint on the inputs ($B(i,j;k,l) = 0 \quad \forall i,j,k,l$).

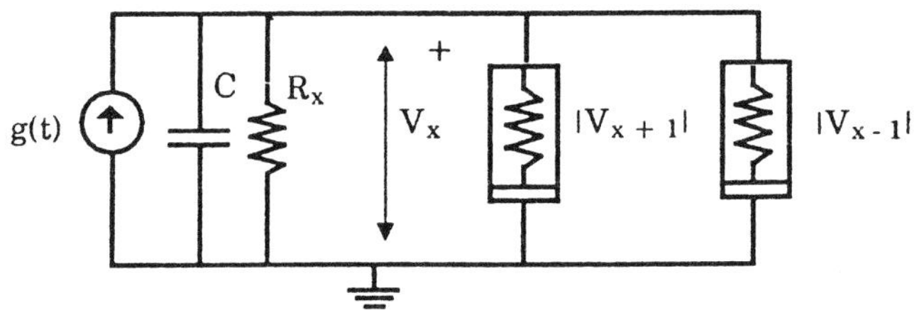

Figure 1. Equivalent circuit

Let us now assume that using switches we can disconnect all the couplings from the neighbours to the cell $C(i,j)$, i.e. $v_{ykl}=0 \quad \forall k,l=i,j$. In this case the equilibrium equation can be written as

$$C\frac{\mathrm{d}v_{xij}}{\mathrm{d}t}=-0\cdot 5A(i,j;i,j)(|v_{xij}+1|-|v_{xij}-1|)+\frac{v_{xij}}{R_x}$$

If we are able to establish an initial condition or, equivalently, to supply the cell (through a switch) with a test input value V_{0xij}, we can restate a theorem in Reference 4 to be used as the basis of any local test strategy.

Local test theorem (for a proof see Theorem 5 in Reference 4)

In an ACNN, if the circuit parameters satisfy

$$A(i,j;k,l)>1/R_{xij}$$

then we have the following properties:

$$\lim_{t\to\infty}|v_{xij}|>1$$

and either

$$\lim_{t\to\infty} v_{yij}=1 \quad \text{if } V_{0xij}>0 \quad \text{for } 1<i<M,\ 1<j<N$$

or

$$\lim_{t\to\infty} v_{yij}=-1 \quad \text{if } V_{0xij}<0 \quad \text{for } 1<i<M,\ 1<j<N$$

Remark In the case where our assumptions apply, there are three equilibrium points in the cell, two of them being stable; the other will be located midway between the two bias voltages and, although never observed, points out a separation between the stable ones. Then the above theorem allows us to test a cell by using two values of V_{0xij}, one positive and one negative, none of them being required to be very precise as follows from the theorem.

Without more precise quantitative models for analogue faults, the only thing we can do is consider the kind of faults which can be masked when a test is applied to an isolated cell. For example, if owing to a fault a cell has more than two stable states and two of them are able to give outputs like the expected ones, then the fault may be masked if we only apply the test as stated in the local test theorem. However, such a fault is very unlikely to be produced and the probability of detecting it might be increased by repeating the test procedure for more than two initial conditions.

5. ARCHITECTURAL STRATEGY

Based on the ability of testing every cell in the array, we can propose different methodologies to apply the test to every cell. We will introduce herein one alternative based on applying a sequence of steps which can be summarized as follows.

1. Reconfiguring the cell mesh to isolate every cell.
2. Fixing initial conditions to 'force' output state $+1$ (in the sense of the local test theorem).
3. Processing/storing (locally) the output value.
4. Fixing initial conditions to 'force' output state -1.
5. Processing/storing (locally) the output value.
6. Comparing (digitally) the two output states.
7. Reading the comparator output.
8. Interpreting (externally) the availability of each cell.

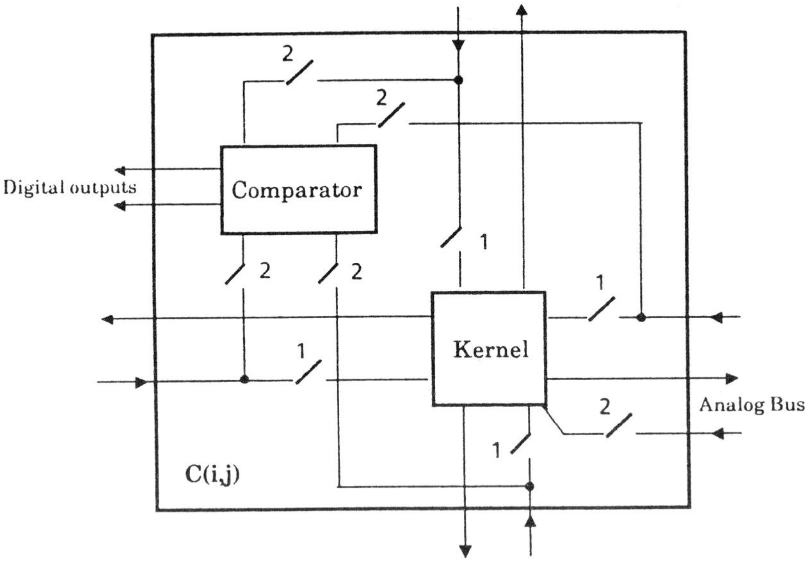

Figure 2. Proposed scheme for a testable cell

Figure 2 shows the basic cell structure for this case. The block labelled kernel is the processing cell itself. Several switches connect (or disconnect) the external inputs in such a way that these inputs are either connected to the kernel (during the normal operating state) or to a comparator (during the testing phase). This comparator is intended for providing a signal indicating whether or not the four signals coming from the neighbours are equal. This is done for both equilibrium states. In this case every cell 'knows' if its surrounding neighbours are all operational. External interpretation will provide a 'map' for excluding the faulty cells.

In this case there exists redundancy in the local test patterns, since the four surrounding cells are checked by every cell. Also, the paths linking the cells are tested in the process.

An actual circuit will depend on the implementation style we select for the cell kernel. Following the known CNN implementations, current is conveying the information. Current comparison is then the basic operation. A simple way to efficiently design the required four-neighbour current comparator is shown in Figure 3. Figure 4 shows a transistor-level representation of the basic block in Figure 3. This figure corresponds to one of the halves of the current comparator. Two complementary current mirrors provide the sum of the input currents with two different polarities. The double output through two inverters gives a digital self-testing output.

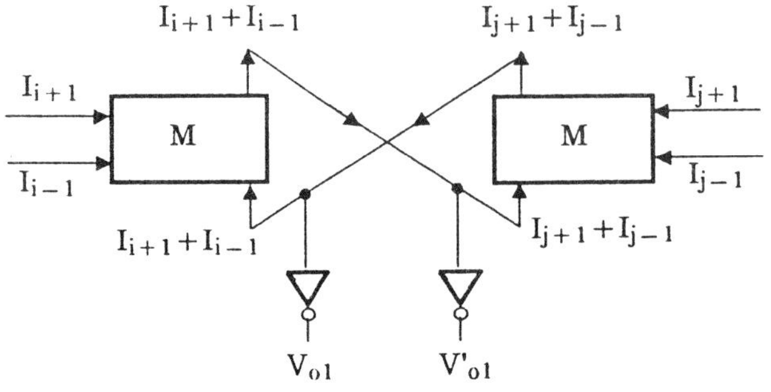

Figure 3. Four-neighbour current comparator (FNCC)

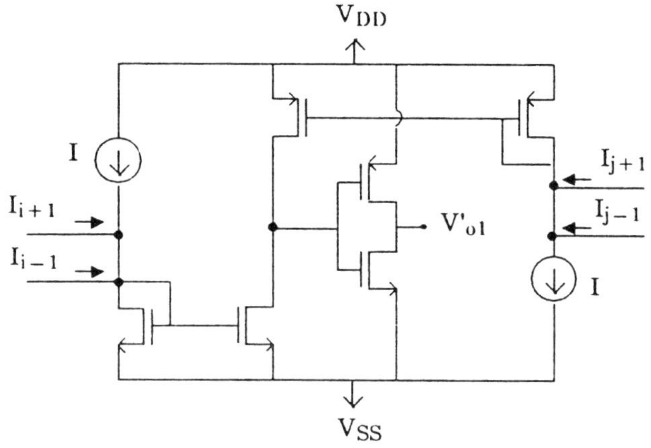

Figure 4. Half-circuit for the FNCC

6. CONCLUSIONS

The main idea drawn from this paper is that developing practical large-scale CNNs relies on integrating design, test and reconfiguration. The first point has been addressed before,[1–3] the second is considered herein and the third is an avenue still unexplored. We have considered an architecture to enhance the testability of ACNNs and to alleviate the verification process. In the work presented here we have dealt with a methodology based on adding some extra circuitry to every cell of a regular ACNN to render feasible a postfabrication test. This methodology is just an initial proposal for taking advantage of the network regularity to perform a global test which can be externally interpreted and hence has potential application for reconfiguring the network. The proposed method is based on a divide-and-conquer strategy since it relies on isolating the cells and testing each one on the basis of the bistable nature of the cells when isolated. Since it requires extra hardware, a trade-off has to be established in terms of testability improvement versus extra silicon area, power consumption and speed.

It is worth finishing this paper by emphasizing again the importance of simultaneously considering design, test and reconfiguration when ACNNs are concerned. We must learn from the digital world that the bottleneck in producing large integrated circuits is mainly due to the separation of these three actions.

REFERENCES

1. L. Yang, L. O. Chua and K. R. Krieg, 'VLSI implementation of cellular neural networks', *Proc. ISCAS 1990*, Vol. 4, pp. 2425–2427, IEEE, New York, 1990.
2. A Rodriguez-Vazquez, R. Dominguez-Castro and J. L. Huertas, 'Accurate design of analog CNN in CMOS digital technologies', *Proc. IEEE Int. Workshop on Cellular Neural Networks and Their Applications*, Budapest, December 1990, IEEE, New York, 1991, pp. 273–280.
3. J. E. Varrientos, J. Ramirez-Angulo and E. Sanchez-Sinencio, 'Cellular neural networks implementation: a current-mode approach', *Proc. IEEE Int. Workshop on Cellular Neural Networks and Their Applications*, Budapest, December 1990, IEEE, New York, 1991, pp. 216–225.
4. L. O. Chua and L. Yang, 'Cellular neural networks: theory', *IEEE Trans. Circuits and Systems*, **CAS-35**, 1257–1272, October (1988).

A DIGITAL MULTIPROCESSOR HARDWARE ACCELERATOR BOARD FOR CELLULAR NEURAL NETWORKS: CNN-HAC†

T. ROSKA, G. BÁRTFAI‡, P. SZOLGAY, T. SZIRÁNYI, A. RADVÁNYI, T. KOZEK§, ZS. UGRAY AND Á. ZARÁNDY§

Computer and Automation Institute of the Hungarian Academy of Sciences (MTA SzTAKI), Úri u. 49, Budapest, Hungary

SUMMARY

Analogue realizations of neural networks are superior in speed. The hardware accelerator boards using catalogue programmable VLSI ICs represent a trade-off having higher reconfigurability and lower cost. This paper presents such a solution for a cellular neural network (CNN).

The architecture of the present design (CNN-HAC) using four standard DSPs to calculate the transient response of a one-layer CNN containing $(0{\cdot}25–0{\cdot}75) \times 10^6$ analogue neural cells (depending on the type of template) is presented. The architecture and also the design principles are independent of the number of processors. The actual design was made in the form of a PC add-on board.

The global control unit, which connects the board to the host firmware and communicates control signals to/from the local control units of the DSPs, was realized mainly with EPLDs.

A special correspondence between the virtual processing elements—calculating the time-discrete models of the analogue neural cells—and the physical ones is discussed in detail. It is realized in an architecture with a simple, two-directional interprocessor communication. This architecture can be 'scaled down' using faster processors, EPLDs and memories. The present version runs with 2 μs/cell/iteration speed.

1. INTRODUCTION

The cellular neural network (CNN) is a special type of analogue non-linear processor array.[1] Its use, including the extension with non-linear and delay-type template elements,[2] provides a rich variety of functional possibilities, especially in real-time image processing.

The fabrication of analogue VLSI CNN arrays is on the way,[3] but in most cases only with a few thousand processing elements (PEs) on a chip and with limited programmability. The present paper describes a hardware accelerator architecture (HAC) for the hardware simulation of a CNN, which provides a much higher speed than software simulation does. On the other hand, it is capable of handling nearly a million PEs with flexible programmability. Hence it represents a useful trade-off in terms of speed–programmability–complexity.

The main points of our architecture[4] are as follows:

(i) transformation of the analogue processors (neurons) of the cellular neural network into virtual fixed point digital ones
(ii) mapping of the array of virtual digital processors onto a realizable set of physical processors
(iii) a simple interprocessor communication method (between the physical processors) providing fully parallel operations
(iv) flexible programming of the various analogue templates.

---

† Part of this research has been reported in the Proceedings of the 1990 International Workshop on Cellular Neural Networks and Their Applications held in Budapest.

‡ Present address: Computer Science Department, Victoria University of Wellington, Wellington, New Zealand.

§ Also with the Technical University of Budapest.

We have designed the accelerator board with four DSPs. The actual parameters and the development environment for specifying the various CNN templates will be discussed as well.

As a result, 2 μs/cell/iteration speed has been achieved.

2. THE ARCHITECTURE

2.1. Transformation of the analogue processors into fixed point digital ones

The CNN processor array with $M \times N$ analogue processing elements is considered (M rows and N columns). Suppose that the processors are connected to their nearest neighbours only, i.e. $r = 1$. The crucial step in calculating the CNN transient response with the forward Euler formula can be summarized[1] as in equation (1). Namely, at the nth time step the new, $(n+1)$th state value of the ijth element, v_{xij}, can be calculated from the step size h and the old, nth state, v_{xij}, and output, v_{yij}, values of the neighbouring cells by

$$v_{xij}(n+1) = (1-h)v_{xij}(n) + h \sum_{C_{kl} \in N_r(ij)} A(ij; kl)v_{ykl}(n) + hI + h \sum_{C_{kl} \in N_r(ij)} B(ij; kl)v_{ukl}(n),$$
$$v_{yij}(n) = f(v_{xij}(n)) \tag{1}$$

The cell amplifier variables are shown in Figure 1, where $\hat{A}$ and $\hat{B}$ are the cloning templates and $N_r(ij)$ is the r-neighbourhood of the ijth element. The function f is a piecewise linear sigmoid of unity slope. In equation (1) linear templates A and B are assumed with $Q = 1$.

The first question is how to determine the representation of the values. Some analysis and extensive software experiments show that a 0·001 relative accuracy is more than enough for the integration. This means that with an input/output voltage between +1 and −1 V and if

$$\Sigma \mid A(ij, kl) \mid + \Sigma \mid B(ij, kl) \mid + I \leqslant 20 \quad \text{and} \quad h \leqslant 1 \cdot 5$$

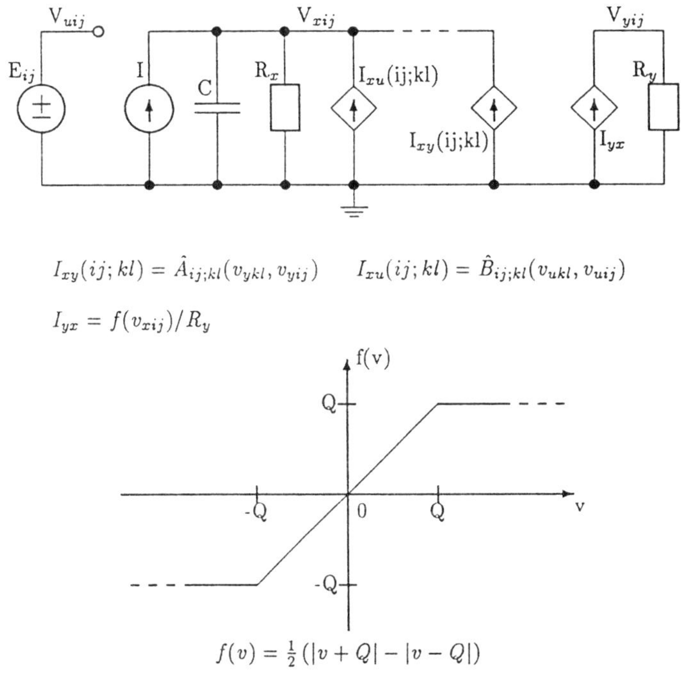

Figure 1. The cell amplifier (neuron)

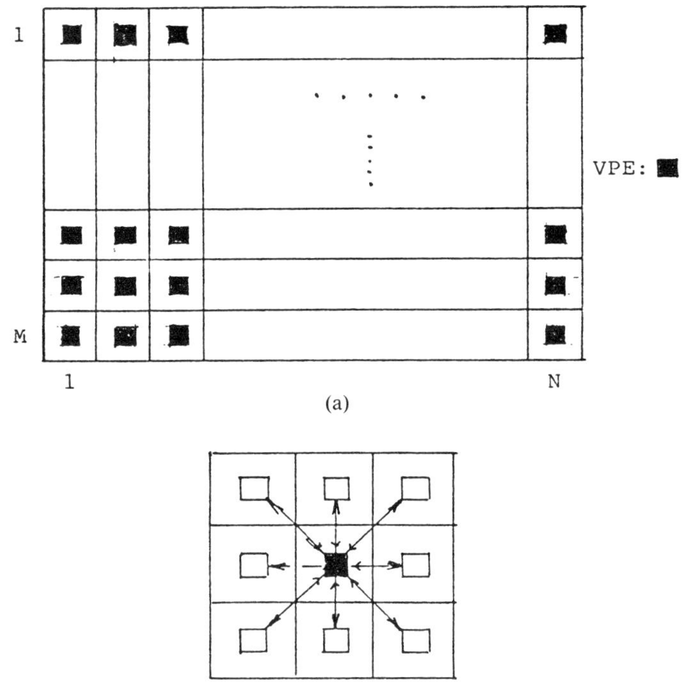

Figure 2. The CPU array and the local communication lines

then a 1 mV resolution is enough. Using a 16 bit fixed point representation, $\pm 32\,000$ units $= \pm 32$ V are the voltage limit values in equation (1).

The next question is how to transform this formula into an architecture of $M \times N$ fixed point digital processors retaining both the structure and the communication lines (Figures 2(a) and 2(b)). For each analogue processor cell (amplifier) a digital processor is assigned. This processor has a local memory to contain the values of the variables v_{uij} and v_{xij} in the N_1-neighbourhood of the *ij*th element in discrete time (present and next values); v_{yij} need not be stored since it can easily be calculated from v_{xij} (see Figure 1). At each time step, before performing the calculations, the processors have to communicate the appropriate data values to their neighbours. In this case all the processors calculate the same expression (equation (1)).

2.2. Decomposition of the virtual processor array into a smaller number of physical processors with a simple communication scheme

Since CNNs can be used also for image processing, large arrays of cells have to be processed. However, the number of digital processors on a board (or chip) is limited. Therefore we are restricted to using much fewer physical processing elements (PPEs) than the number of array cells, called virtual processing elements (VPEs). Figure 3 shows a structure using one PPE for $5 \times 5 = 25$ VPEs.

In order to separate the communication actions between PPEs from the calculation of state equation (1), some additional memory space is sacrificed. In general, in a CNN of an r-neighbourhood template, if each PPE stands for a VPE array of $M_V \times N_V$ size, then the data area of VPEs processed by one PPE has to be enlarged by an r-element-wide border, i.e. a VPE data space of size $(M_V + 2r) \times N_V + 2r)$ is processed by one PPE. This enables independent calculations in each PPE. After each time step the neighbouring VPE data values are communicated in each direction. Figure 4 shows the case for $M_V \times N_V = 6 \times 8$ and $r = 2$.

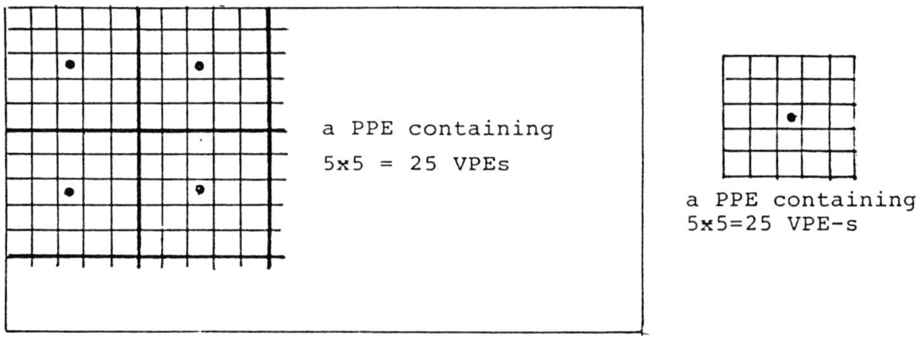

Figure 3. PPEs serving smaller VPE arrays

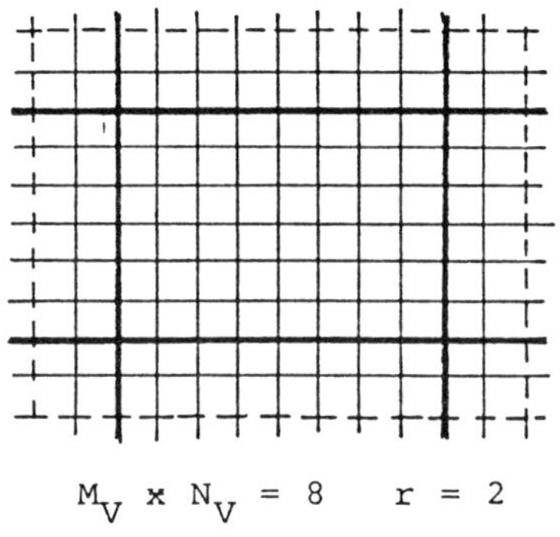

Figure 4. PPE area with margins

However, communicating in eight directions (Figure 2(b)) is time-consuming and may result in synchronization problems. We have chosen the decomposition scheme shown in Figure 5, where a total vertical band of VPEs is assigned in each PPE. The advantage of this scheme is that communication between neighbouring PPEs is required only across the left and right vertical borders between in one direction only at a time.

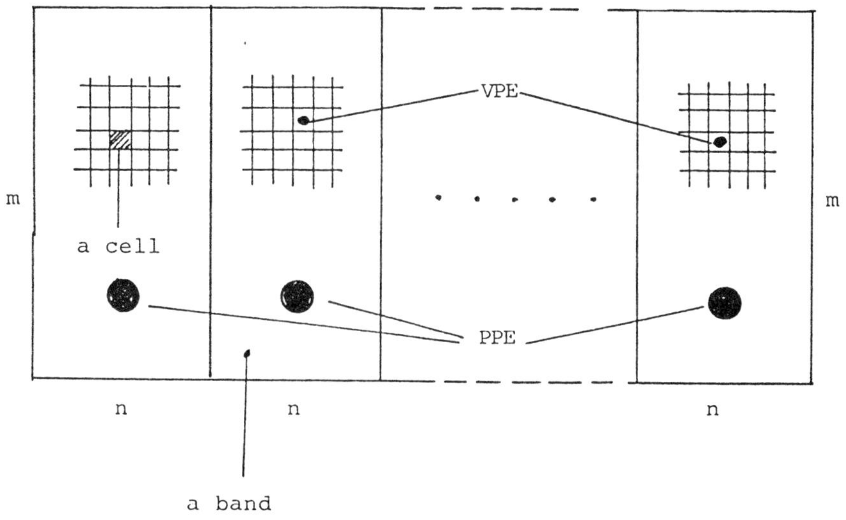

Figure 5. The band decomposition

In vertical band decomposition, where $M_V = M$, i.e. the number of VPE rows assigned to one PPE is equal to that in the original image, for the neighbourhood r we store $N_V + 2r$ elements in a row for each PPE. The first/last r places in each row are for storing the data values belonging to the last/first r VPEs of the left/right neighbouring VPE band. In this architecture the task of the physical processors can be organized fully parallel, with only one synchronization in each time step to perform the comparatively few and simple interprocessor communications.

2.3. *Programmability*

In the calculations of equations (1) we need additions, multiplications and comparisons only. Digital signal processors (DSPs) are ideally suited for this task. Moreover, DSPs can also perform calculations which can realize useful non-linear and delay-type templates too.[2] The programmability of DSPs ensures the change in various cloning templates and their sequencing.

3. THE ACTUAL HARDWARE IMPLEMENTATION

The CNN-HAC has been designed in the form of an IBM PC add-on board. It has some control logic and four essentially identical processor blocks working in parallel when computing the next state equation (1), of a CNN. Each processor block has its own local dynamic memory to store the processor programme code, the cloning templates and the appropriate data values as signed 16 bit numbers.

3.1. *Design considerations*

The CNN-HAC circuitry has been realized on a single IBM PC add-on board with a 265×90 square grid (one grid = 1·27 mm) design area. Preliminary placement calculations showed that besides the common control logic, the four processor blocks including memories and local communication circuitry could be accommodated on this area, assuming that SMD components were applied.

When selecting the components, one basic point was that their total price should be less than $1000. On the basis of extensive time-domain functional simulation of the architecture[5] as well as price/performance, programmability and software support considerations, a Texas TMS320C25 type DSP with 40 MHz clock frequency and 10 MIPS computing power has been chosen for the PPEs.[6] The different control logic sections have been realized by Intel 5C060 (600 gate equivalent) and Intel 5C090 (900 gate equivalent) PLDs programmed with CUPL (Logical Devices Inc.). As external memory to the processors, Hitachi HB56D25608B-10A type $256\text{ k} \times 16$ CMOS dynamic RAM modules have been applied. The interprocessor communication has been organized through AMD AM7202-80JC $1\text{ k} \times 16$ FIFOs.

Figure 6 shows photographs of the CNN-HAC board.

3.2. *The main blocks of the CNN-HAC board*

Figure 7 shows the block diagram with relevant segments of the CNN-HAC architecture. The function of the parts are as follows.

1. PC1 is the central control block supervising the main states and state transitions of the board. It performs the interfacing to the IBM PC bus as well. It is accessed as an I/O periphery in 16 bit CPU transfer mode.

In this solution all the processors are synchronously started in every iteration. The processor blocks have four main states: RESET_&_HOLD, PROCESS, IDLE and HOLD.

2. IT is the interrupt control logic. The board initiates an interrupt to the PC at the end of calculations or when an error has occurred during interprocessor communication.

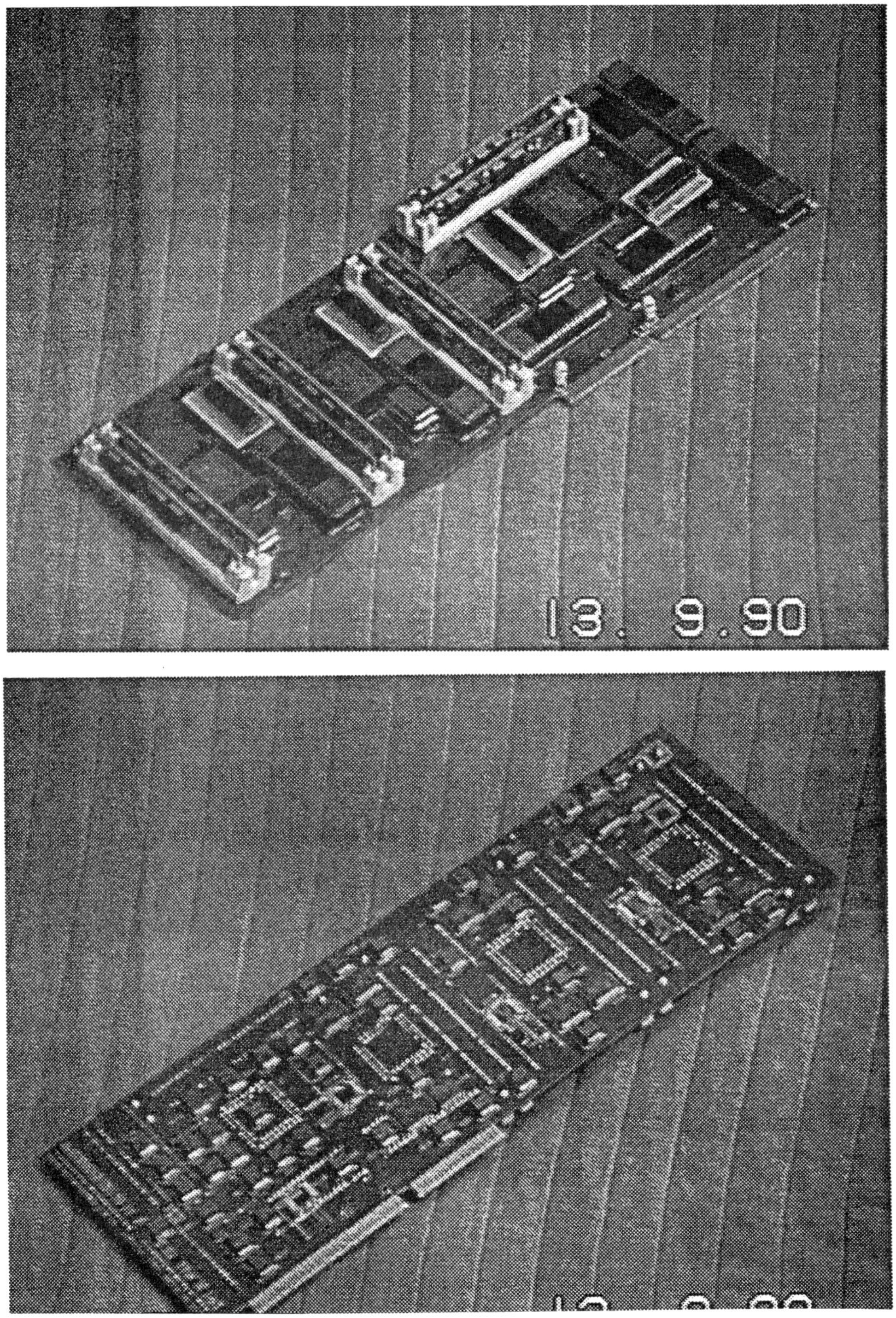

Figure 6. Top and bottom views of the experimental CNN-HAC board

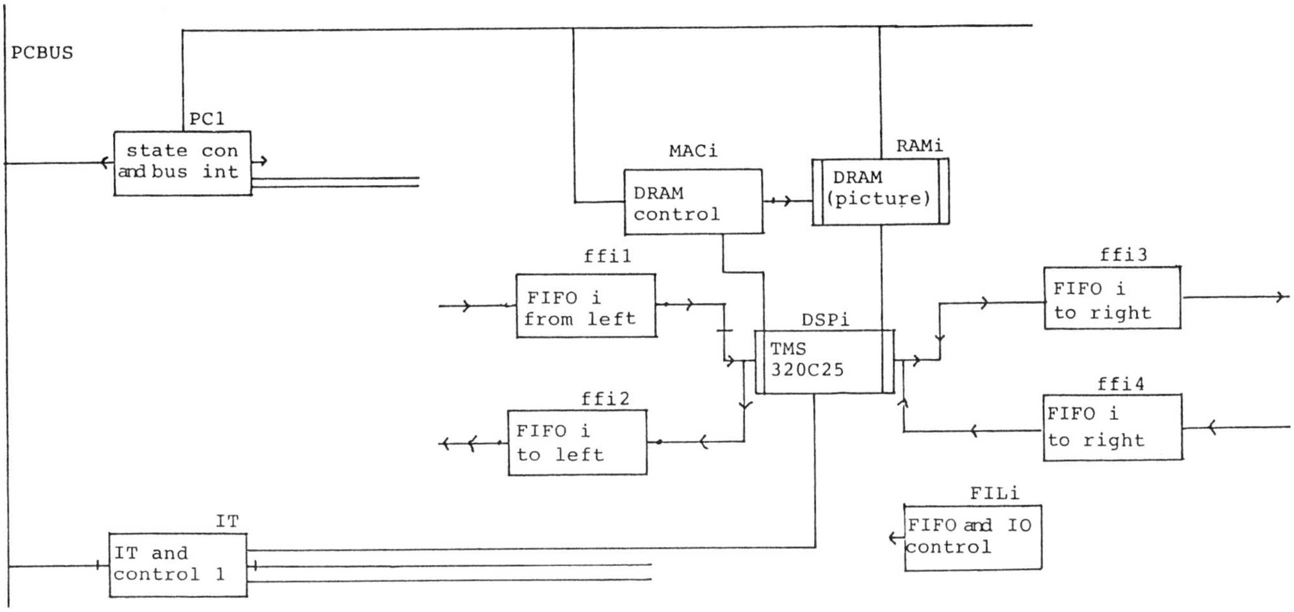

Figure 7. Relevant segments of the CNN-HAC architecture

On the board there are four essentially identical processor blocks, each around a TMS320C25 digital signal processor. The only difference among the blocks is that the right-hand-side/left-hand-side communication circuitry of the rightmost/leftmost block is substituted with a module giving constant zeros.

3. Two FIFO blocks, each having 512 × 16 byte effective storage capacity, provide the communication path between neighbouring processor blocks. The communication is supervised by the FIL module, which functions as the I/O port decoder of processors and also as the interrupt control logic for invalid operation cases when, owing to software malfunction, any of the FIFOs become full or empty.

4. Each processor block has a DRAM memory array of its own to store the data values belonging to the corresponding VPE band and to the margins of the neighbouring ones. The memory operations are controlled by the MAC module, which contains the memory access and read/write controller, the refresh sequencer and clock generator and a combinational logic functioning as the memory address selector.

4. THE CNN-HAC SOFTWARE

The basic task of the CNN-HAC related software is to iterate equation (1) to calculate the discrete-time transient of a CNN structure defined by appropriate template values, starting with given input values from a given initial state. Moreover, some auxiliary tasks such as data acquisition, visualization of the results, etc. are performed. Since the tasks are divided between the host and the CNN-HAC board, their communication has to be organized by additional software.

The allocation of some software segments is self-evident, e.g. the data acquisition and the visualization of results must happen on the host and the calculations related to equation (1) have to be allocated to the CNN-HAC board. Communication routines and interrupt handlers are necessary on both sides of the host–board interface. However, there are tasks which can be performed on either side. Since on one hand we have two computing units running parallel, we have to optimize the allocation of such routines to achieve the best overall performance. The routines which can be on either side are the ones performing the data transformation between the input, internal and output formats. In general, the transformations can run more efficiently on the board, but they do mean an overhead in calculations there. On the other hand, since the length of data blocks of different formats may be considerably different, the allocation of transformation routines may heavily influence the communication demands between the host and the board.

4.1. *The software development environment*

As the design of the CNN board and its software went on in parallel, a software development environment was built up to enable the CNN programme coding and debugging activities to proceed independently of the actual hardware works. The kernel of this tool is the TMS 320 SIMULATOR.[7] It is invoked by a programme called CALLSIM, which besides reading the initial data and template values and displaying the result produces the appropriate data and TMS machine code blocks to be sent to the SIMULATOR and receives the result blocks as well.

To check the mathematical accuracy of the calcualted results, they can be compared with those of the software simulator CNND[8] with implicit integration formula.

4.2. *The host-side CNN-HAC software*

The function of the host-side CNN-HAC software is twofold. On one hand it provides a menu-driven user-friendly framework in which both the selection of input images and template sequences and the visualization needs can be easily stated. A choice of acceptable data formats for black-and-white or grey-scale pictures is available, enabling the programme to input from various sources such as scanners, video cameras, computer graphics, etc. The visualization sampling rate of the CNN transient is selectable in terms of the number of steps of iteration equation (1). In visualizing the results, user-selectable pseudo-colours are used for intermediate grey tones. To scrutinize fine texture pictures, windowing with independent vertical and horizontal magnification is available. Besides, the CNN-HAC host programme package contains a set of utilities for transforming data formats, changing picture size and aspect ratio to compensate distortion of some sources, grey-scale histogram calculation and transformation, grey-scale interpolation during changing picture size, etc.

In addition to managing the user interface, the host-side CNN-HAC software organizes the communication of data blocks and programme modules between the host and the CNN-HAC board. Transient snapshots and steady state results can be uploaded and stored in disk files of selectable format for later use, whenever necessary.

4.3. *The CNN-HAC on-board software*

Using CALLSIM to read and display data and to invoke the TMS 320 SIMULATOR under the control of appropriately prepared journal files, a series of routines to calculate equation (1) and perform the auxiliary tasks has been developed. For special structure sparse templates with one non-zero row or column only, suitably tailored modules are available for downloading to achieve an even higher on-board speed in these practically important special cases. For advanced applications, modules capable of calculating CNN transients with non-linear and delay-type templates[2] are being developed. The routines running on the different processor blocks are identical; they are independent of the position of the processor they run on as well as of the actual number of processor blocks on the board. (Consequently, the same routines will be able to run without modification on future, advanced CNN-HAC boards too.)

In the CNN-HAC architecture each processor block has a local external memory to store the programme routines to execute, the values of the cloning template elements and the v_{uij} and v_{xij} data values of the corresponding VPE band. During a run, depending on the internal memory capacity of the DSP, suitable segments of the programme and data are swapped in and out to achieve the best overall performance.

The six TMS assembler programme modules performing the basic calculations in PROCESS state of the board are as follows.

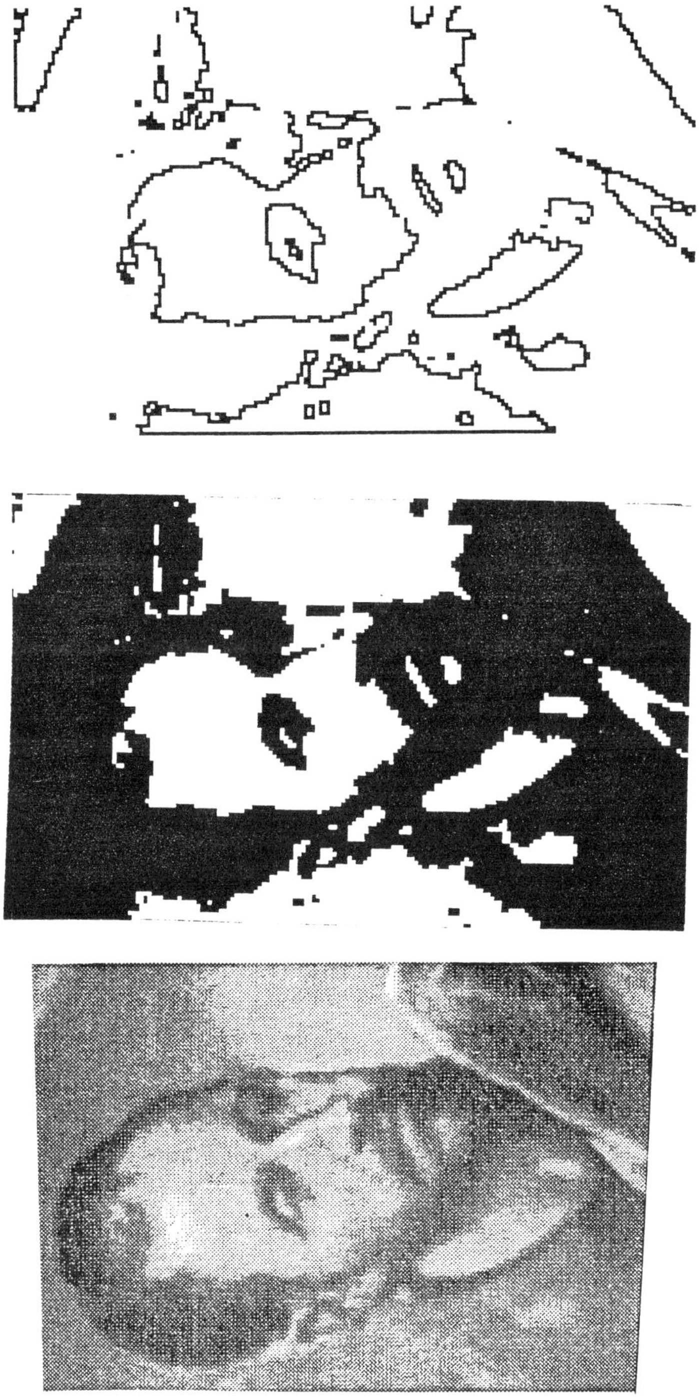

(c) (b) (a)

Figure 8. (a) A grey-scale picture of size $358 \times 224 = 80\,192$ pixels. (b) The resulting black-and-white picture after averaging. (c) The result of edge detection

1. Module INIT initializes the processor and allocates memory for programme variables and data arrays for later use in other modules.
2. Module UCOM sends and receives the v_{uij} input values along the VPE band borders through the interprocessor communication channel via I/O ports and the connecting FIFO circuitry to and from the neighbouring processor blocks. This module is not activated when, owing to special actual template structure, horizontal data flow is unnecessary.
3. Module BTEM, once per transient, calculates the produce–sum related to template B and the v_{uij} input values and adds the constant I to the sum; see equation (1).

After finishing the preparatory calculations made by the previous three modules, the next three modules are loaded into the processor internal programme memory to perform the iteration.

4. Module YCOM sends and receives the v_{yij} output values in just the same way as module UCOM does. Similarly to UCOM, neither is this module activated when, owing to special actual template structure, horizontal data flow is unnecessary.
5. Module ATEM calculates the product–sum related to template A and the v_{yij} output values and adds it to the previously stored result of module BTEM to produce the new iteration of v_{xij} state values; see equation (1).
6. Module SIGM calculates the v_{yij} output values from the v_{xij} state values with a piecewise linear sigmoid function of unity slope.

Having finished the actual iteration step, each processor block assumes IDLE state to wait until an interrupt starts the next iteration or until being sent into HOLD state where the host can read the results. Before HOLD state the processor blocks may optionally be sent to PROCESS state to execute the seventh programme module OTRANS to transform the results into a more compact video format, which, besides requiring much less communication, is directly suitable for displaying on the host.

5. EXAMPLES

To illustrate the CNN-HAC applicability, two examples are presented. The neighbourhood size was $r = 1$ and the step parameter h was set to 0·75. The first example is a grey-scale picture of size 358×244 pixels (VPE) processed with the averaging template[1b] to produce a black-and-white picture. The input picture and the result obtained after eight iterations are given in Figures 8(a) and 8(b) respectively. The measured processing speed was 2 μs/pixel/iteration.

In the second example the black-and-white borders of the previous result are extracted using the edge detector template.[1b] The result is shown in Figure 8(c). To get this result, four iterations were necessary and the processing speed was 2 μs/pixel/iteration.

6. DISCUSSION

This architecture is optimal in the sense that for 16 bit multiply–add–store operations DSPs are ideal and the communication cost is negligible. Keeping the architecture unaltered and using advanced, more powerful components (TMS320C50 DSPs, 1 MB SRAMS, faster PLDs), six to eight times faster operation can be achieved according to our estimation. Moreover, using more compact devices (memories and EPLDs), even more processor blocks can be accommodated on a board of the same size.

ACKNOWLEDGEMENTS

This work has been supported by the National Research Fund in Hungary (OTKA, grant 1212/88) and in part by the National Science Foundation (grant MIP-8614000), the Office of Naval Research (grant N00014-89-5-1402) and a joint grant of the National Science Foundation and the Hungarian Academy of Sciences (INT-90-01336).

REFERENCES

1. L. O. Chua and L. Yang, (a) 'Cellular neural networks: theory' and (b) 'Cellular neural networks: applications', *IEEE Trans. Circuits and Systems*, **CAS-35**, 1257–1272 and 1273–1290 (1988).
2. T. Roska and L. O. Chua, 'Cellular neural networks with nonlinear and delay-type template elements', *Proc. IEEE CNNA-90*, IEEE, New York, 1991, pp. 12–25.
3. (a) J. M. Cruz and L. O. Chua, 'A CNN chip for connected component detection', *IEEE Trans. Circuits and Systems*, **CAS-38**, 812–817 (1991).
   (b) C. Mead and M. Ismail (eds), *Analog VLSI implementation of Neural Systems*, Kluwer, Boston, MA, 1989.
   (c) C. Mead, *Analog VLSI and neural systems*, Addison-Wesley, Reading, MA, 1989.
   (d) K. Halonen, V. Porra, T. Roska and L. O. Chua, 'VLSI implementation of a reconfigurable cellular neural network containing local logic (CNNL)', *Proc. IEEE CNNA-90*, IEEE, New York, 1991, pp. 206–215.
4. T. Roska, G. Bártfai, P. Szolgay, T. Szirányi, A. Radványi, T. Kozek and Zs. Ugray, 'A hardware accelerator board for cellular neural networks: CNN-HAC', *Proc. IEEE CNNA-90*, IEEE, New York, 1991, pp. 160–168.
5. *PC/TWIN Design System, Version 1.0, User's guide*, Computer and Automation Institute of the Hungarian Academy of Sciences (MTA SzTAKI), Budapest, 1989 (in Hungarian).
6. *Second Generation TMS320 User's Guide*, Texas Instruments, Houston, 1989.
7. *TMS320 Family Development Support Reference Guide*, Texas Instruments, Houston, 1989.
8. T. Roska and A. Radványi (eds), *CNND Simulator, User's Guide, Version 3.01*, *Report No. 36/1990*, Computer and Automation Institute of the Hungarian Academy of Sciences (MTA SzTAKI), Budapest, 1990.

A CNN HANDWRITTEN CHARACTER RECOGNIZER†

H. SUZUKI AND T. MATSUMOTO

Department of Electrical Engineering, Waseda University 3-4-1 Okubo, Shinjuku-ku Tokyo 169, Japan

AND

LEON O. CHUA

Department of Electrical Engineering and Computer Sciences, University of California, Berkeley, CA 94720, U.S.A.

SUMMARY

CNNs are used for feature detection in handwritten character recognition. Detected features are fed to a simple classifier network. Performance was tested by using two well-known ETL data base series: (i) ETL3 consisting of numerals, alphabets and several symbols and (ii) ETL8B2 consisting of Japanese Hirakana characters. The average recognition rate for ETL3 is 94·8%, while that for ETL8B2 is 85·7%. Both series include 'hard' characters so distorted that even humans cannot recognize them.

1. INTRODUCTION

This paper reports an attempt to perform *handwritten* character recognition using the CNN paradigm. In any pattern recognition problem, *feature detection* is of utmost importance.[1,2] Once the main features have been extracted, a decision network (classifier network) can be built with relative ease. When the feature detector network performs poorly, however, it is extremely difficult, often impossible, to construct an accurate classifier. Handwritten character recognition is no exception. In particular, features extracted from handwritten characters must have a *reasonable degree of robustness* (invariance) against (i) *shift*, (ii) *rotation*, (iii) *scale* and (iv) other variations due to *individuals*. In addition, the extracted features must have a much lower 'dimension' than the original data. There are two different classes of invariance that need to be distinguished: (a) *geometric* and (b) *statistical.*

Speaking roughly, (a) corresponds to (i)–(iii) above while (b) corresponds to (iv). It should be emphasized that these invariances are much more difficult to achieve than they appear. In addition, sometimes a *complete* invariance must be *avoided*, because if a system is completely invariant under rotation, then '6' and '9' cannot be differentiated, nor can the symbol '(' cannot be distinguished from ')'. Thus feeding raw data into e.g. a three-layer back-propagation network and asking it to classify the character is completely unrealistic. It would not work except for 'toy examples'. There are hundreds of algorithms belonging to class (a) and hundreds more belonging to class (b). This implies that no single algorithm has been identified as the best. The feature extractors proposed in this paper are no exception in this regard. It will simply be claimed that our handwritten character recognition system works.

The test data used in our experiments are taken from the well-known database for handwritten character recognition, the ETL series.[3,4] Each volume in the series contains handwritten characters of at least 100 different persons. Two volumes were tested:

(i) *ETL3*, consisting of the numeral 0–9, the Roman alphabet (capital) A–Z and 12 symbols ' – ', '(', ')', '*', ' + ', ',', ' = ', '¥', etc. (48 classes), several examples of which are given in Figure 1

† Part of this research has been reported in the Proceedings of the 1990 International Workshop on Cellular Neural Networks and Their Applications held in Budapest.

Figure 1. Typical examples of numerals chosen from the database ETL3

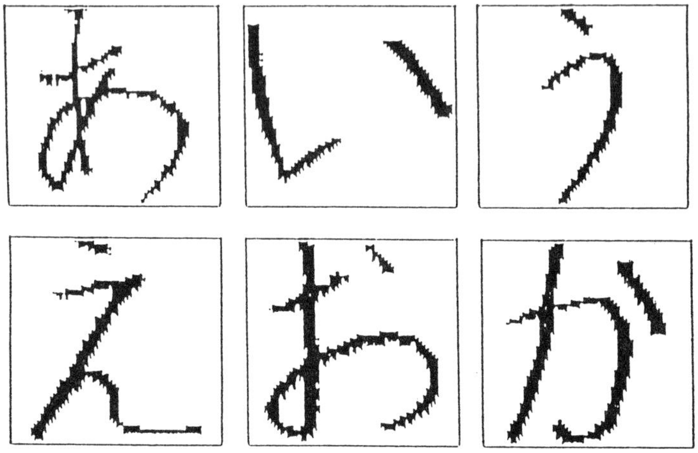

Figure 2. Typical examples of Hirakana characters chosen from the database ETL8B2

(ii) *ETL8B2* consisting of the 71 Japanese *Hirakana* characters, several examples of which are given in Figure 2.

Two CNNs paradigms are used in our character recognition system to extract the essential features:

(i) connected component detectors (CCDs)
(ii) shadow detectors (SDs)

The resulting feature vectors are then fed into a decision network for classification. Out of the 100 data for each character, 30 per cent randomly chosen characters are used to train the decision network. After training, the rest of the data are shown to the network. The average recognition rates are 94·8 per cent for ETL3 and 85·7 per cent for ETL8B2.

2. THE FEATURE EXTRACTORS

2.1. Connected component detectors

Four different CCDs[5] are used to detect the connected components in each character.

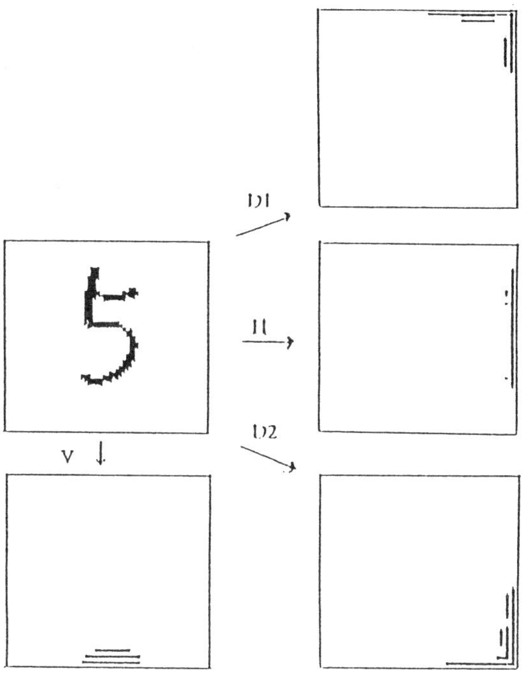

Figure 3. Connected components extracted from the number '5' along four directions

The H- (resp. V-) template detects the number of 'connected components' of a given bipolar (± 1) image in the horizontal (resp. vertical) direction. Similarly, the D1- (resp. D2-) template detects the number of connected components in the 45° (resp. -45°) direction. Figure 3 shows the four CCD results of a typical datum '5'. Observe that all CCDs exhibit complete shift invariance. Our experience tells us that they have a moderate amount of rotation invariance.

2.2. Shadow detectors

The four shadow detectors used in our character recognition system are defined in Reference 6. An example is given in Figure 4, where the LS-template, for instance, gives the 'shadow' which results when a horizontal light beam is applied from the right of the character.

2.3. Data compression

The eight features obtained in Sections 2.1 and 2.2 are not fed directly into a decision network. To see the reason for not doing this, suppose that we are given a 64×64 bipolar image. Then each of the resulting eight features is still a 64×64 bipolar image. The following subsections describe how data compressions is implemented in our system.

2.3.1. CCD data compression Let us first look at Figure 5(a), which gives a typical result obtained by the V-template in Figure 3. Recall that the V-template gives alternating ■s and □s in each column, where the number of ■s gives the number of connected components. Thus the information is contained in how the □s are located. In this particular example there are four 'layers' of ■s. In order to decode these data, let us let Y be the 'width' of the character, i.e. the number of ■s in the 'bottom' layer. We formally define

$$X_{11} = 0, \qquad X_{12} = Y, \qquad X_{13} = 0$$

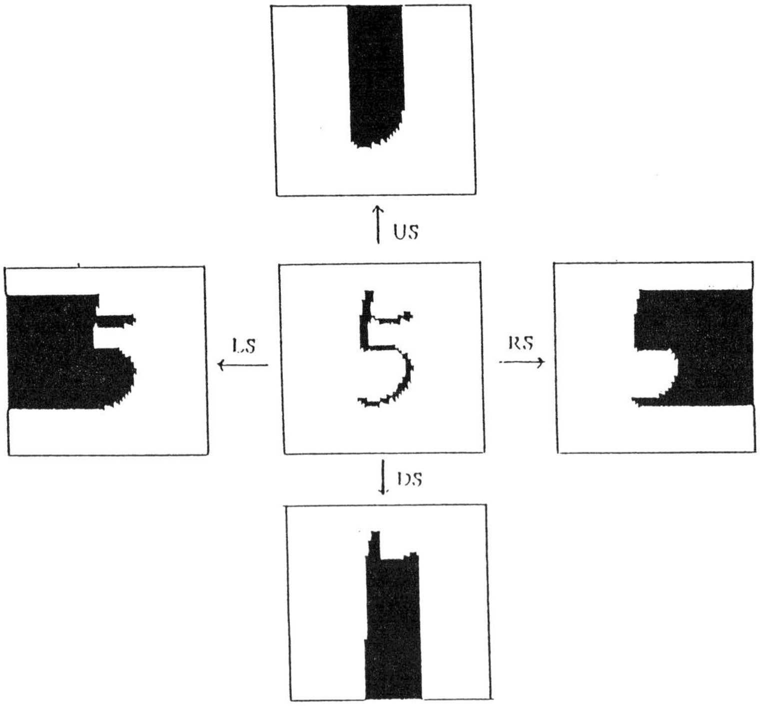

Figure 4. Shadows extracted from the number '5' along four directions

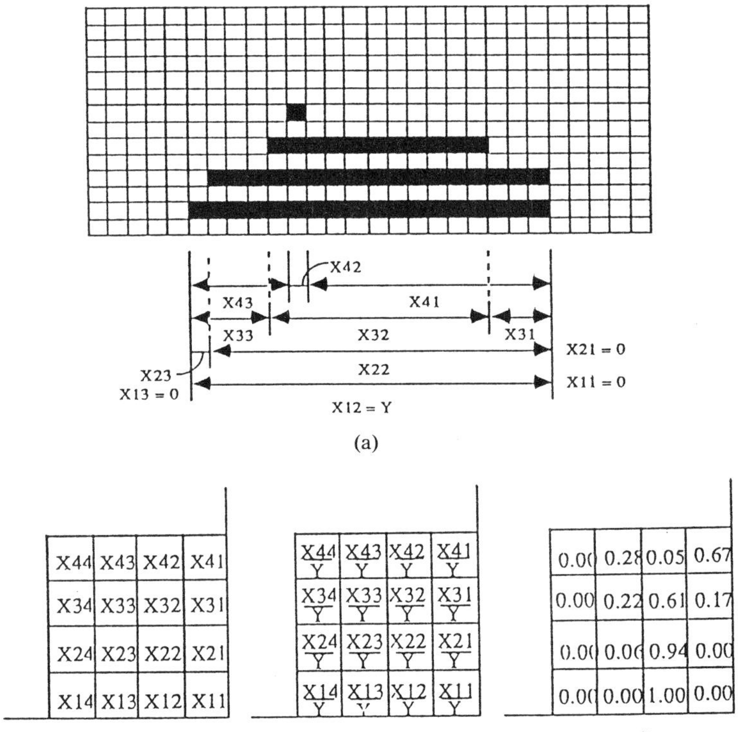

Figure 5. Data compression: (a) a typical example of the CCD result, where X_{ij} is defined as the number of consecutive ■s or □s in the '*i*th layer'; (b) X_{ij}; (c) X_{ij}/Y; (d) values of X_{ij}/Y for (a)

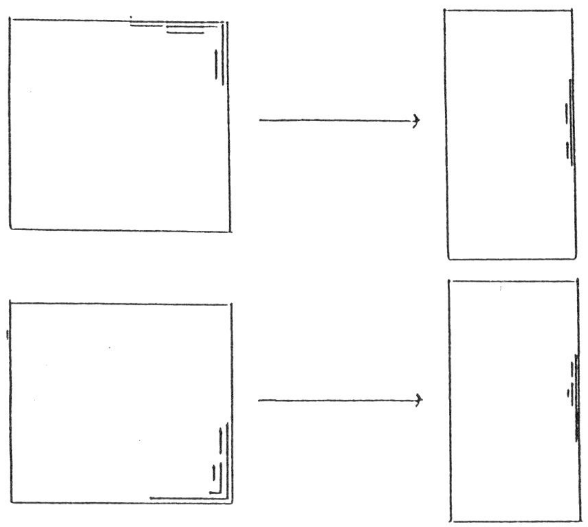

Figure 6. 'Bending' of the D1 and D2 results

Subscript '1' of X_{ij} indicates that the X_{1j} are the data associated with the bottom layer. Similarly, X_{2j} indicate data associated with the ■s and □s of the second (from the bottom) layer, and so on. From these data we obtain Figure 5(b), which is the array of X_{ij}, and Figure 5(c), which is the array of X_{ij}/Y. Figure 5(d) gives the numerical values associated with Figure 5(b).

Since each X_{ij} has been divided by Y, the total size, the resulting data have a great amount of *scale invariance*, which the CCDs themselves do not have. We will call this the *V-data*. A similar operation can be performed on the data resulting from the H-template. The result will be called the *H-data*. In order to obtain the corresponding results for the D1- and D2-templates, we first 'straighten' the data in Figure 3 as shown in Figure 6 (by bending part of the pattern by 90°) and then perform a similar operation. The results will be called the *D1-data* and *D2-data* respectively.

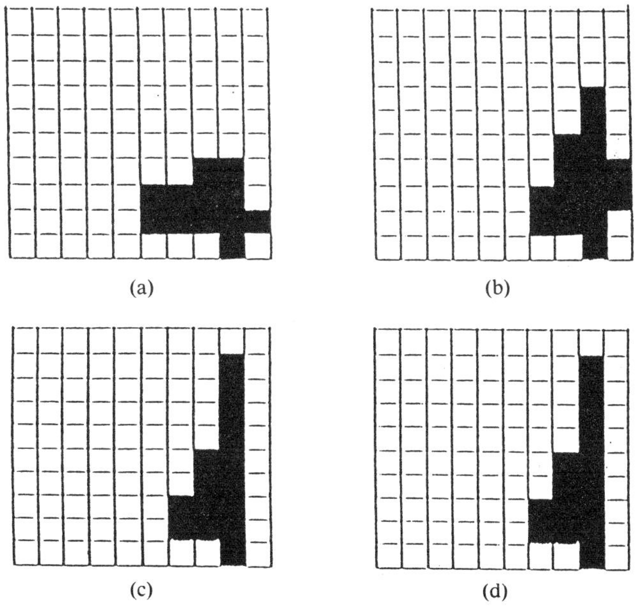

Figure 7. Positions of the X_{ij}/Y on which 99 per cent of the data from ETL3 fall (■ indicates a non-zero X_{ij}/Y): (a) V-data; (b) H-data; (c) D1-data; (d) D2-data

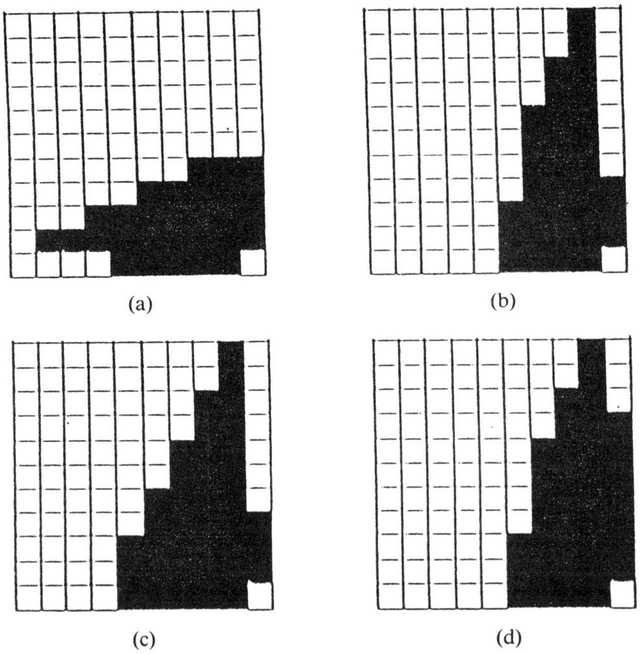

Figure 8. Positions of the X_{ij}/Y on which 99 per cent of the data from ETL8B2 fall (■ indicates a non-zero X_{ij}/Y): (a) V-data; (b) H-data; (c) D1-data; (d) D2-data

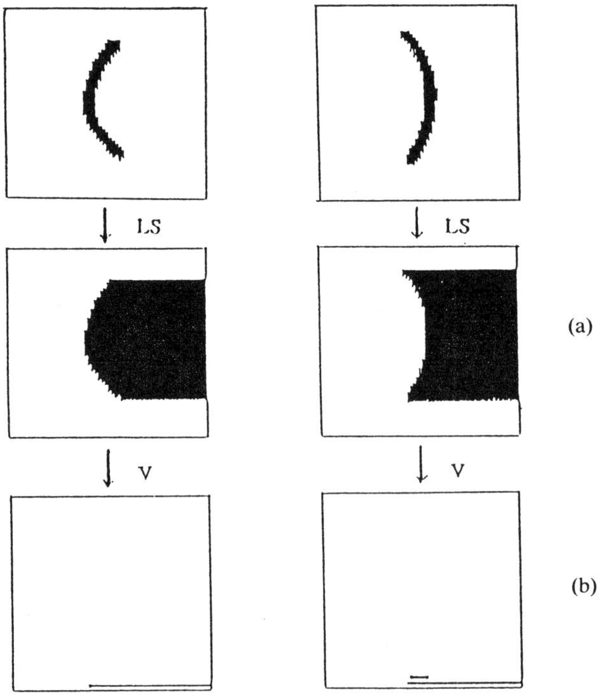

Figure 9. SD templates differentiating '(' from ')': (a) results of LS to '(' and ')'; (b) results obtained by applying the V-template to (a)

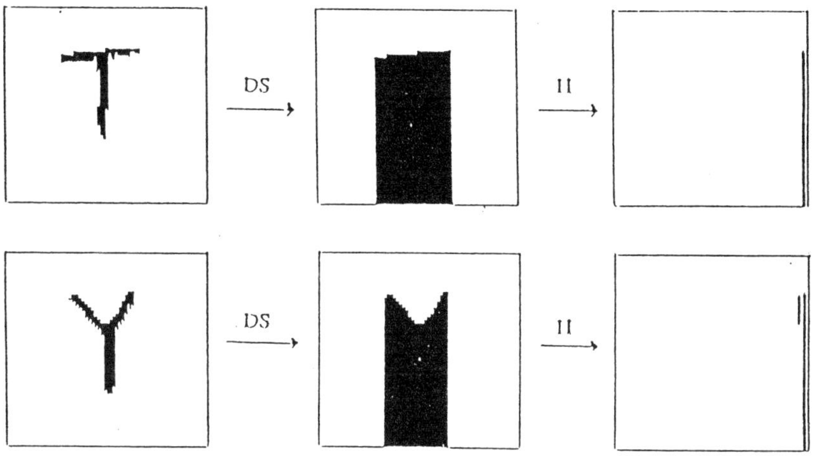

Figure 10. SE templates differentiating 'T' from 'Y'

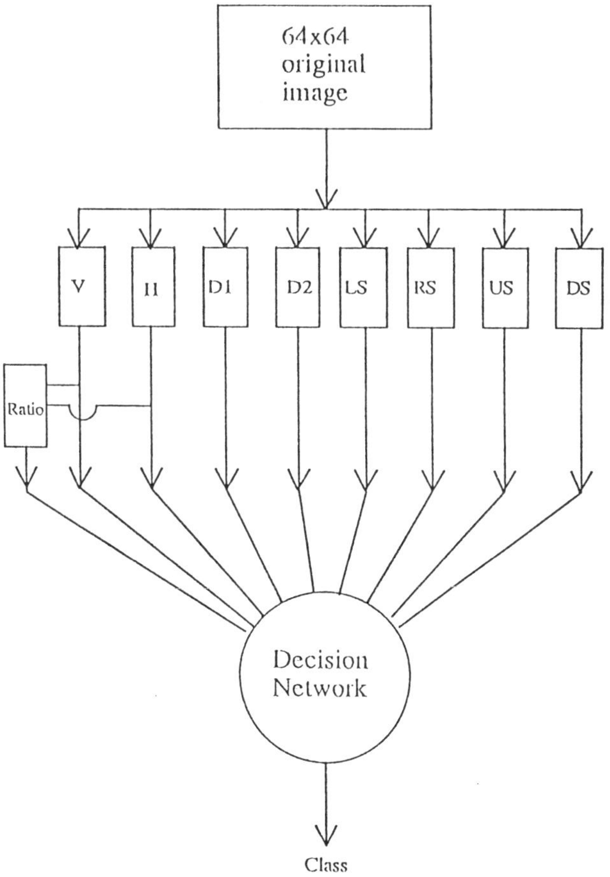

Figure 11. Overall schematic diagram of our handwritten character recognition system

Now a question arises. How many different variations of X_{ij}/Y can result? If there are as many as 64×64, then these transformed data are not any better. It turns out, however, that there are very few compared with 64×64. In fact, the dark squares in Figure 7(a) show the positions of the non-zero X_{ij}/Y-values on which 99% of the V-data fall for ETL3 (■ indicates a non-zero X_{ij}/Y). Only the 10×10 lower right corner of the 64×64 array is shown. The rest are all □s. Note that the connected ■s do not necessarily mean that the original image has only one connected component. A ■ indicates that X_{ij}/Y is non-zero. Figures 7(b)–7(d) show the corresponding results for the H-data, D1-data and D2-data respectively.

Figure 8 shows the results for ETL8B2. Naturally, Figure 8 requires more information than Figure 7 because the Hirakana characters are more complex than the numerals and the Roman alphabet.

2.3.2. SD data compression. From our earlier experiments we have observed that the CCD features alone are already very efficient for handwritten character recognition. One of the factors which limits the effectiveness of the CCD feature detectors, however, is a difficulty in distinguishing between, for instance, '(' and ')' or 'T' and 'Y' because they give rise to very similar CCD data. Let us now apply the LS-template to '(' and ')' (Figure 9). Then one immediately sees the distinction. Of course, the shadow data (Figure 9(a)) do not include any data compression. However, if we apply the V-template to these data, we will obtain Figure 9(b), which clearly distinguishes the symbol '(' from ')'. Similarly, Figure 10 demonstrates how the DS-data can be used to differentiate the letter 'T' from the letter 'Y'.

Speaking roughly, the shadow features detect *directional convexity* or *directional concavity* of a given image. Figure 11 shows the four different shadow features. We can then apply the CCD templates followed by operations similar to that of Figure 5. The resulting data will be called the LS-, RS-, US- and DS-data respectively.

2.3.3. The ratio feature In order to further strengthen our feature detection capabilities, we add one more feature, called the *ratio*, which is defined as the ratio between the length of the 'first layer' of the H-data and that of the V-data.

Table I summarizes the dimensions of the feature vector which incorporates the basic features used in our character recognition system, namely CCD, SD and ratio features.

Table I. Components of the feature vector

	CCD	SD	Ratio	Total
ETL3	57	8	1	66
ETL8B2	137	16	1	154

3. THE DECISION NETWORK

A three-layer back-propagation network (simulated on a digital computer) was chosen as our decision network simply because it was readily available in our environment. There is nothing special about our choice and other networks would do as well. The number of inputs units is the dimension of the feature vector shown in Table I (66 for ETL3 and 154 for ETL8B2). The number of output units is equal to the number of categories (characters) for classification (48 for ETL3 and 71 for ETL8B2). On the basis of our earlier experiments, the number of hidden units was chosen to be equal to the number of output units.

4. RESULTS

The overall schematic diagram used in our experiments is shown in Figure 11. The data supplied by the ETL series are of array size 64×64, some of which are extremely ragged, e.g. see Figure 12.

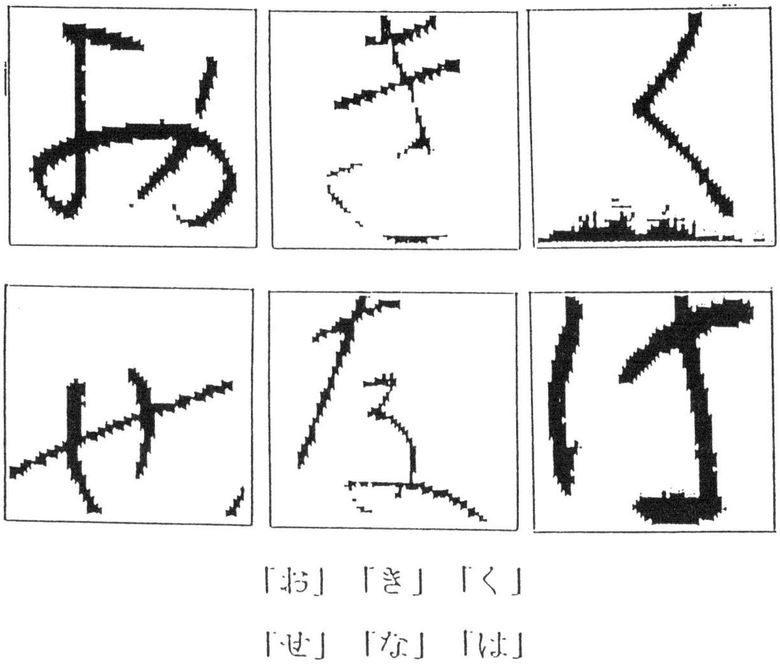

Figure 12. A sample of some raw ETL raw data which are hard to classify even by humans

4.1. Experiment 1

From the 100 data for each character, 30 per cent were chosen randomly and used for training the decision network. The network was trained to output +1 for a desired category. It took an Alliant FX parallel mini-supercomputer approximately 2 weeks for ETL3 and 6 weeks for ETL8B2. After training, the rest of the data were shown to the network. The recognition rate was examined with two different output 'thresholds':

(i) threshold = 0·0, which means that the output unit which outputs the largest value, whatever it is, is considered to be the decision
(ii) threshold = 0·9, where a decision is considered to be made *only when* it exceeds 0·9, otherwise it is considered as a *reject*.

Table II shows the results of this experiment. A reject was counted as an error.

Table II. Effect of threshold on recognition rate

	ETL3		ETL8B2	
Threshold	0·0	0·9	0·0	0·9
Average recognition rate	94·8%	84·0%	85·7%	80·3%

4.2. Experiment 2

In order to test the quality of our feature extractors, we fed all (100) data into the network for training. After training, the data were again shown to the system. If the recognition rate were 100 per cent, our feature detectors would have been perfect. However, there is one *important* factor involved: several

Table III. Feature detection capability after 'bad' samples have been excluded

	ETL3		ETL8B2	
Threshold	0·0	0·9	0·0	0·9
Feature detection capability	99·9%	99·4%	99·9%	99·9%

original data are so bad that even a human cannot classify them. Figure 12 provides a sample of some of them. Thus we tested the network after removing these 'bad' data. (There were 60 out of 100×48 for ETL3 and 391 out of 100×71 for ETL8B2.) Table III gives the results. This, we believe, shows that our feature detectors are functioning correctly, almost as well as a human.

4.3. *Execution time*

All our experiments described above have been done on a digital computer with double precision.

(i) The feature extraction (CNN) simulated on a serial digital computer takes a long time because it is a stable limit point of a 4096-dimensional ODE. If the CNN is fabricated on an analogue chip, the convergence is determined by the capacitance and resistance of the implementation and typically takes several microseconds. This part does not appear to demand high accuracy, though verification will be needed to draw a definite conclusion.

(ii) The encoding of X_{ij} and the forward part of the back-propagation network are simple but may need reasonable precision. Thus this stage should be done by a digital system. If a dedicated DSP were used, it would take a matter of milliseconds.

(iii) Learning would require a fair amount of precision and would take a long time as described, which is true in most back-propagation applications. Therefore this part should be done by a digital computer.

4.4. *Brief comparative comments*

In this subsection we try to make brief comparative studies of our method with other paradigms. In Reference 7, ETL9 Hirakana character (similar to ETL8B2) recognition is performed. Given a 64×64 raw image, the data are divided into 8×8 subimages. By applying four different templates (vertical, horizontal, $+45^\circ$ and -45°), four different directional components are computed. These are used as feature vectors of dimension $4 \times 8 \times 8 = 256$. For the classifier network a hierarchical architecture is proposed. The system consists of two hierarchies: (i) eight 'subnets' and (ii) one 'supernet'. Since there are a total of 71 Hirakana characters, each subnet is responsible for nine Hirakana characters. All of them are back-propagation networks with 256 input units (the dimension of the feature vector), eight hidden units and 10 output units. The number of output units is 10 instead of nine. The last output unit is to tell that the input is *not* within a particular subnet's 'jurisdiction' if a given feature vector does not match any of the nine. After all the subnets have been well trained, the supernet is trained, again by back propagation. The training of the eight subnets is done in parallel on an NCUBE. For training, 100 data were used, and after the training, another 100 were shown to the network. The recognition rate is 82 per cent.

In Reference 8 an experiment is performed on the numerals 0–9 of the ETL1 data, which are similar to 0–9 of ETL3. In addition to line segment features, edge features are detected by applying simple templates. Furthermore, 'holes' are detected by a combination of the line and edge features. An unsupervised learning is utilized and 151-dimensional feature vectors are obtained. The details will be omitted, however. The classifier is based on the correspondence detection idea from binocular stereo vision. After learning, the *same* data were fed into the network. The capability is 95·5 per cent, which, however, is not the recognition rate.

The paradigm shown in Reference 9 is very well known and we will not redescribe it here. Experiments conducted in Reference 9 do not use the ETL data. Furthermore, recognition rates are not reported.

The system reported in Reference 10 utilizes a non-neural net algorithm. It first normalizes a given data set by scale, rotation and shift so that every character is located at the centre with full scale on 64×64 pixels. Then, by dividing the 64×64 data into 8×8 subimages, vertical, horizontal, $+45°$ and $-45°$ features are detected. Reference patterns (code words) are generated by averaging the feature vectors over half of the given data. The classification is done by computing the *generalized distance* between a reference pattern and the feature vector of an input. Speaking roughly, the generalized distance takes the inner product of two vectors while maintaining a reasonable amount of robustness against shift. The data set used is ETL8. There are two free parameters in the generalized distance. The best recognition rate with the two parameters tuned is 83 per cent on the Hirankana characters. The result is better (85 per cent) on the 881 Kanji characters. This appears to be attributable to the fact that Hirakana characters have many curves while Kanji characters are basically unions of line segments.

All the systems described so far are entirely digital, while the system described in Reference 11, which is an extension of Reference 12, is partly analogue and partly digital; namely, of the whole net with five layers, the first three are feature detectors using template convolutions and are executed by an analogue chip, while the last two layers are done by a digital system because more accuracy is needed as one goes up the hierarchy of networks. The system is successfully applied to handwritten zip code recognition.

5. CONCLUDING REMARKS

An attempt has been made to utilize CNN connected component detectors and CNN shadow detectors to obtain features of handwritten Japanese characters. The detected features are fed into a standard back-propagation classifier. The test data are taken from the ETL database series. The experiments indicate that the CNN feature detectors capture good features for recognition tasks. For implementation the CNN part should be done by an analogue chip[13] because it is slow by digital but very fast by analogue and because high accuracy is unnecessary. The classifier as well as the learning should be done by digital because they both demand greater accuracy.

ACKNOWLEDGEMENTS

The authors would like to thank Dr K. Yamamoto and Dr T. Saito of the Electro Technical Laboratory for providing us with the database and References 3, 4 and 7–9.

This research was supported in part by the Japanese Ministry of Education, the Casio Foundation and the Science and Engineering Laboratory, Waseda University.

REFERENCES

1. C. Y. Suen, 'Distinctive features in automatic recognition in handprinted characters', *Signal Process.*, **4**, 193–207 (1982).
2. M. D. Levine, 'Feature extraction: a survey', *Proc. IEEE*, **57**, 1391–1407 (1969).
3. T. Saito, H. Yamada and S. Mori, 'An analysis of hand-printed character data base (III)', *Bull. Electrotech. Lab.*, **42**, 385–434 (1978).
4. T. Saito, H. Yamada, K. Yamamoto and S. Mori, 'An analysis of hand-printed character data base (V)', *Bull. Electrotech. Lab., Waseda, Univ.*, **45**, 49–77 (1981).
5. T. Matsumoto, L. O. Chua and H. Suzuki, 'CNN cloning template: connected component detector', *IEEE Trans. Circuits and Systems*, **37**, 633–635 (1990).
6. T. Matsumoto, L. O. Chua and H. Suzuki, 'CNN cloning template: shadow detector', *IEEE Trans. Circuits and Systems*, **37**, 1070–1073 (1990).
7. Y. Mori, 'Aiming at a large-scale neural network', *IEICE Technical Report PRU-88-59*, 1988.
8. Y. Tsukui and Y. Hirai, 'Character recognition by neuro-pattern matching', *IEICE Technical Report PRU-88-56*, 1988.
9. K. Fukushima, S. Miyake and T. Ito, 'Neocognitron: a natural neural network model for a mechanism of visual pattern recognition', *IEEE Trans. Syst., Man Cyber.*, **13**, 826–834 (1983).
10. T. Saito, H. Yamada and K. Yamamoto, 'An analysis of hand-printed character data base (VI)', *Bull. Electrotech. Lab., Waseda Univ.*, **46**, 695–725 (1982).

11. B. E. Boser, E. Säckinger, J. Bromley, Y. LaCun, R. E. Howard and L. O. Jackel, 'An analog neural network processor and its application to high-speed character recognition', *Proc. IJCNN*, Vol I, 415–420 (1991).
12. H. P. Graf, L. D. Jackel and W. E. Hubbard, 'VLSI implementation of a neural network model', *IEEE Comput. J.*, (March), **21**, 41–49 (1988).
13. J. M. Cruz and L. O. Chua, 'A CNN chip for connected component detection', *IEEE Trans. Circuits and Systems*, **38**, 812–817 (1991).

DETECTING MOVING AND STANDING OBJECTS USING CELLULAR NEURAL NETWORKS

T. ROSKA, T. BOROS AND A. RADVÁNYI

Computer and Automation Institute of the Hungarian Academy of Sciences, Úri-u 49, H-1014 Budapest, Hungary

AND

P. THIRAN AND L. O. CHUA

Department of Electrical Engineering and Computer Sciences and the Electronics Research Laboratory, University of California, Berkeley, CA 94720, U.S.A.

SUMMARY

The general framework of motion detection based on discrete time samples of the moving image is defined. Four types of motion detection problem are studied. The simplest one is a model resembling the famous Hubel-Wiesel experiment with a cat's retina for detecting the motion of an object having a given speed in a given direction. The most complicated case is the determination of the vertical and horizontal velocity components of a moving image.

Various cloning template sequences are proposed for detecting different types of motion. In the sampled mode the consecutive black-and-white snapshots are fed to the input and to the initial state nodes of the cellular neural network respectively. After the transients have decayed, the output gives the information necessary for detecting the presence or absence of a specific motion as well as for estimating the direction and magnitude of the velocity vector. In continuous mode the sampling process is eliminated by the use of delay-type templates.

Conditions are analysed under which the detection is correct. The circuit realization of some motion detectors is discussed and the use of a programmable dual-CNN structure is proposed

1. INTRODUCTION

Motion detection and estimation is an evergreen problem. Several digital computer algorithms and more recently regular analogue 'neural' computing structures have been used for various classes of motion problems. Living systems solve these problems quickly and elegantly. A famous discovery by Hubel and Wiesel[1] showed that if a bar was moving across a particular region of a cat's visual field in a certain direction and if the speed was around a given value, then some specific cortical neurons would fire (detection). The impressive results in Reference 2 show that some simple, analogue regular VLSI computing structures are capable of motion detection and elementary retina functions.

The cellular neural network[3,4] (CNN) is a novel and natural framework for geometrically regular, analogue, non-linear and dynamic computing structures especially suited for image-processing applications. In this paper we will present some CNN algorithms (sequences of CNN templates) which are able to detect and estimate some types of motion. Let us emphasize that in this paper we are not intending to give any biological model, though biological motivation helped us in posing the abstract problem.

The four problem classes, the two types of circumstances and the two modes of operation which are relevant to CNN are as follows.

Problem 1

Let us consider a simple object (e.g. a bar) moving with a constant speed. Detect the object if it moves with a given speed in a particular (say horizontal) direction. This problem is similar to the Hubel–Wiesel experiment.

Problem 2

In this problem the direction of the motion is not taken into account. Detect a simple moving object if the absolute value of the speed is within a given range (e.g. one pixel drift per sampling period).

These two problems will henceforth be referred to as *motion detection* problems.

Problem 3

Determine the magnitude of the velocity if the object is moving in a given (horizontal) direction.

Problem 4

Determine the components v_x and v_y of the velocity vector of a simple object moving in an arbitrary direction.

These two problems are referred to as *speed estimations*.
Two types of circumstances can be considered.

Type A

The single moving object is black while the background is white. The object is restricted to be rigid and fully connected; this is the *ideal* circumstance.

Type B

Disturbing effects are taken into account, e.g. grey-scale images, connectedness of the picture, blurring and noise, small perturbations in the direction of motion, multiple objects, shadowing effects and uncertainty due to the finite resolution; this is the *real* circumstance.

The two types of modes of operation are as follows.

Sampled mode

The consecutive image sequences are sampled in equal time steps.

Continuous mode

The pixel intensities are varying continuously in time, thereby requiring the use of delay-type templates.

In this paper we are not considering how to transmit the changing pixel intensity values from the photosensor layer (camera) to the cells (these questions are extensively discussed in other papers, e.g. Reference 2). Many of the presented results are case studies showing working templates for some useful problems. In our solutions the whole object is considered and not just a part (e.g. an edge) of it. Under these circumstances the aperture problem does not enter the picture.

The general framework, basic principles and notations are given in Section 2. The sampled mode approach to motion detection problems is presented in Section 3. Examples of continuous mode operation are provided in Section 4. Type B solutions are shown in Section 5. Finally, conclusions and further possibilities are discussed in Section 6. The interested reader should refer to Reference 5 for solutions of speed estimation problems.

2. NOTATIONS, GENERAL FRAMEWORK AND BASIC PRINCIPLES

2.1. Sampled mode

In the sampled mode the moving object is represented by a series of image samples P_0, P_1, P_2, ..., P_q, ... The sampling time is Δt; hence $P_q = P(q\Delta t)$. The images are quantized in space and are represented by pixels of size $\Delta x \times \Delta y$. Usually $\Delta y = \Delta x$, i.e. the resolution of the screen is Δx both vertically and horizontally. The intensity of a pixel varies between $+1$ (black) and -1 (white). Consider two consecutive images $P(q\Delta t)$ and $P((q+1)\Delta t)$. An image is applied (input) to a CNN by assigning its pixel intensity values either to v_{uij} or to $v_{xij}(0)$, i.e. to the input nodes or to the initial states of the network.

Using the standard CNN model,[3] the dynamics of a cell is described by the equations

$$C\dot{v}_{xij}(t) = -R^{-1}v_{xij}(t) + \sum_{C(k,l)\in N_1(i,j)} A(i,j;k,l)v_{ykl}(t) + \sum_{C(k,l)\in N_1(i,j)} B(i,j;k,l)v_{ukl} + I, \quad 1 \leqslant i,$$

$$1 \leqslant i,k \leqslant M, \quad 1 \leqslant j,l \leqslant N \tag{1a}$$

$$v_{yij}(t) = 0{\cdot}5(|v_{xij}(t)+1| - |v_{xij}(t)-1|), \quad 1 \leqslant i \leqslant M, \quad 1 \leqslant j \leqslant N \tag{1b}$$

$$|v_{xij}(0)| \leqslant 1 \quad \text{and} \quad |v_{uij}| \leqslant 1, \quad 1 \leqslant i \leqslant M, \quad 1 \leqslant j \leqslant N \tag{1c}$$

where $N_1(i,j)$ denotes the $r=1$ neighbourhood of the cell. Usually $P(q\Delta t)$ is applied to the input voltages while $P((q+1)\Delta t)$ is applied to the initial states. With the assumption $q=0$, we will often write $v_{uij} = P_0$ and $v_{xij}(0) = P_1$. The settling time of the CNN is denoted by t_s. We always suppose that $\Delta t \gg t_s$. In Section 3 we will denote the difference picture by P_d, the result of the speed detection by P_s, the result of the improved speed detection by P_{is} and the outcome of the motion detection algorithm by P_{det}.

2.2. Continuous mode

In the continuous mode the moving object is represented by the image $P = P(t)$, the pixel intensities of which are permanently varying in time. This image is fed to the input nodes of the CNN: $v_{uij} = P(t)$. The initial state of the CNN is usually identical to the input image at $t=0$: $v_{xij}(0) = P(0)$.

Processing of time-varying images requires the use of delay-type templates.[4] The CNN dynamics is described by the state equation

$$C\dot{v}_{xij}(t) = -R^{-1}v_{xij}(t) + \sum_{C(k,l)\in N_1(i,j)} A(i,j;k,l)v_{ykl}(t) + \sum_{C(k,l)\in N_1(i,j)} B(i,j;k,l)v_{ukl}(t)$$

$$+ \sum_{C(k,l)\in N_1(i,j)} A^{\tau}(i,j;k,l)v_{ykl}(t-\tau) + \sum_{C(k,l)\in N_1(i,j)} B^{\tau}(i,j;k,l)v_{ukl}(t-\tau) + I,1$$

$$1 \leqslant \mathrm{i},k \leqslant M, \quad 1 \leqslant j,l \leqslant N \tag{2}$$

where $A^{\tau}(i,j;k,l)$ and $B^{\tau}(i,j;k,l)$ are called the *delayed feedback and delayed control operators* respectively.

The relation between the delay time τ and the transient time of the circuit, t_s, is $\tau \gg t_s$, i.e. the CNN should be much faster than the represented motion itself. the delay time τ has to be adjusted according to the motion speed. The output equation (1b) and the constraint (1c) remain unchanged.

2.3. Basic principles of the solution strategy

P1. The difference principle. This means that two consecutive images are combined by exploiting the unique feature of the CNN paradigm that provides the possibility of assigning two independent image samples to the input nodes (v_{uij}) and to the initial state nodes ($v_{xij}(0)$). In many cases the logical difference of these images is generated as an intermediate step of a more complex algorithm.

P2. The delay principle. When the input image is changing continuously and is assigned to the input nodes of the CNN, the consecutive phases of the motion are generated by a delay-type template. There is a threefold correspondence between (i) the actual input image and the current sample, (ii) the delayed input image and the previous sample and (iii) the delay time τ and the sampling time Δt.

P3. The multiple-template principle. The solution is performed by applying several different templates, one after another, to solve sophisticated motion detection and estimation tasks. From an algorithmic point of view the role of the templates is similar to that of the *macro instructions* used in a digital computer programme. In this manner an analogue algorithm is created by the use of consecutive *analogue macro instructions* (templates).[4] Possibly, several simple templates can be unified into one complex template, i.e. the analogue algorithm can be shortened by installing more complex macro instructions.

3. MOTION DETECTION IN SAMPLED MODE

3.1. Solution of Problem 1 (direction-sensitive case)

Our main purpose in this subsection is to propose a template sequence which is able to detect an object if it is moving in a certain (horizontal) direction and if the absolute value of the velocity vector lies in a small neighbourhood of $V = \Delta x/\Delta t$. If the motion does not meet one of these two requirements, then the screen will become completely white, i.e. no detection will occur. In the affirmative case the difference of P_1 and P_0, defined by $P_d \triangleq P_1 \backslash P_0$,† will remain on the screen. This problem statement implies that the given speed limit can be adjusted appropriately by varying the sampling rate. The CNN algorithm consists of three consecutive steps.

Step 1. Taking the difference. In order to take the difference of the two snapshots P_1 and P_0, the truth table given in Table I should be realized. This logical operation can be performed by means of the simple template

$$
\begin{aligned}
A &= [1 \cdot 1], \qquad & B &= [-1], \qquad & I &= -1 \\
v_{uij} &= P_0, \qquad & v_{xij}(0) &= P_1, \qquad & v_{yij}(\infty) &= P_d
\end{aligned}
\tag{3}
$$

The result of the subtraction is depicted in Figure 1. Obviously, if there is no motion or if the motion is too slow, then the difference picture will contain no black pixel, i.e. the problem is solved after one step. Now the question is: what happens if the motion is too fast?

Step 2. Speed detection. In this step we are going to determine whether the displacement of an object during the sampling period is equal to or larger than one pixel. For convenience, suppose that the object is moving horizontally. From Figure 1 one can easily see that if the object moved exactly one pixel to the right, then each black pixel in P_d would have a left neighbour in P_0. This means that we have to delete

Table I. Truth table for the subtraction

P_0 v_{uij}	P_1 $v_{xij}(0)$	P_d $v_{yij}(\infty)$
−1	−1	−1
−1	1	1
1	−1	−1
1	1	−1

† This is the usual notation for the difference between two sets.

(a) (b) (c)

Figure 1. The previous sample P_0. (b) The current sample P_1. (c) The difference picture defined by $P_d \triangleq \mathbf{P}_1 \backslash \mathbf{P}_0$

every black pixel in P_d which has no left neighbour in P_0. This is the basic idea of direct neighbour detection. The CNN template corresponding to this operation is

$$\begin{gathered} A = [0\ \ 2 \cdot 1\ \ 0], \qquad B = [2\ \ 0\ \ 0], \qquad I = -2\ v_{uij} = P_0, \\ v_{xij}(0) = P_d, \qquad v_{yij}(\infty) = P_s \end{gathered} \tag{4}$$

The action of this template can be seen in Figure 2. In this case the motion was too fast; consequently the screen has been completely erased.

This template seems to be excellent at performing speed detection. However, if we consider the example sketched in Figure 3, it turns out that undesirable effects might arise if the size of the object is larger than one pixel. The grey image in Figure 3(a) represents the previous sample, which cannot be seen on the screen. In spite of the large horizontal displacement, some pixels in the difference picture will not be eliminated because they still have adjacent black pixels in P_0. Under these circumstances we have to modify our template. We are going to delete a black pixel in P_d if it has no left neighbour in P_0 at all

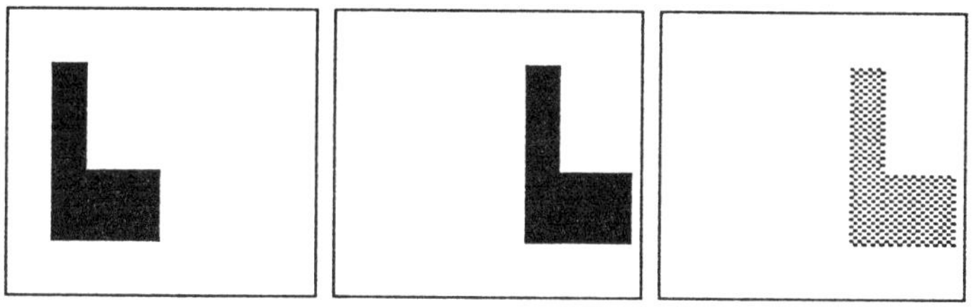

(a) (b) (c)

Figure 2. (a) The previous sample P_0. (b) The current sample which is the same as the difference picture $P_1 = P_0$. (c) direct neighbour detection. The displacement is too large, hence the screen will become white by the end of the transient, i.e. no detection will occur. The grey pixels represent the evolving transient process

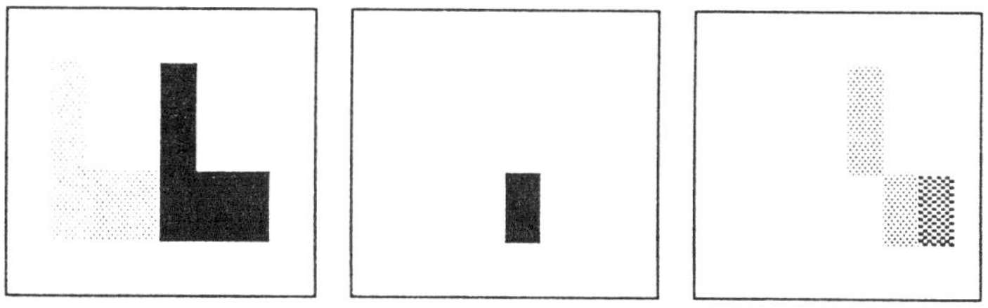

(a) (b) (c)

Figure 3. (a) The previous sample P_0 (printed with grey pixels) and the current sample $P_1 = P_d$ (printed with black pixels). (b) Inaccurate speed detection due to the close edges. The displacement was larger than one pixel; nevertheless, the screen did not become empty. (c) Improved speed detection. Template (5) first deletes the crucial pixels near the edges of the previous sample. The object completely disappears by the end of the transient

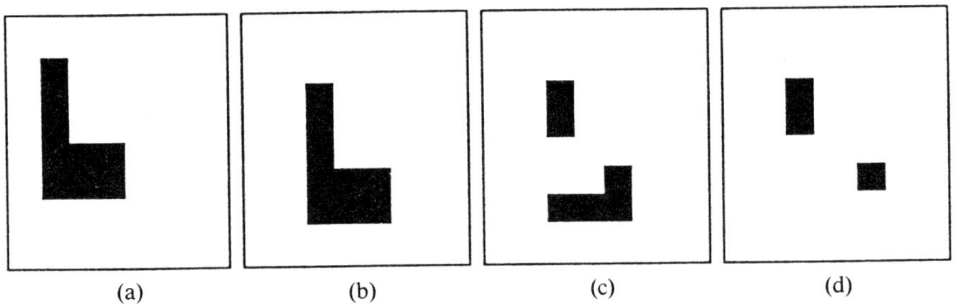

Figure 4. (a) The previous sample P_0. (b) The current sample P_1. (c) The difference picture P_d. (d) The disturbing effect of the vertical drift. The object moves in a diagonal direction, but the speed detector template is not able to eliminate this type of motion. Erroneous detection occurs

or if it has one as well as having a right neighbour in P_0. Our next template is able to perform this task:

$$A = [1 \;\; 2\cdot 1 \;\; -2], \qquad B = [2 \;\; 0 \;\; 0], \qquad I = -4$$
$$v_{uij} = P_0, \qquad v_{xij}(0) = P_d, \qquad v_{yij}(\infty) = P_{is} \tag{5}$$

The behaviour of this template is illustrated in Figure 3(c).

Step 3. Direction detection. So far we have managed to delete the object if the horizontal drift is smaller or greater than one pixel per sampling period. Of course, erroneous detection will occur if the object displaces itself vertically as well (Figure 4). Therefore we have to perform the third step of our algorithm with the following template, which erases the screen in the case of vertical displacement:

$$A = \begin{bmatrix} 0 & 1 & 0 \\ 0 & 4\cdot 1 & 0 \\ 0 & 1 & 0 \end{bmatrix}, \qquad B = \begin{bmatrix} -1 & 1 & 0 \\ 0 & 0 & 0 \\ -1 & 1 & 0 \end{bmatrix}, \qquad I = -1 \tag{6}$$
$$v_{uij} = P_0, \qquad v_{xij}(0) = P_{is}, \qquad v_{yij}(\infty) = P_{det}$$

As a result of our investigation we have a three-step analogue algorithm which deletes the image either if the motion is not perfectly horizontal or if the magnitude of the velocity is outside the given range.

Remarks 1

(i) it is interesting to notice that by combining templates (4)–(6) we can construct a new template which is able to perform all three steps of the analogue algorithm at the same time. We have merely to scale and add up the elements $A(i, j; k, l)$ and $B(i, j; k, l)$ in an appropriate order. A suitable combination of

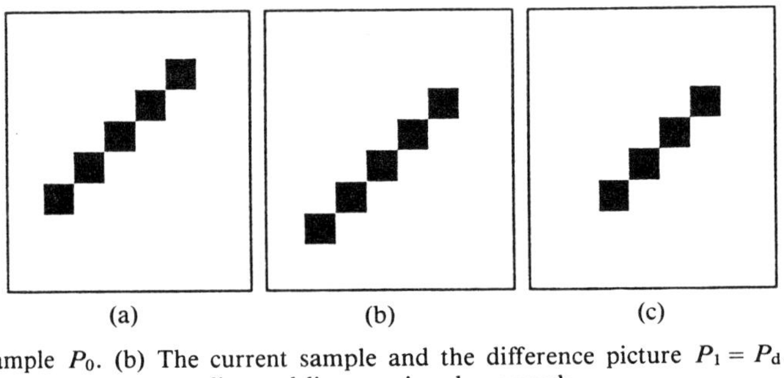

Figure 5. (a) The previous sample P_0. (b) The current sample and the difference picture $P_1 = P_d$. (c) Incorrect detection of a diagonal line moving downwards

the three templates is

$$A=\begin{bmatrix}0 & 0\cdot 5 & 0\\ 1 & 2\cdot 1 & -2\\ 0 & 0\cdot 5 & 0\end{bmatrix},\qquad B=\begin{bmatrix}-0\cdot 5 & 0\cdot 5 & 0\\ 2 & -8 & 0\\ -0\cdot 5 & 0\cdot 5 & 0\end{bmatrix},\qquad I=-11 \tag{7}$$

$$v_{uij}=P_0,\qquad v_{xij}(0)=P_1,\qquad v_{yij}(\infty)=P_{\text{det}}$$

An interesting question for the following investigations is: under what conditions may several simple templates be unified into one complex template?

(ii) After performing several tests, we found only one special, distorted horizontal motion for which the direction detector template (6) did not work properly (see Figure 5). The additional template which is able to detect this unique motion is

$$A=\begin{bmatrix}0 & 0 & 1\\ 0 & 4\cdot 1 & 0\\ 0 & 0 & 1\end{bmatrix},\qquad B=\begin{bmatrix}0 & -1 & 1\\ 0 & 0 & 0\\ 0 & -1 & 1\end{bmatrix},\qquad I=-1 \tag{8}$$

$$v_{uij}=P_0,\qquad v_{xij}(0)=P_{\text{det}},\qquad v_{yij}(\infty)=P_{\text{out}}$$

This means that one more step will need to be added to the previous algorithm.

(iii) The correct functions of the above templates can be proved analytically in the same way as shown in Section 4.

(iv) Another approach to solving Problem 1 is studied in Reference 6.

3.2. Solution of Problem 2 (direction-independent case)

In this subsection we will consider another type of motion detection problem; namely, whether the absolute value of the velocity vector is appropriate (i.e. whether it lies in a certain range around $V=\Delta x/\Delta t$) or not. We do not care about the direction of motion. If the motion is slower or faster than some prescribed V, then the screen will become totally white, i.e. the moving object will not be detected. In the affirmative case the *current image sample* will remain on the screen. A convex† object is restricted to be larger than one pixel in every direction; furthermore, it is not allowed to contain holes. The method consists of four steps. The two examples presented in Figure 6 will help us in understanding the analogue algorithm.

Step 1. Taking the difference This step is exactly the same as was shown in Section 3.1. The results after the subtraction are depicted in Figure 7.

Step 2. Speed detection. It is obvious that we cannot use the speed detector template (5) introduced in the previous subsection because it produces the wrong answer if the motion is not perfectly horizontal.

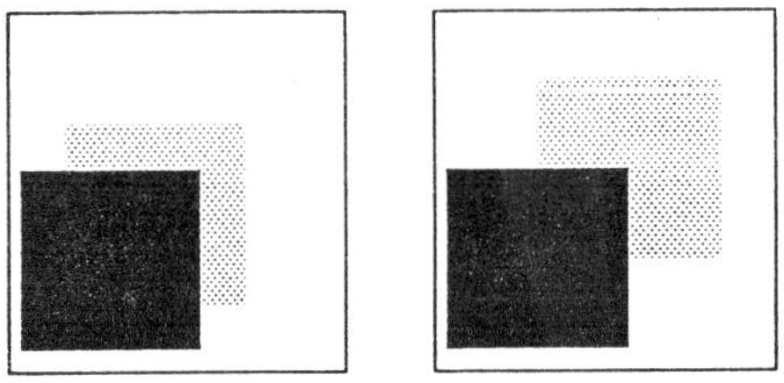

(a) (b)

Figure 6. (a) Motion with correct speed. The previous sample is printed in black while the current sample is printed in grey. (b) Motion with high speed

† In this paper an object is called *convex* is any line segment determined by two of its points lies inside the object.

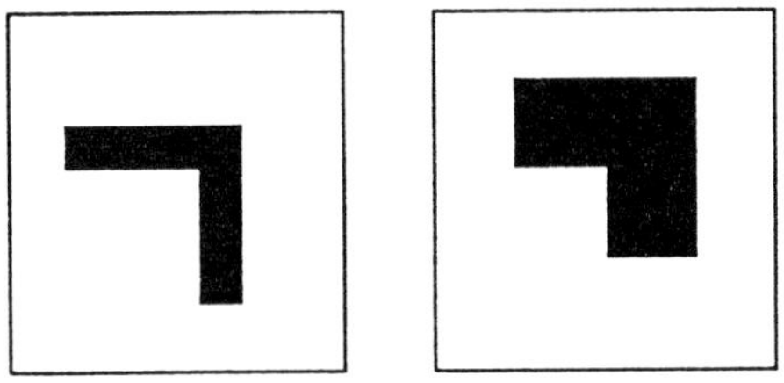

Figure 7. The difference picture in the case of (a) favoured and (b) disfavoured motion

However, from Figure 7 one can easily see that the moving object is of the correct speed only if the dark regions of the difference picture are 'one pixel wide'. This means that in the affirmative case every black pixel has at most three black neighbours, while in the opposite case there definitely exist some pixels which have at least four neighbours. Consequently, we have to determine whether the black pixels in P_d have more than three neighbours or not. This task can be achieved by the symmetric template

$$A = \begin{bmatrix} 0 & 0 & 0 \\ 0 & 2 & 0 \\ 0 & 0 & 0 \end{bmatrix}, \qquad B = \begin{bmatrix} -0\cdot25 & -0\cdot25 & -0\cdot25 \\ -0\cdot25 & 0\cdot75 & -0\cdot25 \\ -0\cdot25 & -0\cdot25 & -0\cdot25 \end{bmatrix}, \qquad I = -2 \tag{9}$$

$$v_{uij} = P_d, \qquad v_{xij}(0) = P_d, \qquad v_{yij}(\infty) = P_s$$

Notice that every template used for solving this kind of problem must necessarily show perfect symmetry, since none of the horizontal, vertical or diagonal directions is preferred.

Step 3. Improved speed detection. After performing Step 2 of the algorithm, the difference picture will remain on the screen when the speed is favourable, while some pixels of it will disappear if the speed is not favourable. In Step 3 we complete the speed detection. As a result, the difference picture remains entirely on the screen if the speed is correct, while it completely disappears if the speed is too large. The following template will accomplish the improved speed detection:

$$A = \begin{bmatrix} 1 & 1 & 1 \\ 1 & 2 & 1 \\ 1 & 1 & 1 \end{bmatrix}, \qquad B = \begin{bmatrix} -1 & -1 & -1 \\ -1 & 0 & -1 \\ -1 & -1 & -1 \end{bmatrix}, \qquad I = 0 \tag{10}$$

$$v_{uij} = P_d, \qquad v_{xij}(0) = P_s, \qquad v_{yij}(\infty) = P_{is}$$

This template eliminates a pixel if it has a black neighbour in P_d but there is no black pixel at the same place in P_s. It is interesting to observe the propagation effect which occurs when two or more consecutive pixels are to be deleted on the screen (see Figure 8).

Step 4. Image reconstruction. So far the difference picture will remain on the screen if the motion is in the correct range of speed. In order to finish our task, we use the following template which

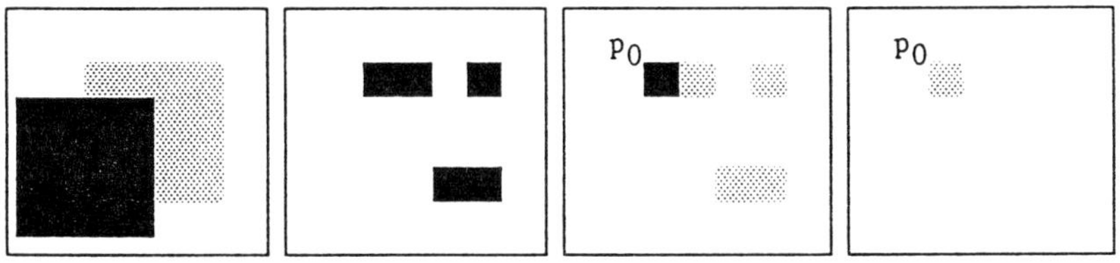

Figure 8. (a) The previous sample P_0 (black) and the current sample P_1 (grey). (b) The difference picture P_d. (c) The transient process of template (10) which accomplishes the speed detection, i.e. deletes the remaining pixels of the difference picture. In the first part of the transient these pixels disappear which have one or more black neighbours deleted in Step 2 of the algorithm. (4) In the second part of the transient a so-called propagation effect occurs

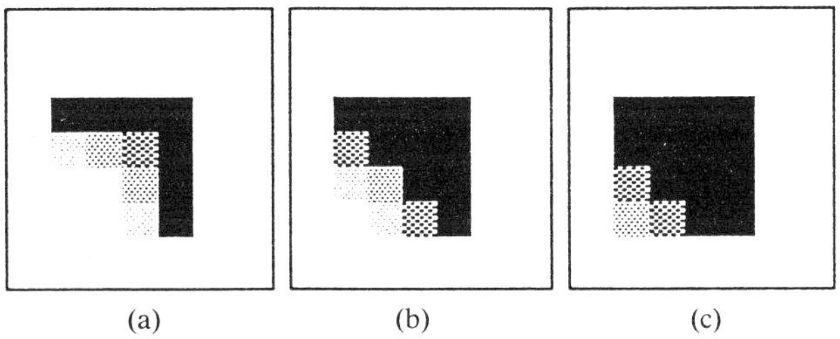

Figure 9. The reconstruction of the current sample of Figure 6(a) using the output of the improved speed detector as a framework. If the output of the speed detector is empty, then the reconstruction will be suppressed

reconstructs the current image of the moving object from the difference picture:†

$$A=\begin{bmatrix}0\cdot2 & 0\cdot2 & 0\cdot2\\ 0\cdot2 & 1\cdot5 & 0\cdot2\\ 0\cdot2 & 0\cdot2 & 0\cdot2\end{bmatrix},\qquad B=\begin{bmatrix}0 & 0 & 0\\ 0 & 2 & 0\\ 0 & 0 & 0\end{bmatrix},\qquad I=0 \tag{11}$$
$$v_{uij}=P_1,\qquad v_{xij}(0)=P_4,\qquad v_{yij}(\infty)=P_{\text{det}}$$

After performing this step, the original image of the object will appear on the screen instead of the difference picture, assuming that the motion satisfies the requirements (Figure 9).

4. MOTION DETECTION IN CONTINUOUS MODE

4.1. Tracing of moving objects

The simplest problem in the field of continuous time motion detection is the tracking of small objects moving across the vision field. This task can be solved by means of the simple template

$$A=[1\cdot5],\qquad B=[1\cdot1],\qquad I=1$$
$$v_{uij}=P(t),\qquad v_{xij}(0) \tag{12}$$

This is a very simple, perfectly symmetric, completely stable template which actually performs the binary OR function. We assume that the transient time t_s is smaller than the time interval between two characteristic events (changes) in the input picture. Template (12) 'adds' each new input pattern to the actual state of the CNN (Figure 10).

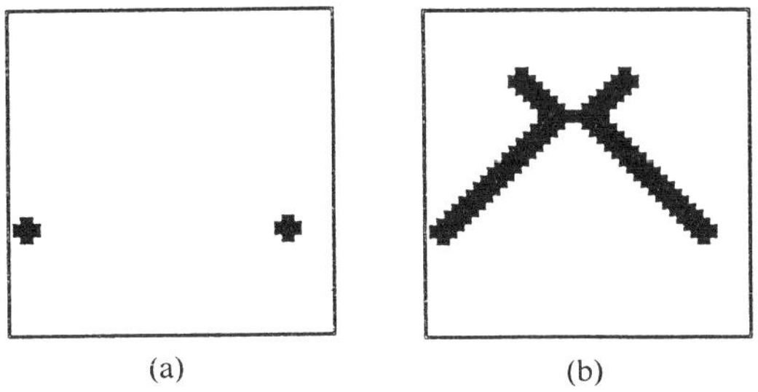

Figure 10. (a) Two particles moving across the screen. (b) The trajectories of the particles. Template (12) adds each new snapshot coming from the input to the trajectory represented by the state variables

† This template can be called an 'analogue latch'. If the initial state is completely white, then the output will be totally white as well. If a fragment of the input pattern is already present at the initial state nodes, then the *whole* input pattern will become visible at the output after the transient. The propagation effect occurs in this case as well.

4.2. Generating the moving part (difference picture)

In several motion detection problems the basic question is how to separate the moving parts of the image from those which are not changing during the experiment. Template (3) is able to solve this problem by taking the logical difference of the consecutive image samples. In the case of continuous input pictures, however, we cannot use the initial state nodes as independent input ports to the CNN. In order to take the difference of the actual phase and some preceding phase of the motion, we have to use the delay-type template

$$\begin{array}{lllll} A=[1], & B=[3], & I=-3, & As=[0], & Bs=[-3] \\ \tau \geqslant 5RC, & v_{uij}=P(t), & v_{xij}(0)=P(0), & v_{yij}(t)=P_{\mathrm{d}}(t) \end{array} \tag{13}$$

This simple template subtracts the delayed input image $P(t-\tau)$ from the actual input image $P(t)$ and provides the difference picture continuously in time. The rate of change of the input picture is required to be much smaller than the speed of the CNN. Template (13) has a fast transient: $t_s \approx RS$.

In a real situation the extracted difference picture is corrupted by noise due to various light reflections and disturbances in the input picture (Figure 11). In order to overcome this difficulty, a new delay-type template has been devised which is able to perform subtraction and noise filtering simultaneously. We will now proceed in providing the design method.

The template we are looking for is direction-independent, i.e. it shows perfect central symmetry. Furthermore, suppose that the basic operation of subtraction is performed by the control operator and the delayed control operator. This means that the expected structure of the template is

$$A=\begin{bmatrix} 0 & 0 & 0 \\ 0 & 1 & 0 \\ 0 & 0 & 0 \end{bmatrix}, \quad B=\begin{bmatrix} a & a & a \\ a & b & a \\ a & a & a \end{bmatrix}, \quad I, \quad A^{\tau}=[0], \quad B^{\tau}=\begin{bmatrix} c & c & c \\ & & \\ c & c & c \end{bmatrix} \tag{14}$$

where a, b, c, d and I are constants. This template is completely stable and the trajectory of the circuit will always reach the saturation regions.† Besides, it can be proved that the circuit transients are monotonous in time, i.e. each state variable is either strictly increasing or strictly decreasing in the interval $0 \leqslant t \leqslant t_s$ (see Appendix). Now the question is: how do we determine the constants a, b, c, d and I?

The truth table of the logical subtraction is depicted in Table I. In this case, however, the first column belongs to the delayed input image, the second column belongs to the actual input image, while the third column provides the result of the transient. For example, the first row of Table I states that if a certain pixel is white in both in the actual and delayed input pictures, then the steady state of the corresponding state variable should be -1 independently of the neighbouring cells. In this case the following inequality

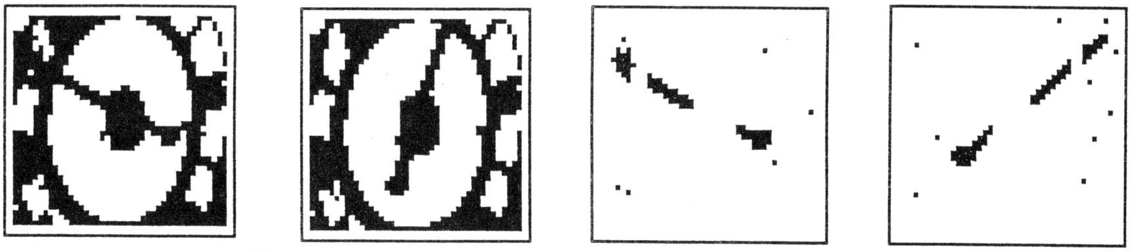

(a) (b)

Figure 11. (a) Different snapshots of a running clock. The hand turns around while other parts of the picture stand still. The snapshots were made with the help of a camera. The resolution of the screen is 44×44. (b) Template (13) provides the difference picture continuously in time. The output image is corrupted by noise due to light reflections and small perturbations in the preprocessing system

† It is worth mentioning that there exists an additional equilibrium point in the linear region of the output characteristics as well, but this equilibrium point is not stable; consequently, the circuit will always leave it owing to the noise and small perturbations.

holds:

$$\dot{v}_{xij}(t) = -v_{xij}(t) + v_{yij}(t) + I - b \pm 8a - d \pm 8c < 0 \quad (0 < t < t_s) \tag{15}$$

The ± sign denotes that this inequality must be fulfilled independently of the neighbouring input variables (either if the neighbouring pixels in the input image are white or if they are black). Similarly, the state equations corresponding to the third and fourth rows of Table I are

$$\dot{v}_{xij}(t) = -v_{xij}(t) + v_{yij}(t) + I - b \pm 8a + d \pm 8c < 0 \quad (0 < t < t_s) \tag{16}$$
$$\dot{v}_{xij}(t) = -v_{xij}(t) + v_{yij}(t) + I + b \pm 8a + d \pm 8c < 0 \quad (0 < t < t_s) \tag{17}$$

The second row of the truth table requires further investigation. It states that if a specific pixel is black in the actual input picture and the same pixel is white in the delayed input picture, then the pixel should remain black at the end of the transient. This would be the normal operation of subtraction. At the same time, however, we wish to perform noise filtering on the difference picture. This means that if a pixel has less than two black neighbours, then it is considered to be noise and should remain white:

$$\dot{v}_{xij}(t) = -v_{xij}(t) + v_{yij}(t) + I + b + 8a - d + 8c < 0 \quad (0 < t < t_s) \tag{18}$$
$$\dot{v}_{xij}(t) = -v_{xij}(t) + v_{yij}(t) + I + b + 8a - d + 6c < 0 \quad (0 < t < t_s) \tag{19}$$
$$\dot{v}_{xij}(t) = -v_{xij}(t) + v_{yij}(t) + I + b + 8a - d + 4c > 0 \quad (0 < t < t_s) \tag{20}$$

According to the presented method, we have to write every possible statement of the truth table in the form of an inequality. Of course, the set of inequalities might be either contradictory, in which case no CNN can be devised to solve the task, or redundant, in which case the superfluous statements can be neglected.

If we consider the inequalities (15)–(20) at $t = 0$, we have $v_{xij}(0) = v_{yij}(0)$. If the inequalities hold at $t = 0$, they will be fulfilled at $0 < t < t_s$ as well, according to the monotonousness of the transients. Thus we obtain the following set of inequalities:

$$\begin{aligned} I - b \pm 8a - d \pm 8c &< 0 \\ I - b \pm 8a + d \pm 8c &< 0 \\ I + b \pm 8a + d \pm 8c &< 0 \\ I + b + 8a - d + 8c &< 0 \\ I + b + 8a - d + 6c &< 0 \\ I + b + 8a - d + 4c &> 0 \end{aligned} \tag{21}$$

this set determines a convex polyhedron in the five-dimensional parameter space (in this case there is no contradiction between the inequalities). We have to choose the values of a, b, c, d and I so that the corresponding point in the parameter space will be located inside the polyhedron. A suitable choice is $a = 0{\cdot}25$, $b = 2$, $c = -0{\cdot}25$, $d = -2$ and $I = -4{\cdot}75$. The delay-type template consisting of these values is

$$\begin{aligned} A &= \begin{bmatrix} 0 & 0 & 0 \\ 0 & 1 & 0 \\ 0 & 0 & 0 \end{bmatrix}, \quad B = \begin{bmatrix} 0{\cdot}25 & 0{\cdot}25 & 0{\cdot}25 \\ 0{\cdot}25 & 2 & 0{\cdot}25 \\ 0{\cdot}25 & 0{\cdot}25 & 0{\cdot}25 \end{bmatrix}, \quad I = -4{\cdot}75 \\ A^{\tau} &= \begin{bmatrix} 0 & 0 & 0 \\ 0 & 0 & 0 \\ 0 & 0 & 0 \end{bmatrix}, \quad B^{\tau} = \begin{bmatrix} -0{\cdot}25 & -0{\cdot}25 & -0{\cdot}25 \\ -0{\cdot}25 & -2 & -0{\cdot}25 \\ -0{\cdot}25 & -0{\cdot}25 & -0{\cdot}25 \end{bmatrix}, \quad \tau \geqslant 10RC \end{aligned} \tag{22}$$

The performance of the template can be seen in Figure 12.

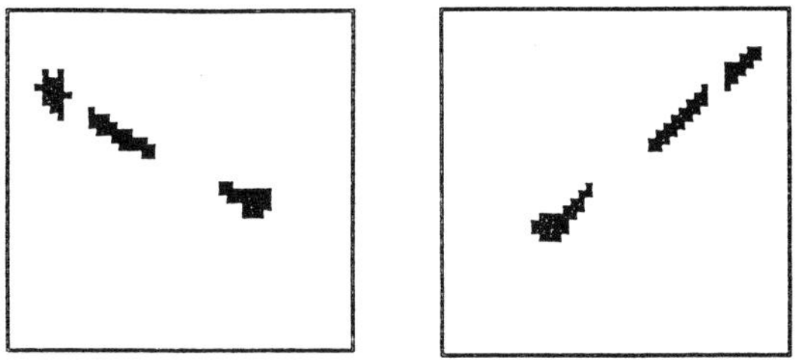

Figure 12. Template (22) is able to perform subtraction and noise filtering at the same time. The figure shows a snapshot of the continuously changing output

Remarks 2

(i) Further steps of the motion analysis can be performed by consecutive CNN layers using the difference picture as an input image.

(ii) Recently, a variant of this design method has been successfully used for creating a delay-type template which is able to solve Problem 2 (direction-independent motion detection) in continuous mode:

$$A=\begin{bmatrix}0&0&0\\0&1&0\\0&0&0\end{bmatrix},\qquad B=\begin{bmatrix}0&0&0\\0&6&0\\0&0&0\end{bmatrix},\qquad I=-2$$
$$A^{\tau}=\begin{bmatrix}0\cdot68&0\cdot68&0\cdot68\\0\cdot68&0\cdot68&0\cdot68\\0\cdot68&0\cdot68&0\cdot68\end{bmatrix},\qquad B^{\tau}=\begin{bmatrix}0&0&0\\0&0&0\\0&0&0\end{bmatrix},\qquad \tau\geqslant 10RC \tag{23}$$

As can be seen, in this case $A^{\tau}\neq 0$, so Proposition 1 does not apply directly.

5. MOTION DRASTICALLY CORRUPTED BY NOISE (MIMICRY)

One of the most fascinating features of some animal species (e.g. chameleons, rays, octopods, etc.) is the ability to change their coloration in order to blend in with the surroundings, since in the highly competitive animal world camouflage is one of the commonest forms of protection against predators. In most cases the animal mimicking its surrounding becomes absolutely imperceptible until everything is still. Should the surroundings stir, however, the motionless mimicry will reveal the concealed animal.

Inspired by camouflage in nature, we define a new class of problems—called *mimicry*—to challenge CNNs. Let us consider a hiding animal that stands in front of a noisy background. Since the statistical property of the animal's fuzzy pattern is the same as that of the noisy background, no one is able to distinguish the animal from the environment. However, if the background noise changes statistically while the pattern of the animal remains constant, then the animal can be extracted from the background by simply averaging (integrating) the images in time over a sufficiently long period.

For this purpose we use the integrating template (24) of zero neighbourhood, neglecting any local, spatial information of the picture. We assume that the time constant of the integration is adequately large:

$$A=[1],\qquad B=[\varepsilon],\qquad I=0$$
$$v_{uij}=P(t),\qquad v_{xij}(0)=0,\qquad \varepsilon=0\cdot06 \tag{24}$$

Experiments for testing the template have been carried out by generating an image sequence which represents random noise that changes continuously in time and setting the pixels of the central area to arbitrary but *constant* intensity values. The central area with a constant pattern will represent the 'animal', while the randomly changing outer areas will exemplify the noisy background. Two snapshots of the input image sequence are shown in Figure 13. It can be seen that no single snapshot has any

Figure 13. Two consecutive snapshots of the continuous input image of mimicry. An 'animal' with constant pattern is hiding in the permanently changing background. Owing to the uniform statistics of the pixel intensities, the animal remains imperceptible

particularly *meaningful* portion. However, if we use template (24) for averaging the images in time, then the animal will become visible at the centre of th image (Figure 14).

Notice that if a randomly varying pixel intensity is fed to the input of an individual cell, then the output of the cell will remain close to zero (average grey). In contrast, if the pixels belonging to the animal have constant intensities, then the outputs of the corresponding cells will tend to +1 (black) or −1 (white) (see Figure 14). In our experiments random noise with zero expected value has been used. In the case of different expected values the bias I of the CNN should be adjusted suitably.

In a more complex experiment the convergence of the output variables which belong to the animal can be speeded up by using a non-linear feedback template to measure how close the output variables are to the saturation regions. The appropriate non-linear template is

$$A = \begin{bmatrix} f(x) & f(x) & f(x) \\ f(x) & 0\cdot97 & f(x) \\ f(x) & f(x) & f(x) \end{bmatrix}, \qquad B = \begin{bmatrix} 0 & 0 & 0 \\ 0 & \varepsilon & 0 \\ 0 & 0 & 0 \end{bmatrix}, \qquad I = 0 \qquad (25)$$

$$v_{uij} = P(t), \qquad v_{xij}(0) = 0, \qquad \varepsilon = 0\cdot06$$

where $f(x)$ denotes a non-linear function of $x = v_{xij} + v_{xkl}$. The non-linear characteristic can be seen in Figure 15:
This template fully explores the integral, spatial information extracted from the local neighbourhood.

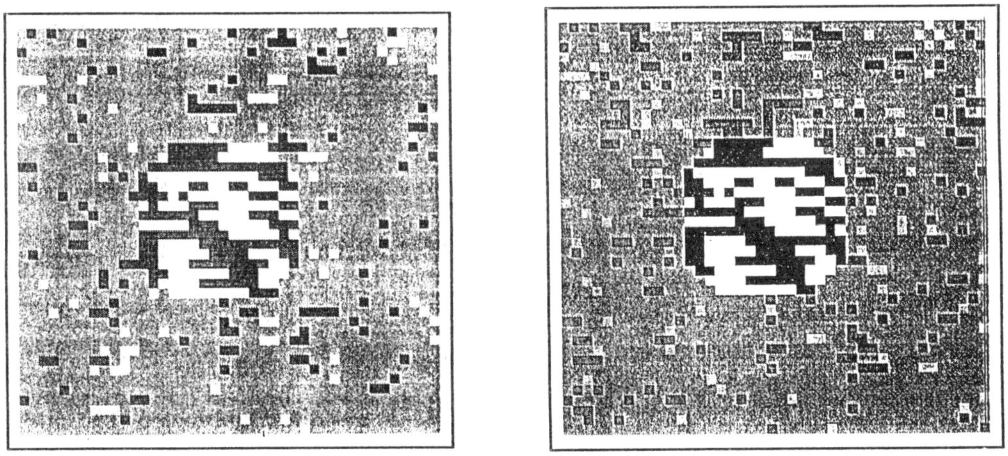

Figure 14. Two consecutive snapshots of the continuous output picture of template (24). The expected value of the background noise is zero; consequently the background remains grey after the integration. Input pixel intensities belonging to the animal are constant; therefore the corresponding output variables tend to either +1 (black) or −1 (white)

6. DISCUSSION

The results of this paper can be applied in many fields, e.g. robotics. The circuit realization could be either a hardware accelerator[7] or a VLSI circuit. In the latter case the programmability is of crucial importance. Moreover, in some cases a local logic is advisable. Such an architecture is under fabrication.[8]

The methods presented in this paper are partly analytical and partly heuristic. Although the four classes of problems require different template sequences, some standard steps are obvious. In our examples mostly black-and-white images are presented, although the circuits work on grey-level images as well. Grey-scale problems require further experiments which are beyond the scope of the present paper.

ACKNOWLEDGEMENTS

This work has been supported by National Research Fund of Hungary grant 1212/1987, National Science Foundation grant MIP-86 14000, the Belgian American Educational Foundation and the joint grant INT 90-01336 of the National Science Foundation and the Hungarian Academy of Sciences.

APPENDIX

Consider the template

$$A=\begin{bmatrix}0&0&0\\0&A_0&0\\0&0&0\end{bmatrix},\qquad B=\begin{bmatrix}a&a&a\\a&b&a\\a&a&a\end{bmatrix},\qquad I,\qquad A^{\tau}=0,\qquad B^{\tau}=\begin{bmatrix}c&c&c\\ & & \\c&c&c\end{bmatrix}\tag{26}$$

where A_0, a, b, c, d and I are constants and $A_0>1$. The circuit state equation and the output equation with the assumptions $C=1\ \mu\text{F}$ and $R=1\ \text{k}\Omega$ are

$$\dot{v}_{xij}(t)=-v_{xij}(t)+A_0v_{yij}(t)+b\sum_{C(k,l)\in N_1\backslash C_{ij}}v_{ukl}(t)+av_{uij}(t)+d\sum_{C(k,l)\in N_1\backslash C_{ij}}v_{ukl}(t-\tau)+cv_{uij}(t-\tau)+I,$$

$$1\leqslant i,k\leqslant M,\quad 1\leqslant j,l\leqslant N\tag{27a}$$

$$v_{yij}(t)=f(v_{xij}(t))=0\cdot5(|v_{xij}(t)+1|-|v_{xij}(t)-1|),\quad 1\leqslant i\leqslant M,\quad 1\leqslant j\leqslant N\tag{27b}$$

Proposition 1

If the input of the circuit changes *much slower* than the transient time, then the CNN with template (26) is not only *completely stable* (positive cell-linking template[4]), but in addition the state variables $v_{xij}(t)$ are *monotonous* in time, i.e.

$$\text{if } \dot{v}_{xij}(t_0)\leqslant 0,\quad \text{then } \dot{v}_{xij}(t)\leqslant 0\quad (t>t_0),\quad 1\leqslant i\leqslant M,\quad 1\leqslant j\leqslant N$$
$$\text{if } \dot{v}_{xij}(t_0)\geqslant 0,\quad \text{then } \dot{v}_{xij}(t)\geqslant 0\quad (t>t_0),\quad 1\leqslant i\leqslant M,\quad 1\leqslant j\leqslant N$$

Proof. Let us write the state equation in the form

$$\dot{v}_{xij}(t)=-v_{xij}(t)+A_0v_{yij}(t)+g_{ij},\quad 1\leqslant i\leqslant M,\quad 1\leqslant j\leqslant N\tag{28}$$

where

$$g_{ij}=b\sum_{C(k,l)\in N_1\backslash C_{ij}}v_{ukl}(t)+av_{uij}(t)+d\sum_{C(k,l)\in N_1\backslash C_{ij}}v_{ukl}(t-\tau)+cv_{uij}(t-\tau)+I\tag{29}$$

is independent of time because we assume that the input image does not change significantly during the transient (i.e. the speed of the motion is much smaller than the speed of the circuit). The steady state response of cell C_{ij} is determined by

$$-x_{ij}+A_0f(x_{ij})+g_{ij}=0,\quad \text{where } x_{ij}=\lim_{t\to\infty}v_{xij}(t)\tag{30}$$

Let us represent the equation $A_0f(x_{ij})=x_{ij}-g_{ij}$ graphically (Figure 15). Now we have three cases.

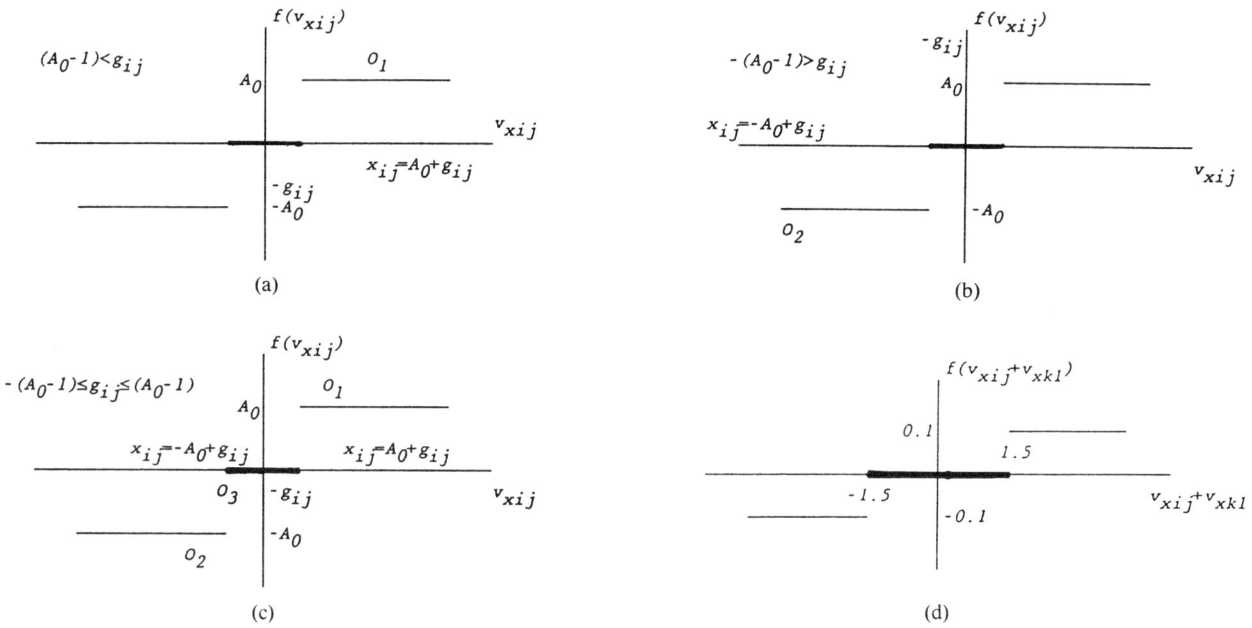

Figure 15. The operating points of cell C_{ij} (a, b and c) and the off-centre non-linear function in the mimicry template (d)

(a) If $A_0 - 1 < g_{ij}$, then we have one single equilibrium point at O_1 (Figure 15(a)). The eigenvalue of the linearized equivalent circuit at O_1 is $\lambda = -1$; hence the equilibrium point is stable. It is obvious that during the transient $v_{xij}(t) \leqslant x_{ij} = A_0 + g_{ij}$; therefore from (28) it follows that

$$\dot{v}_{xij}(t) \geqslant 0, \quad t \geqslant t_0 \tag{31}$$

This means that the state variable of cell C_{ij} will monotonously increase from $v_{xij}(0)$ until x_{ij} where it asymptotically reaches the stable equilibrium point.

(b) Similarly, if $-(A_0 - 1) > g_{ij}$, then a stable equilibrium point exists at O_2 (Figure 15(b)). Now during the transient $v_{xij}(t) \geqslant x_{ij} = -A_0 + g_{ij}$; therefore from (28) we obtain

$$\dot{v}_{xij}(t) \leqslant 0, \quad t \geqslant t_0 \tag{32}$$

This means that the state variable $v_{xij}(t)$ will monotonously decrease and asymptotically settle down to the stable equilibrium point at O_2.

(c) If $-(A_0 - 1) \leqslant g_{ij} \leqslant A_0 - 1$, then there exist three equilibrium points (Figure 15(c)). Two of them (at O_1 and O_2) are located in the horizontal region of $f(x_{ij})$; they are locally stable. In addition, there exists an equilibrium point at O_3, which is not stable because the eigenvalue of the linearized equivalent circuit at that point is $\lambda = 1/(A_0 - 1) > 0$. It is easy to show that (i) if $\dot{v}_{xij}(t_0) = (A_0 - 1)x_{ij}(t_0) + g_{ij} > 0$, then $v_{xij}(t)$ will increase until it reaches O_1; (ii) if $\dot{v}_{xij}(t_0) = (A_0 - 1)x_{ij}(t_0) + g_{ij} < 0$, then $v_{xij}(t)$ will increase until it reaches O_2. Q.E.D.

From the proof presented above it is obvious that the state variables are strictly increasing or decreasing for $t_0 < t < t_s$ and that the circuit will always find a stable equilibrium point in the horizontal region of the piecewise linear characteristic $f(v_{xij})$. This means that in the steady state $|v_{yij}(t)| = 1$ $(1 \leqslant i \leqslant M, 1 \leqslant j \leqslant N)$ holds. Problems similar to Proposition 1 are profoundly investigated in Reference 9.

REFERENCES

1. (a) D. H. Hubel and T. N. Wiesel, 'Receptive fields, binocular interaction and functional architecture in the cat's visual cortex,' *J. Physiol.,* **160**, 106–154 (1962). (b) R. M. J. Cotteril, 'A possible role for coherence in neural networks' in R. M. J. Cotteril (ed.), *Computer Simulation in Brain Science*, Cambridge University Press, Cambridge, MA, 1988, pp. 164–168.
2. (a) C. Mead, *Analog VLSI and Neural Systems*, Addison-Wesley, Reading, MA, 1989, Chap. 14. (b) C. mead and M. Ismail (eds), *Analog VLSI Implementation of Neural Systems*, Kluwer, Boston, MA, 1989.
3. L. O. Chua and L. Yang, (a) 'Cellular neural networks: theory,' and (b) 'Cellular neural networks: applications', *IEEE Trans. Circuits and Systems*, **CAS-35**, 1257–1272 and 1273–1290 (1988).
4. T. Roska and L. O. Chua, 'Cellular neural networks with nonlinear and delay-type elements', *Proc. Int. Workshop on Cellular Neural Networks and Their Applications, CNNA '90*, IEEE, New York, 1991, pp. 12–25.
5. T. Roska, T. Boros, P. Thiran and L. O. Chua, 'Detecting simple motion using cellular neural networks', *Proc. Int. Workshop on Cellular Neural Networks and Their Applications, CNNA '90* IEEE, New York, 1991, pp. 127–138.
6. L. O. Chua and P. Thiran, 'An analytic method for designing simple cellular neural networks', *ERL Memo UCB/ERL M91/3*, UC Berkeley, January 1991.
7. T. Roska, G. Bártfai, P. Szolgay, T. Szirányi, A. Radványi and Zs. Ugray, 'A hardware accelerator board for cellular neural networks: CNN-HAC', *Proc. Int. Workshop on Cellular Neural Networks and Their Applications, CNNA '90*, IEEE, New York, 1991, pp. 160–168.
8. K. Halonen, V. Porra, T. Roska and L. O. Chua, 'VLSI implementation of a reconfigurable cellular neural network containing local logic (CNNL)', *Proc. Int. Workshop on Cellular Neural Networks and Their Applications, CNNA '90* IEEE, New York, 1991, pp. 206–215.
9. L. O. Chua and B. E. Shi, 'Exploiting cellular automata in the design of cellular neural networks for binary image processing', *ERL Memo UCB/ERL M89/130*, UC Berkeley, November 1989.
10. T. Roska, 'Analog events and a dual computing structure using analog and digital circuits and operators', in *Discrete Event Systems: Models and Applications*, P. Varaiya and A. B. Kurzhanski (eds), Springer, Berlin, 1987, pp. 225–238.

CELLULAR NEURAL NETWORK DESIGN FOR SOLVING SPECIFIC IMAGE-PROCESSING PROBLEMS

KRZYSZTOF ŚLOT

Institute of Electronics, Technical University of Łódź, Stefanowskiego 18/22, 90-924 Łódź, Poland

SUMMARY

This paper is concerned with determining simple CNN-paradigm-based circuits for solving specific binary image-processing problems. Single-layer CNN templates for performing specified shape extraction, extraction of shapes which contain a desired feature and modified feature detection are derived using a formulated design strategy. The concept of CNNs composed of modified cells is developed and circuits designed for realizing simultaneous detection of distinct features and for extracting horizontal line midpoints are presented.

1. INTRODUCTION

Of the various neural network structures currently available, the cellular neural network[1,2] (CNN) paradigm possesses a very attractive and important property, namely the relative ease of VLSI circuit realization owing to its locally interconnected architecture. The processing capabilities of CNNs have been intensively studied and many useful applications have been described,[3-7] especially for the simplest, single-layer CNN structure.

The goal of this paper is to determine simple circuits in the CNN paradigm for solving specific signal-processing problems. The main results of the paper are the following.

1. A methodology of CNN design for realizing specific shape extraction tasks is presented.
2. A pattern-detecting CNN featuring simple realization of a processing result read-out is proposed.
3. Applications of CNNs composed of modified cells[8] in realizing simultaneous detection of distinct shapes and in extracting horizontal line midpoints are described.

The paper has the following structure. A strategy for single-layer CNN design presented in Section 2 is used in deriving shape-extracting and feature-detecting CNN templates (described in Section 3). The concept of using a CNN composed of modified cells is developed in Section 4 and two processing examples for which this structure seems to be optimal in the sense of required circuit complexity are presented in Section 5.

Throughout the paper a binary representation of the input information is assumed. Furthermore, the relation that ensures a binary output voltage from every cell in the equilibrium state, i.e. $A_{ij,ij} > R_x^{-1}$, is satisfied. All the presented applications concern binary image processing. The terms *feature*, *desired feature*, *feature detection* and *shape* that are used are understood as follows.

(a) *Feature*—any binary function defined over a 2D rectangular discrete grid.
(b) *Desired feature*—an element of interest in a feature set.
(c) *Feature detection*—the operation of producing information concerning the presence or absence of a feature. This information is represented by a binary-valued $\{-1, +1\}$ equilibrium cell output voltage. A voltage is referred to as *high* if it equals '+1' and *low* if it equals '−1'. The output voltage of a cell c_{ij} (Vy_{ij}) after processing is high if and only if the cell corresponds to the central pixel of a desired feature.

(d) *Shape*—a set of connected *black* pixels. An image pixel is referred to as *black* if it corresponds to high output or high input voltage. Connectedness here means that in the 1-neighbourhood of any of the pixels of a considered set there is at least one other element that also belongs to this set.

The dynamics of each network cell is described by the state equation[1]

$$C\frac{\mathrm{d}Vx_{ij}}{\mathrm{d}t} = -\frac{1}{R_x}Vx_{ij} + I + \sum_{kl} A_{ij,kl}Vy_{kl} + \sum_{kl} B_{ij,kl}Vu_{kl}$$

In all the presented examples it is assumed that the values of $A_{ij,kl}$ and $B_{ij,kl}$ are of the same order as $1/R_x$ and that $R_x = 1$.

2. A DESIGN STRATEGY FOR THE SINGLE-LAYER CNN

The first step in the CNN design procedure is to estimate the required level of circuit complexity which is necessary to realize the given task. In cases where the signal-processing function can be viewed as non-recursive filtering with spatially local operators, we may expect a single-layer CNN structure to be sufficient. The implementation task can be simplified if the function to be performed by the circuit can be expressed as follows.

Propagate certain signals through the network if a certain condition with respect to the input information is satisfied.

This statement suggests that a design strategy for the single-layer CNN should exploit the distinct roles of the feedback and feedforward operators. The condition on the input cell voltages should be tested by means of the elements of **B**, while propagation should be controlled by suitable connection of cell output voltages by means of the elements $A_{ij,kl}$. Using this approach, an appropriate form for the operators can be derived.

3. APPLICATION OF DESIGN STRATEGY FOR BINARY IMAGE PROCESSING

Some applications for the single-layer CNN obtained as a result of appropriate formulation of the processing problem are introduced in this section. In all the examples presented it is assumed that binary input information (image) is represented as cell input voltages.

3.1. Global information recovery in a locally interconnected circuit

The goal in this example is to extract the specified shapes from an input image. The initial image, which is composed of the pixels corresponding to the cell's initial states, contains isolated black pixels called *key* pixels. If any of the key pixels overlaps with any of the shapes present in the input image, this shape is extracted. The circuit arrangement proposed by Chua and Shi[3] that realizes this task requires a three-layer CNN structure. The algorithm of this problem solution, however, can be formulated according to the pattern introduced, so we may expect that the task can be realized in a single-layer CNN. Let us assume that all initial cell states ($Vx_{ij}(0)$) are set to '-1' except for the cells that correspond to the key pixels, for which $Vx_{ij}(0) =$ '$+1$'.

Specified shape recovery can be described as a process of propagation of an increase in the cell state values for those cells which correspond to the shape chosen by high initial state. The template design reflects the distinct role of the operators: the feedback operator is responsible for the propagation ('diffusion') of high cell states so its entries are positive. Furthermore, since there is no preferred propagation direction, the operator is symmetric. The feedforward operator tests if an input voltage of the cell is high and only then allows for an increase in the state so that it contains a single non-zero element in the centre, which is positive. Templates thus have the form

$$\mathbf{A} = \begin{bmatrix} \alpha & \alpha & \alpha \\ \alpha & \alpha_0 & \alpha \\ \alpha & \alpha & \alpha \end{bmatrix}, \qquad \mathbf{B} = \begin{bmatrix} 0 & 0 & 0 \\ 0 & \beta_0 & 0 \\ 0 & 0 & 0 \end{bmatrix}, \qquad \alpha, \beta_0 > 0$$

To obtain appropriate values for α, α_0 and β_0, several requirements which must be satisfied to ensure desired circuit properties are formulated. These are based on the cell state equation and determine the signs of derivative for various values of controlling voltages. Since it is assumed that voltages are binary-valued, the requirements have the form of linear combinations of the unknown parameters and are formulated for the worst cases, i.e. for the cases where the inequalities are least well satisfied. The requirements are as follows (a_0 denotes $\alpha_0 - R^{-1}$).

1. The cell state voltage is increased only when

(a) high initial voltage (key) corresponds to high input voltage, which occurs only when the chosen pixel belongs to the stored shape:

$$\beta_0 + a_0 - 8\alpha + I > 0$$

(b) the initial voltage is low but the pixel belongs to the shape chosen, so its state should increase during the transient:

$$\beta_0 - a_0 - 6\alpha + I > 0$$

2. The cell state is decreased when

(a) the pixel does not belong to the chosen shape:

$$-\beta_0 - a_0 + 8\alpha + I < 0 \quad \text{or} \quad \beta_0 - a_0 - 8\alpha + I < 0$$

(b) high initial state corresponds to low input voltage:

$$-\beta_0 + a_0 - 8\alpha + I < 0$$

This set of inequalities can be solved using linear programming techniques. The parameters used in the processing example presented in Figure 1 are as follows:

$$\alpha_0 = 1\cdot 5, \qquad \alpha = 0\cdot 3, \qquad \beta_0 = 2\cdot 5, \qquad I = 0$$

It must be stressed that the global information recovery described in this subsection can be realized using a single-layer CNN only when key pixels are isolated. In the general case appropriate requirements constitute a contradictory set of inequalities.

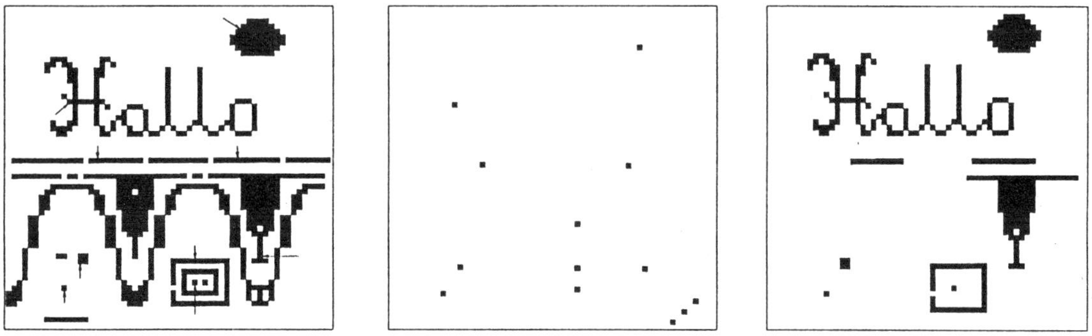

(a) (b) (c)

Figure 1. Recovery of specified shapes. (a) Information stored in CNN inputs. Arrows point to those pixels which overlap with key pixels. (b) Initial states of the CNN (initial image with key pixels). (c) The processing result

3.2. Extraction of the shape containing a desired feature

The processing goal can be described as an extraction of all of the pixels that belong to the shape that contains the desired feature. If we assume that for all the cells $Vx_{ij}(0) = -1$, then the circuit should operate according to the following rule: the cells that encounter the desired feature should increase their states and this increase in the state should be propagated among other elements of the shape. Once again the feedback operator is assumed to be symmetric. The feedforward operator should test whether a cell corresponds to the desired feature or to a shape in an input image. An appropriate form of the operators is

$$\mathbf{A} = \begin{bmatrix} \alpha & \alpha & \alpha \\ \alpha & \alpha_0 & \alpha \\ \alpha & \alpha & \alpha \end{bmatrix}, \qquad \mathbf{B} = \begin{bmatrix} \beta' & \beta' & \beta' \\ \beta' & \beta_0 & \beta' \\ \beta' & \beta' & \beta' \end{bmatrix}, \qquad \beta' \neq 0, \qquad \alpha > 0$$

$\beta' = \beta > 0$ if the cell corresponding to the central pixel of the desired feature expects a black pixel (i.e. high output voltage) in the considered position and $\beta' = -\beta$ otherwise. Parameter values are chosen to satisfy the following set of requirements (again $a_0 = \alpha_0 - R^{-1} > 0$ is assumed).

1. The cell state is increased when

(a) a cell encounters in the analysed neighbourhood a configuration of input voltages that corresponds to the desired feature:

$$\beta_0 + 8\beta - a_0 - 8\alpha + I > 0$$

(b) a cell corresponds to the shape in the input image and has at least one neighbour in a high state (propagation):

$$\beta_0 - 8\beta - a_0 - 6\alpha + I > 0$$

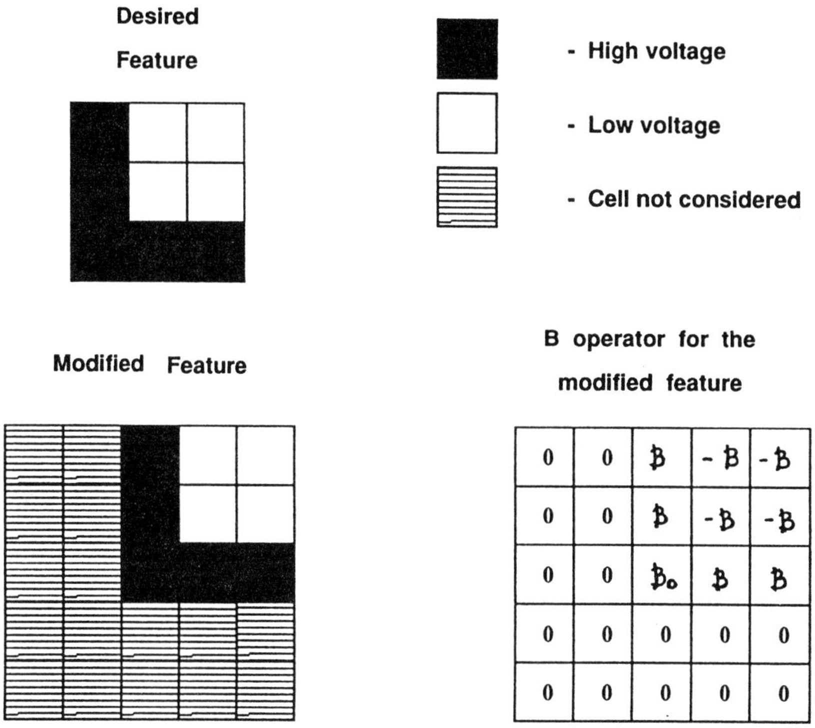

Figure 2. Redefinition of the desired feature

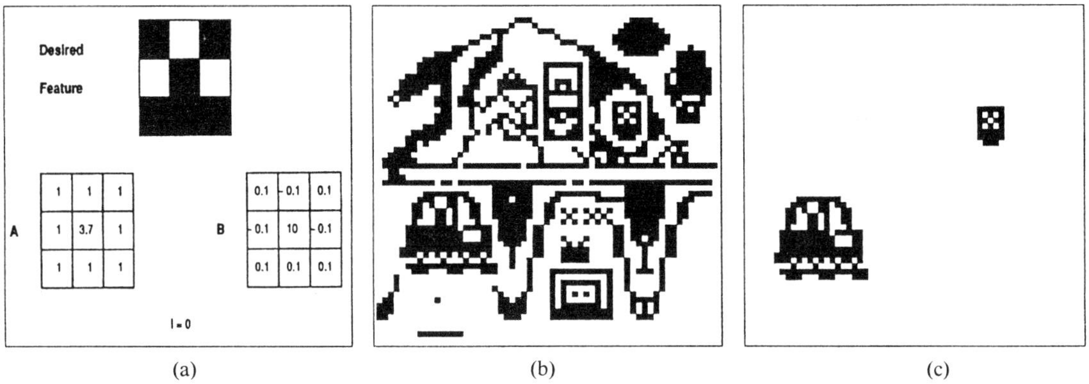

Figure 3. Extraction of shapes containing a desired feature. (a) The feature of interest and the template. (b) Input image. (c) The processing result

2. The cell state must not be increased when

(a) a cell encounters a feature similar but not identical to the desired one:

$$\beta_0 + 6\beta - a_0 - 8\alpha + I < 0$$

(b) a pixel does not belong to any shape:

$$-\beta_0 + 8\beta - a_0 + 8\alpha + I < 0$$

These requirements are not contradictory if we assume that the central pixel of the desired feature is black. However, we can always modify the desired feature definition to make the central pixel black by enlarging the cell neighbourhood. This is shown in Figure 2 together with the corresponding **B**-matrix. The processing example together with the chosen set of parameters is presented in Figure 3.

3.3. Simplification of information read-out

One of the very important problems to be solved in the VLSI implementation of a CNN is the realization of the input/output interface. The analysis of the result is typically performed in discrete time steps after the transient decays. This involves accessing every cell in the network and takes many time steps. The procedure would be simplified considerably if the output information could be concentrated in a small number of cells, e.g. those belonging to one of the boundary rows (or columns), rather than dispersed throughout the entire array. Since this requires that the information be compressed, one can expect that this approach could be used only for a subset of processing problems, e.g. to detect the presence of a feature. This leads to the concept of a circuit which performs two tasks: it detects the desired feature and transports this information to the boundary cells. Assuming that all initial states are set to be low, the circuit should be designed to ensure an increase in the state voltages in the following two cases:

(a) for the cells that encounter the desired feature in the analysed neighbourhood—this can be seen as a test of the condition related to the input, so it can be accomplished by an appropriate choice of the **B**-operator, which is obtained in the same way as in the previous example.
(b) for the cells that propagate information, i.e. that repeat an increase in the state after their neighbours in the same row (or column)—this process is determined by the **A**-operator, which has positive entries only in the central row (column) and zero entries elsewhere.

An example of this type of processing together with the chosen parameter set is shown in Figure 4 (a propagation along the rows is assumed).

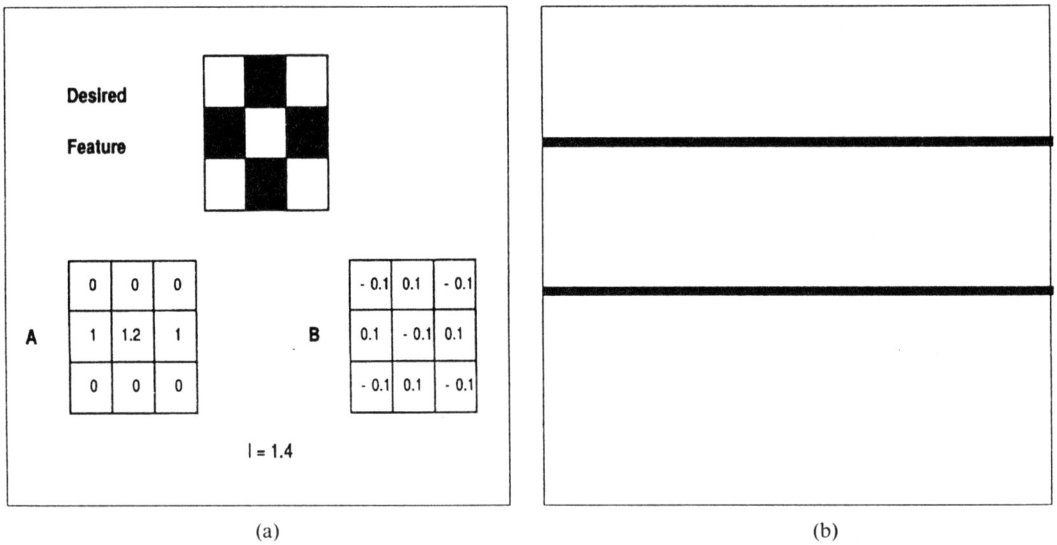

Figure 4. Modified feature detection process. (a) The feature of interest and the template. (b) The processing result (input image given in Figure 3(b))

4. CNN CELL MODEL MODIFICATION

There exist many signal-processing problems which cannot be solved with a single-layer CNN and where we have to use more complex circuits, e.g. multilayer CNNs.[2,3] However, sometimes the problem to be solved can be reformulated as a set of simple tasks, where each task can be solved using a simple single-layer structure. The tasks can be performed simultaneously and their results undergo certain logical operations. The circuit architecture that seems to be appropriate to realize these problems consists of several CNNs arranged in parallel. Parallel CNN connection is understood as a circuit composed of several CNNs, each of them designed to perform a different task, where every CNN is fed with the same input information and the corresponding output voltages of every network are passed through appropriate logical functors. Taking into consideration the severe difficulties in implementing such a structure in VLSI[10], an idea for simplifying the required circuit has been proposed.[8] According to this idea, several parallel connected CNNs are replaced by a single one composed of modified cells. The modified circuit permits complex parallel signal processing, featuring simultaneous realization of different tasks and logical operations on their results. The basic processing element of the proposed structure differs from the original cell in its state-producing part. Namely, the state voltage of every cell is controlled by several input subcircuits through an appropriate diode-resistive structure which implements the desired quasi-logical (we deal with continuous values of arguments) operation.

Network design is the process of choosing the template for each input subcircuit and the architecture of the diode-resistive part of the cell appropriate for the implementation of the specified quasi-logical function. Any γth input subcircuit is described by parameters $\mathbf{A}_\gamma$, $\mathbf{B}_\gamma$ and I_γ and these parameters are determined as if to describe a basic CNN that would realize each partial task separately. Partial state voltages ($Vx\gamma_{ij}$) produced by these subcircuits are the basis for the cell state (Vx_{ij}) determination. Despite the functional differences between the proposed structure and parallel connected CNNs, the processing result in many cases is the same. The modified circuit structure also offers some useful new processing capabilities.

5. BINARY IMAGE-PROCESSING APPLICATIONS OF CNNS COMPOSED OF MODIFIED CELLS

One of the problems for which the modified structure seems appropriate is the detection of a specified feature regardless of the feature's orientation in the input image (or, more generally, the simultaneous detection of distinct features). This problem can be easily formulated in a form that contains simple operations linked by logical expressions, namely: indicate the presence of feature if it has an orientation ϕ OR indicate its presence if it has an orientation ψ OR .... The cell circuit arrangement, parameter values and processing example are shown in Figure 5. The first input subcircuit of the modified cell acts as the detector of the feature in its first orientation (i.e. increases its state ($+Vx1$) when the appropriate pattern of cell output voltages is encountered in the given cell neighbourhood), while another one detects another feature orientation. A diode-resistive subcircuit produces the cell state voltage '$-Vx$' as the lower one of the voltages of input subcircuits, thus performing the function similar to logical OR over '$+Vx1$' and '$+Vx2$' voltages.

One of the image-processing problems that are difficult to realize in the parallel CNN connection is presented in Figure 6. The processing goal—to extract the midpoints of horizontal lines—is obtained by

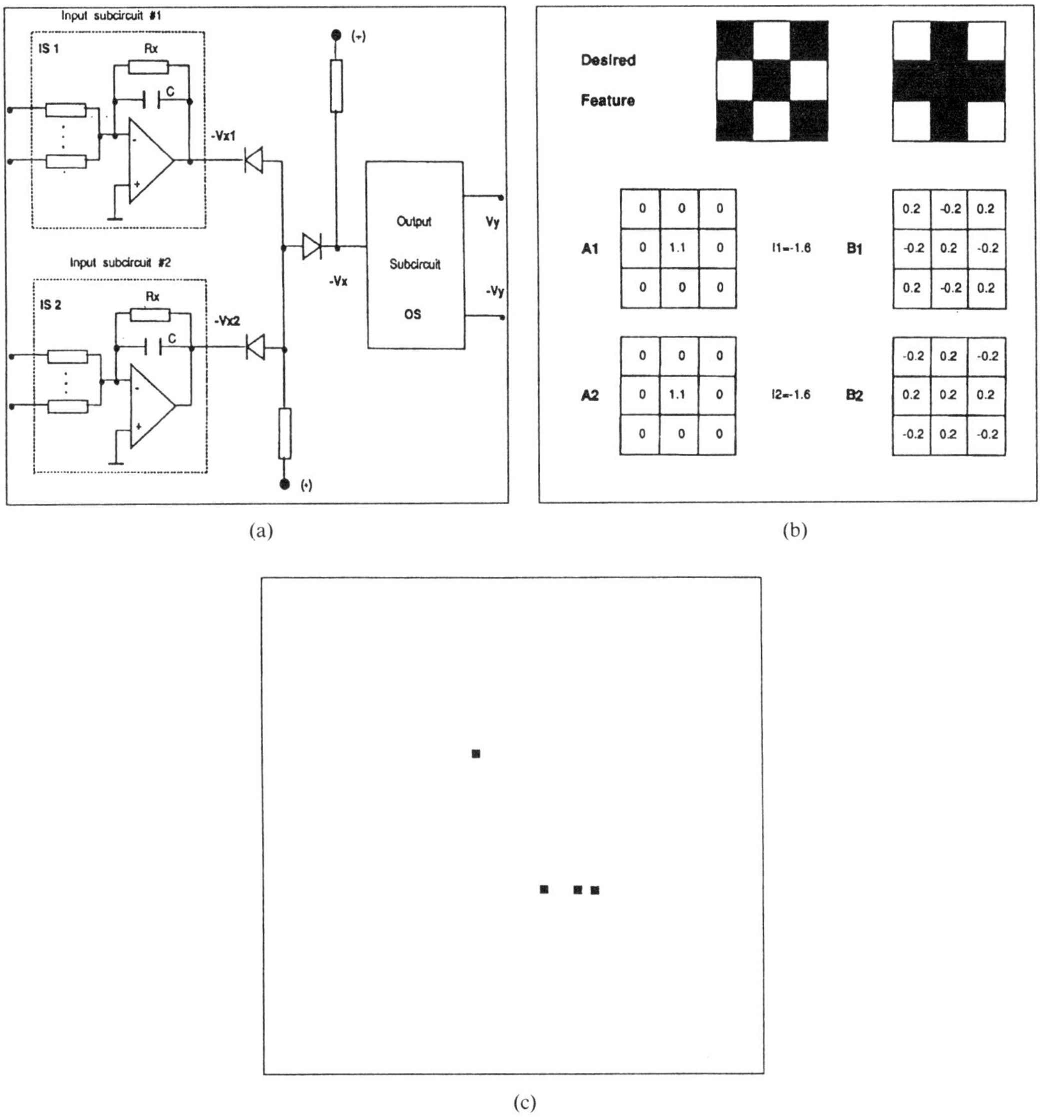

Figure 5. Feature detection using a CNN composed of modified cells. (a) Structure of the cell (output subcircuit is presented in Reference 8). (b) The desired feature in its two orientations and appropriate templates. (c) The processing result (input image given in Figure 3(b))

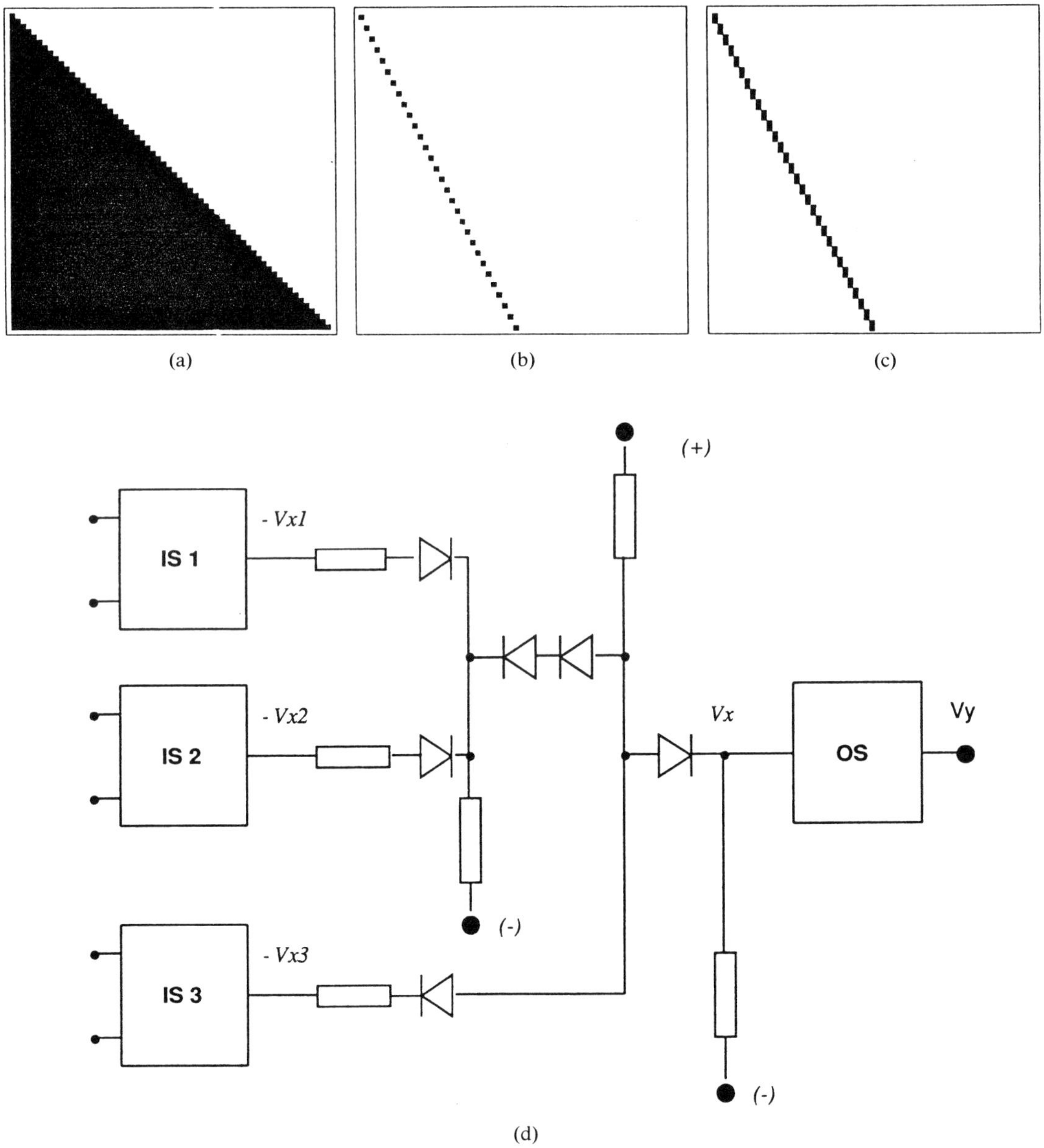

Figure 6. Extraction of midpoints of horizontal lines. (a) Input image. (b) Processing result—each CNN cell contains two input subcircuits. (c) Processing result for the circuit with the cells given in (d). (d) Structure of the cell with three input subcircuits

performing the quasi-logical 'AND' operation over the voltages produced by two input subcircuits in every network cell. The first subcircuit acts as a connected component detector[4] (CCD) which translates the image towards the right network edge, while the second is a CCD acting in the opposite direction. Their parameters are as follows:

$$\mathbf{A1} = [1 \;\; 2 \;\; -1], \qquad \mathbf{A2} = [-1 \;\; 2 \;\; 1], \qquad \mathbf{B1} = 0, \qquad \mathbf{B2} = 0, \qquad I1 = 0, \qquad I2 = 0$$

The circuit works correctly only if there are an odd number of black elements in the horizontal line (see Figure 6(b)). To obtain a horizontal midpoint detector without this restriction, we have to introduce a third input subcircuit which should break the symmetry of the problem. An appropriate choice of parameters for the additional part is

$$\mathbf{A3} = [0 \ \ -1 \ \ 2 \ \ 0 \ \ -1], \qquad \mathbf{B3} = 0, \qquad I3 = -2 \cdot 1$$

If we denote partial state voltages produced on the outputs of cell input subcircuits by $Vx1_{ij}$, $Vx2_{ij}$ and $Vx3_{ij}$ respectively, then the state voltage Vx_{ij} of the complex cell is produced according to the relation

$$Vx_{ij} = \max\{Vx3_{ij}, \min\{Vx1_{ij}, Vx2_{ij}\}\}$$

which corresponds to the logical operation

$$Vx_{ij} = Vx3_{ij} + (Vx1_{ij} * Vx2_{ij})$$

where '+' denotes logical OR and '*' denotes AND.

6. CONCLUSIONS

The CNN design methodology proposed in Section 2, which emphasizes the different roles of the operators **A** and **B**, can be useful in developing novel applications for the single-layer CNN. Furthermore, by describing a problem in terms of the operations to be performed on inputs and input conditions, we can use the design strategy to understand better some existing CNN applications, e.g. hole filling[4] (as presented in Reference 9).

Some signal-processing capabilities of a network composed of modified cells were illustrated in Section 5. By using more complex templates for the input subcircuits, we can increase the number of applications for this structure, e.g. we can extract the shapes that contain one of several desired features, which is an extension of the application presented in Section 3.2.

ACKNOWLEDGEMENTS

I would like to thank Professor T. Roska, Professor T. Kacprzak and M. P. Kennedy for their helpful comments on this paper.

This research is sponsored in part by the Polish Science Committee under Grant 180/ONS-T/OS/026/90-2 and by the American National Science Foundation under Grant MIP 89-12639.

REFERENCES

1. L. O. Chua and L. Yang, 'Cellular neural networks: theory', *IEEE Trans. Circuits and Systems*, **CAS-35**, 1257–1272 (1988).
2. L. O. Chua, and L. Yang, 'Cellular neural networks: applications', *IEEE Trans. Circuits and Systems*, **CAS-35**, 1273–1290 (1988).
3. L. O. Chua and B. Shi, 'Multiple layer cellular neural networks: a tutorial', *Memo UCB/ERLM90/113*, University of California, Berkeley, December 1990.
4. T. Matsumoto, T. Yokohama, H. Suzuki and R. Furukawa, 'Several image processing examples by CNN', *Proc. IEEE CNNA-90*, IEEE, New York, 1991, pp. 100–111.
5. T. Roska and L. Chua, 'Cellular neural networks with nonlinear and delay-type template elements', 12–25.
6. F. Zou, S. Schwarz and J. A. Nossek, 'Cellular neural network design using a learning algorithm', 73–81.
7. T. Boros, K. Lotz, A. Radvanyi and T. Roska, 'Some useful, new, nonlinear and delay-type templates', *Report DNS-1-1991*, Hungarian Academy of Sciences, 1991.
8. K. Ślot, 'Determination of cellular neural network parameters for feature detection of two-dimensional images', *Proc. IEEE CNNA-90*, IEEE, New York, 1991, pp. 82–91.
9. L. O. Chua and P. Thiran, 'An analytic method for designing simple cellular neural networks', *Memo UCB/ERLM91*, University of California, Berkeley, 1991.
10. K. Halonen, V. Porra, T. Roska and L. O. Chua, 'VLSI implementation of a reconfigurable cellular neural network containing local logic (CNNL)', *Proc. IEEE CNNA-90*, IEEE, New York, 1991, pp. 206–215.

ON BIOLOGICAL SENSORY INFORMATION PROCESSING PRINCIPLES RELEVANT TO CELLULAR NEURAL NETWORKS

WALTER HEILIGENBERG
Neurobiology Unit, Scripps Institution of Oceanography, University of California at San Diego, La Jolla, CA 92023, USA

AND

TAMÁS ROSKA
Computer and Automation Institute of the Hungarian Academy of Sciences, Budapest, H-1111 Hungary

1. INTRODUCTION

Cellular neural networks, CNNs,[1] and their extensions[3] provide a natural and universal model of analog processor arrays on a multi-layer geometrical grid. Their local connectivity and regular structure appear most efficient and parsimonious for electronic and biological implementations alike.

Their simple architecture, combining analog dynamics and logic,[2] and thus referred to as 'analogic or dual computing', represents a very powerful and appealing design. We have used this concept by introducing the dual computing CNN (e.g. References 13 and 3c). The 'programmable' CNNs[3c] provide a new quality: a simple way to solve CNN algorithms, a kind of analog software. Qualitatively different functions will thus be reproduced by the same machine or model using different ('templates' or template sequences, called programs.

Studies of the neural basis of behavior in various 'champion' species[4a] or anatomical structures have revealed prototype solutions (e.g. References 4b, 5, 6a, 6b) that nature adopted for well defined tasks faced by different species or by different systems within the same animal.

From the study of task-specific models of specific biological phenomena and their silicon implementations,[9,10,12] as well as from our initial work on vision-related CNN models,[11] it has become clear that the programmable CNN paradigm offers a unique challenge: to serve as a universal model for a multitude of diverse biological phenomena. Moreover, it became apparent during our recent discussions that the dual CNN model contains a number of properties that reflect basic principles found in the neural design of behavior. In this short report we will summarize a few of these principles that appear both inspiring and instructive.

2. THE PROGRAMMABLE ANALOGIC (DUAL) CNN PARADIGM, A UNIVERSAL 3D ANALOG-AND-LOGIC MACHINE

Detailed descriptions of this paradigm can be found in References 1 and 3, and see especially the first paper of this book and its references. Here, we only summarize its main aspects. Suppose the 2D layers of neurons are placed on horizontal grids (Figure 1). Each neuron has three continuous characteristic signals: the input (u), the state (x), and the output (y), and different layers are stacked upon each other in topographic register. A vertical cross section of a single layer is shown in Figure 2. Neurons are interacting locally, within a neighborhood or receptive field. The radius of this neighborhood may vary from one to a few dozen elements. In Figure 2, $r = 1$ and the feedforward and feedback interaction weights are represented in the cloning template matrices. These template matrices reflect the geometry of the grid. If successive layers have fewer grid points we imply 'convergence' (i.e. several neurons in a layer converge to a single one on the next layer).

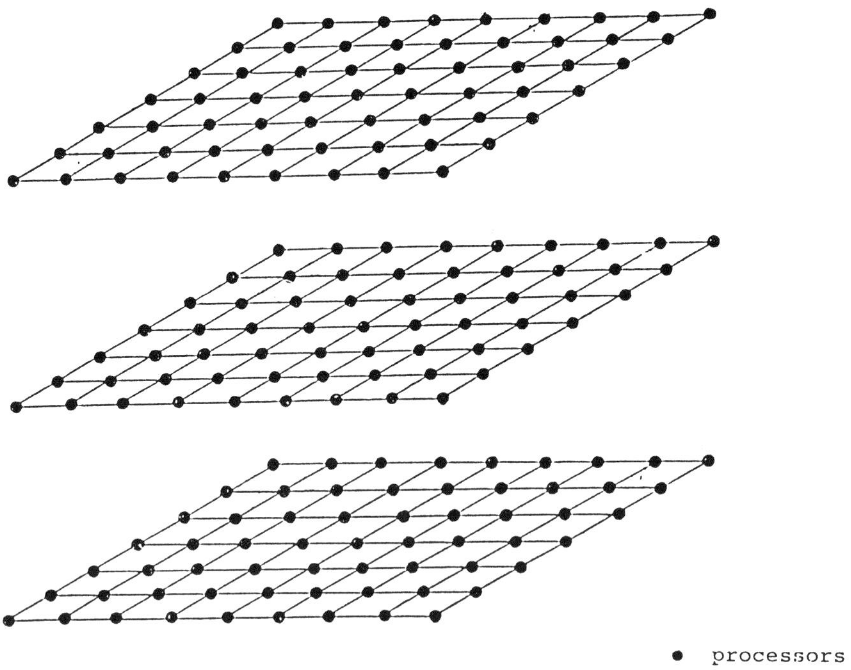

Figure 1. Multiple layers of neurons (processors) placed on geometric grids

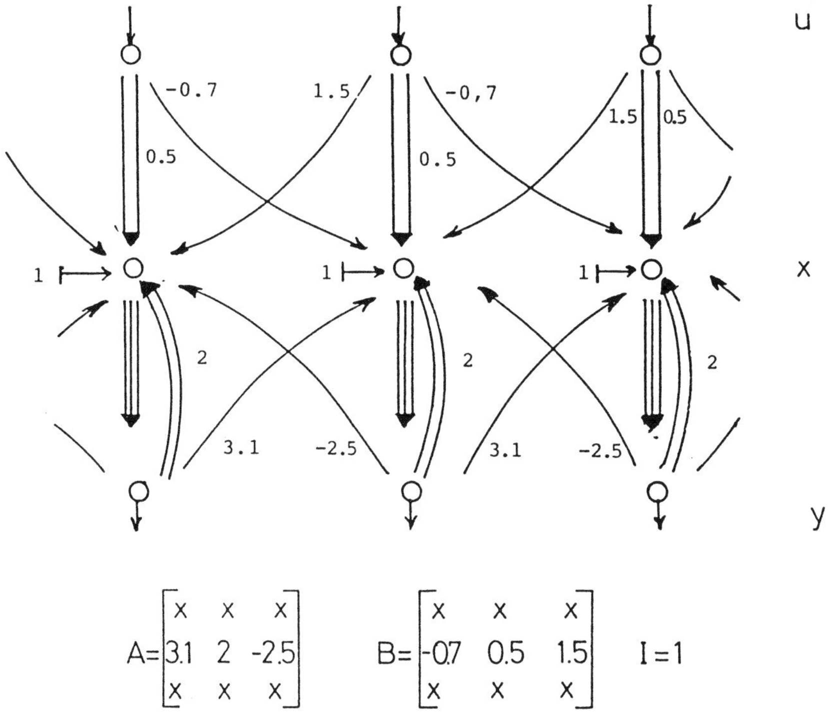

Figure 2. A vertical cross section of a layer consisting of three sub-layers. The lateral interactions are represented in the off-center terms of the feedback (A) and feedforward (B) templates

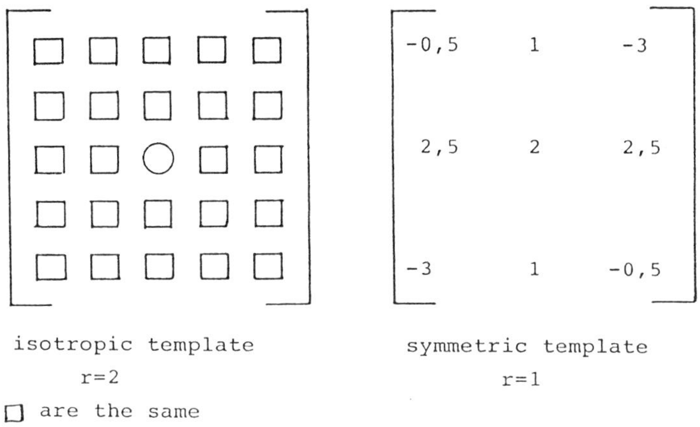

Figure 3. Isotropic and symmetric templates

The weight patterns of the interactions represented in the template reflect not only the anatomical (structural) properties of the synapses but their functional characteristics as well. For simplicity, we assume that the weight patterns are locally invariant, i.e. all elements within the plane have the same local interaction pattern, and their organization is isomorphic; hence their function is translation-invariant. The template patterns of Figure 3, called isotropic and symmetric, perform rotational invariant and symmetric functions, respectively.

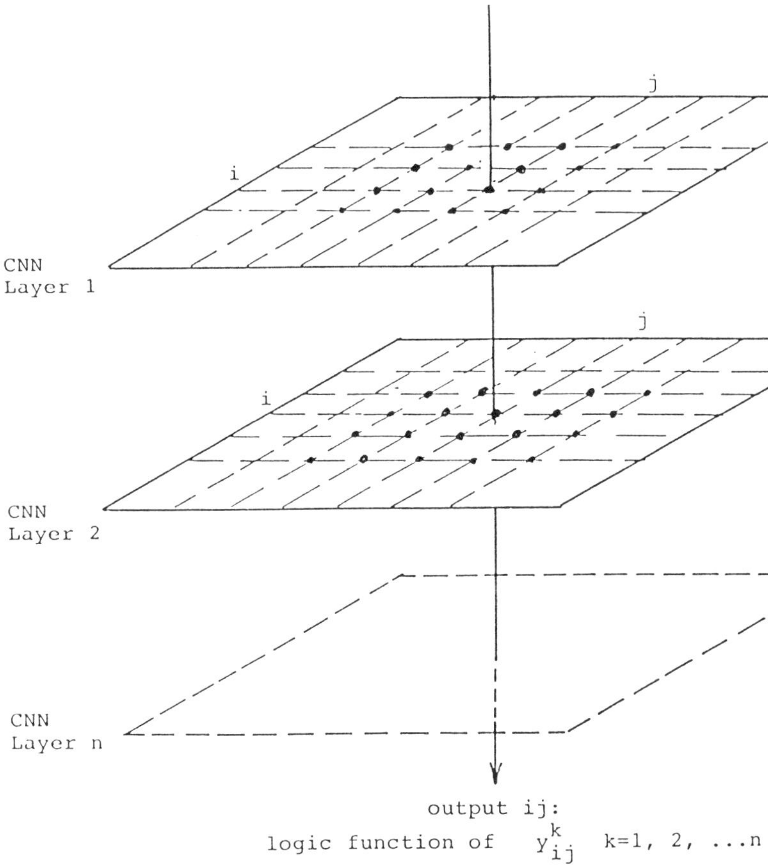

Figure 4. A simple analogic (dual) CNN structure

The template represents the interaction of each cell within its receptive field. All center-surround effects are of this kind. The template elements (atoms) represent the synapse strengths (weights) in different geometric directions and distances. A positive value implies an excitatory and a negative value implies an inhibitory effect. Thus, due to the geometric aspects of the templates, the different center-surround effects can be modeled as functions of relative space.

The basic function of a CNN layer is that, due to its locally interactive nonlinear array dynamics, an output map is generated from an input map. The function is represented by the cloning template patterns (feedforward template, B, feedback template, A, and the off-set term, I). By changing these patterns, i.e. by reprogramming the CNN layer, different functions can be implemented (e.g. noise reduction, motion detection, edge detection, corner detection, etc.). Several CNN layers, stacked upon each other, can interact in a similar manner, i.e. neurons can influence each other within their receptive fields in three dimensions.

In a multilayer analogic (dual) CNN, the results of many locally interacting analog layers can be combined globally by simple local logic. In Figure 4 this 'vertical' logic line is shown for one cell only, but all cells have this kind of vertical local logic lines. Thus, analog dynamics and logic are combined in a special way.

This 3D analog-and-logic machine is so natural that it is not surprising that many biological sensory information processing organs can be represented by this model. It appears as a unifying model in a wide area. In the following, we will try to support this view.

3. PARALLEL TOPOGRAPHIC MAPS AND 'VERTICAL' LOGIC DECISIONS

Continuous variables in a stimulus pattern, such as the frequency of a tone, the location of a stimulus source in space, or the distance to a target, are commonly represented topographically within strata of neurons, referred to as 'neuronal maps'. Often, these maps are multi-layered, and different local computations may extract different aspects of the stimulus pattern within different layers, for example, the phase information and amplitude information, respectively, associated with a periodic signal emanating from the spatial location represented at a given point of the map. Since different layers are stacked upon each other in topographic register, 'vertical' connections (perpendicular to the layer) between these layers can readily serve to perform functional or logical operations between different aspects of the stimulus pattern. A given type of vertical connections could, for example, perform an 'AND'-gating between phase and amplitude representations and detect, for example, the joint occurrence of a phase lead and a rise in amplitude for a signal at a given point in space.

Topographic junctions between maps containing ordered representations of various spatial and temporal aspects of stimulus patterns can perform very complex feature extractions. Even different sensory modalities, such as vision and audition, can be combined in this manner to yield, for example, multimodal spatial stimulus representations as we find them in the tectum or superior colliculus of vertebrate animals.

Stacking different laminae (2D layers) of neurons on top of each other in the third dimension provides space–time efficient processing with minimum communication overhead. To use the same information in several places, nature employs multiple neuronal representations, and individual neurons may send collaterals to several targets for the sharing of their information.

The minimization of communication costs is one of the most crucial tasks in building supercomputers and array computers. The relatively local connectivity in a CNN model ensures high space–time efficiency not only in the biological but in the silicon implementation.

Figures 5, 6 and 7 show some structures and organizations of multi-layered torus semicircularis of electric fish (see the details in Reference 4b).

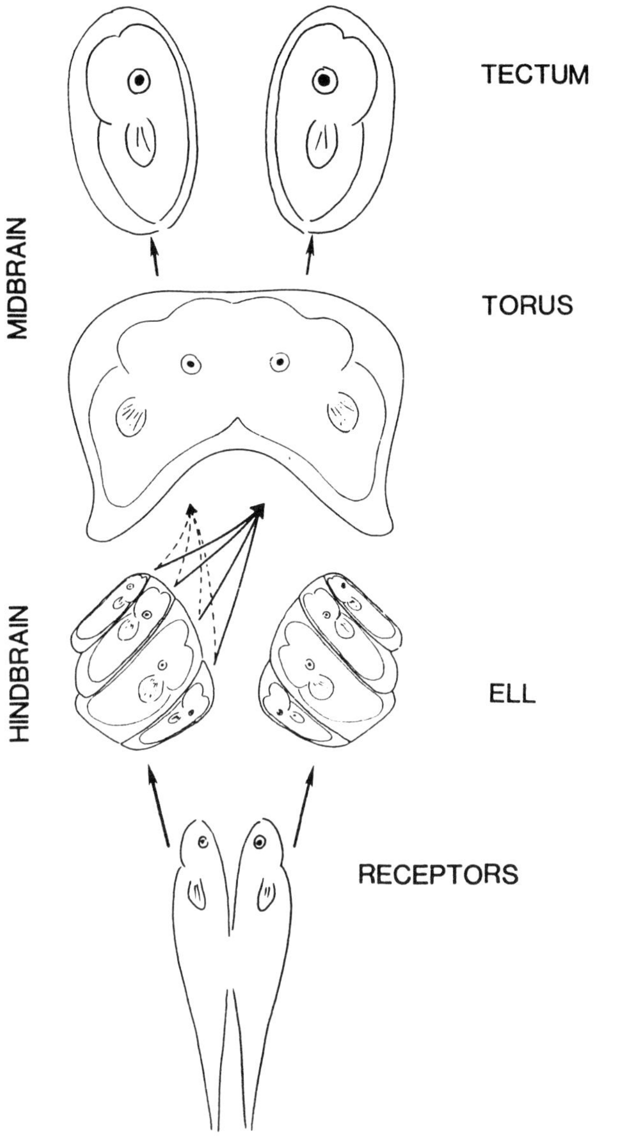

Figure 5. The spatial order of electrosensory information is preserved in layers of higher-order neurons in the hindbrain and midbrain due to their somatotopic organization, i.e. the topographic projections from a lower-order surface to a higher-order surface. Electroreceptors on the fish's body surface project ipsilaterally, via primary afferents, to four somatotopically ordered maps in the electrosensory lateral line lobe (ELL) of the hindbrain. Whereas ampullary receptors (tuned to low-frequency signals) project only to the most medial map, each tuberous receptor (tuned to high-frequency signals) projects to the three remaining maps via collaterals of its primary afferent. The maps of the ELL project predominantly to the contralateral side of a single, multilayered map in the torus semicircularis of the midbrain. An ipsilateral topographic projection of the torus to the optic tectum leads to a convergence of visual and electrosensory spatial information. The enlarged representation of the head in all maps reflects the higher density of electroreceptors in this region of the body surface, which is comparable to a fovea in a visual system

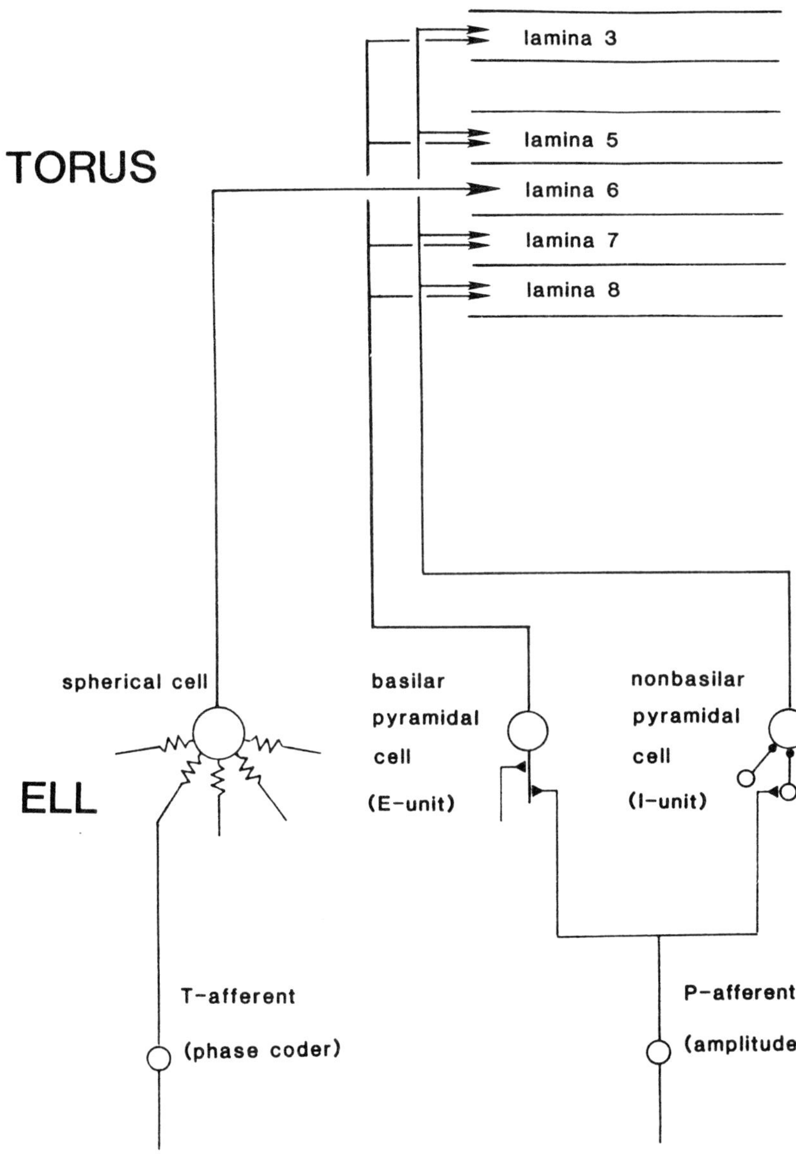

Figure 6. The organization of the ELL provides separate substrates for the processing of phase information and amplitude information conveyed by tuberous electroreceptors. Primary afferents from T-type receptors, or **phase coders**, form electrotonic synapses with spherical cells, and a single spherical cell may receive inputs from several T-afferents originating within its receptive field on the body surface. Primary afferents from P-type receptors, or **amplitude coders**, form excitatory synapses on the basilar dendrites of basilar pyramidal cells, or **E-units**, and also excite granule cells which, in turn, inhibit nonbasilar pyramidal cells, or **I-units**. As a consequence of these connections, E-units are excited by a rise in stimulus amplitude, while I-units fire in response to a fall in stimulus amplitude. These units are thus comparable to on- and off-units in a visual system. Whereas the spherical cells project exclusively to lamina 6 of the torus semicircularis, the pyramidal cells project to various laminae above and below lamina 6. All projections are topographic so that the somatotopic order of electrosensory information is maintained to the level of the torus. And further, the laminae of the torus are stacked in topographic register so that any vertical connection across different laminae would link representations of the same point on the body surface. Such links, for example, could connect phase information in lamina 6 with the topographically corresponding amplitude information represented in higher or lower laminae. Filled triangles indicate excitatory synapses, and filled circles indicate inhibitory synapses.

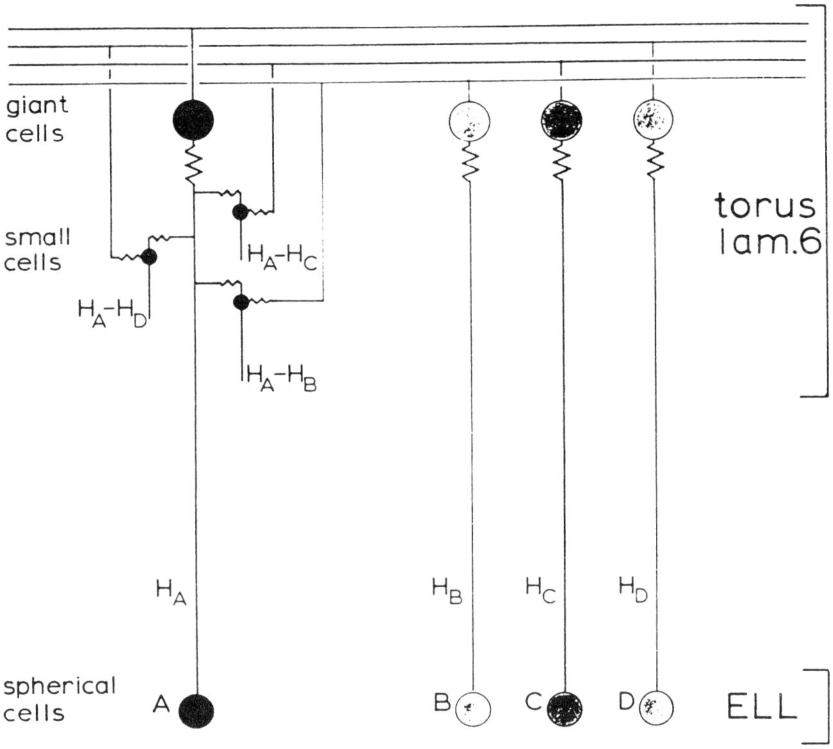

Figure 7. Lamina 6 of the torus semicircularis contains a network for the computation of differential phase between any two areas on the body surface. The spherical cells in the ELL (see Figure 6) receive somatotopically ordered information about the phase, i.e., the timing of zero-crossings, of electric signals on the body surface, and they relay this information, H_A H_B, ..., in a topographic projection to lamina 6 of the torus by synapsing on the somata of giant cells and the dendrites of nearby small cells. Giant cells, in turn, relay the phase-locked spikes of spherical cells through their large axonal arbors to wide areas of lamina 6, and their axonal collaterals synapse on the somata of small cells, which apparently can accommodate only a single large synapse of this kind. Therefore, a small cell is controlled by inputs from two sources: the input at its dendrites is derived from the area on the body surface represented at its location in lamina 6 (area A in this example), while the input at its soma originates from the area represented at the location of the giant cell soma whose axon collateral forms this contact (area B, for example). Small cells fire intermittently, and their rate of firing is controlled by the differential timing of their two inputs. Giant cells receive inputs from several spherical cells within a receptive field, and only fire in response to a nearly synchronous arrival of spikes from several spherical cells. For clarity, only a single spherical cell is shown to contact a giant cell in this diagram. A local module, as shown for area A in this figure, exists for the representation of any area on the fish's body surface.

4. ONTOGENETIC ASPECTS, AND THE MINIMAL CODING OF AN ANALOG PROCESSING ARRAY—THE ROLE OF THE CLONING TEMPLATE

Whenever the computation of some aspect within a stimulus pattern can be formulated as a topographic operation, such as the center-surround organization of a receptive field for the detection of objects of a given size, then the same local network structure and synapse-strength (weight) pattern can be employed everywhere in the map. This 'isomorphy' results in the translation-invariant processing capability inherent in the cloning template of the CNN layer. As a consequence, the ontogeny of this map does not require individual instructions for every synapse to be formed and only calls for general instructions for the wiring of the local pattern of connections, and these instructions can be coded as functions of relative distances and directions within the map. This, of course, does not exclude that parameters of these functions may change gradually across longer distances within the map. For example, receptive fields could be larger in the periphery than in the center of the map. A non-uniform grid size and locally varying weights in CNN connections are examples of these phenomena.

The importance of this structural modularity in artificial (silicon or optical) implementations is manifested in their regular geometric structure, which is of great advantage in manufacturing as well.

5. PROTOTYPE BIOLOGICAL SOLUTIONS—CNN PROGRAMS

Apart from the traditional center-surround organization of receptive fields, which serves for the detection of local contrasts, other modular network structures have recently been identified which serve more complicated operations. These include a network within lamina 6 of the torus semicircularis of gymnotiform electric fish that determines phase differences between periodic electric signals impinging upon different points of the fish's body surface.[4a] Similarly, the nucleus laminaris of the auditory system of owls contains a network for the detection of interaural phase differences that can be computed independently at any sound frequency.[6b]

The programmable analogic (dual) CNN model implemented on silicon reflects an efficient property of programmability. Using the same prototype machine for several purposes, the economy of scale becomes dominant. Moreover, several biological CNN algorithms can be implemented by CNN analog software.[3c] The recently discovered CNN Universal Machine[17] represents the truly universal solution: the unified theoretical framework as well as the practical implementation of the stored program analog- and logic array computer.

Figure 8 shows a model of the joint computation of amplitude and phase information (see the details in Reference 4b).

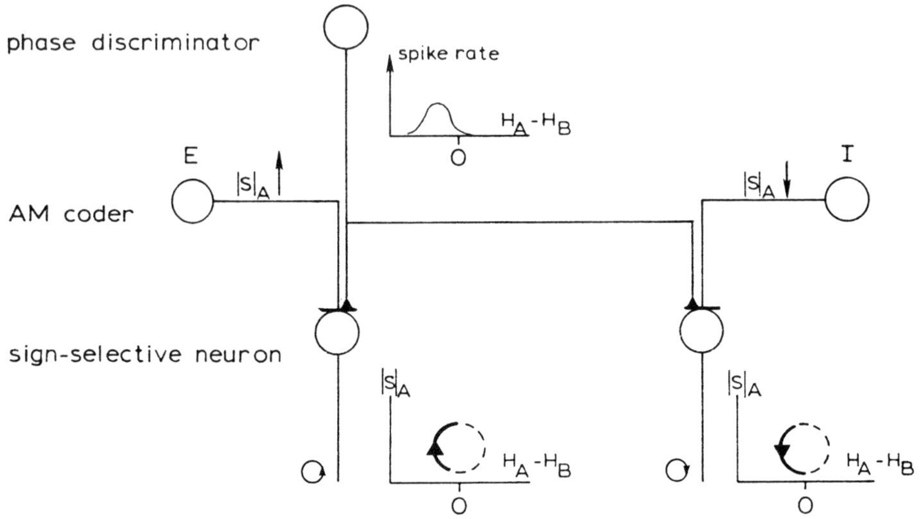

Figure 8. A model for the joint computation of amplitude and phase information by a sign-selective neuron. Information is provided from two areas, A and B, on the body surface. A phase-discriminator neuron that is excited by small negative values of the differential phase, $H_A - H_B$ (top), i.e., small phase leads of area A over B, is assumed to facilitate input from amplitude modulation (AM) coders for area A to sign-selective neurons. Two classes of AM coders, E-neurons and I-neurons, are respectively excited by a rise and a fall in the local stimulus amplitude ($|S|_A$). As a consequence of facilitation, the sign-selective neuron on the left prefers a clockwise rotation centered at phase zero, i.e. a combination of a rise with a phase advance, while the neuron on the right prefers a counterclockwise rotation. The establishment of this circuitry requires vertical connections between the phase-coding lamina 6 of the torus (see Figure 7) and amplitude-coding laminae above or below lamina 6 (see Figure 6).

6 HYPERACUITY EMERGING FROM CNN MODELS

The topographic representation of a stimulus variable within a neuronal map not only allows for the establishment of modular, local networks for the extraction of specific stimulus features. A further 'benefit' of this organization is that it offers a simple way of determining the value of a stimulus variable with much larger resolution than would be expected from the tuning of single receptors or a higher-order neuron may contain several simple neurons. Given an array of broadly tuned sensory elements that cover a wide range of a continuous stimulus variable, x, then pooling of the responses of these elements, weighted by their spatial rank within the array, will yield a function $G(x)$ that is linear in x, provided that the tuning curves of neighboring elements overlap sufficiently.[4d] This form of weighted pooling could readily be obtained by increasing the synaptic weights of the elements continually across the array in the direction of the representation of larger values of x. The fact that this mechanism of weighted pooling

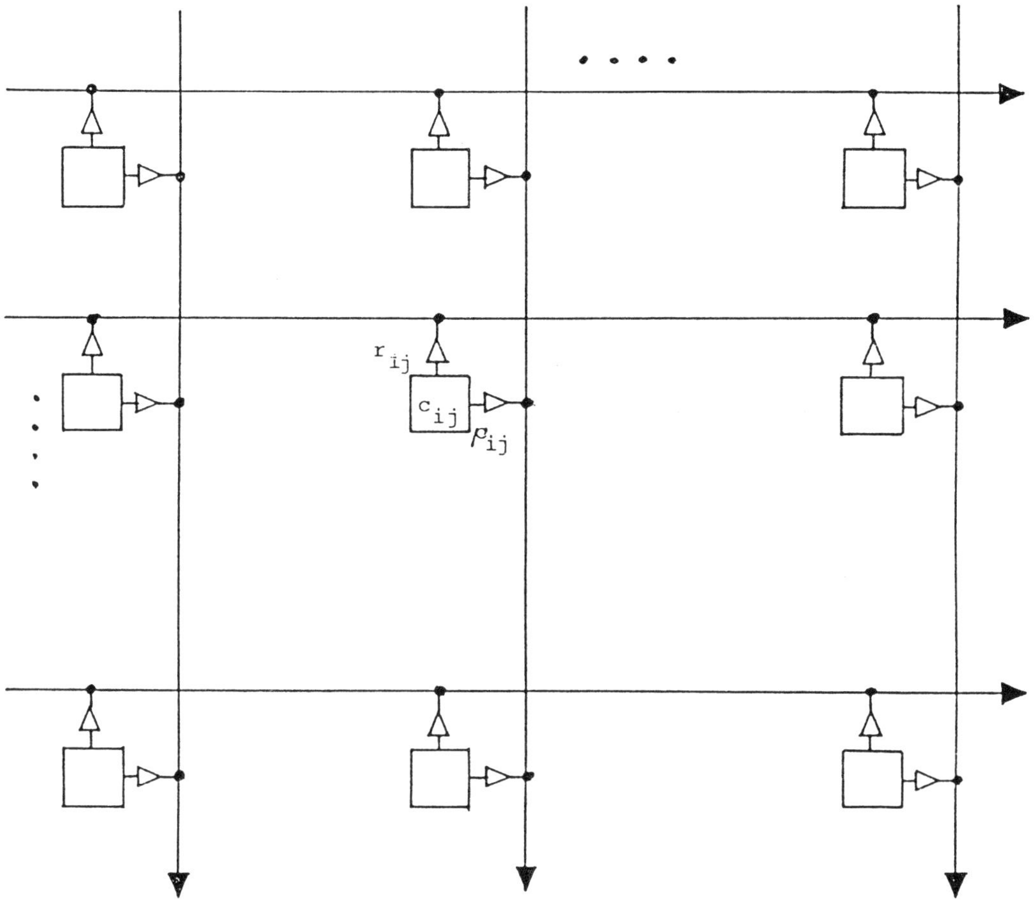

r_{ij}; ρ_{ij}: voltage controlled current sources (controlled by output)

r_{ij}: weight to the row i

ρ_{ij}: weight to the column j

$r_{i(j+1)} = r_{ij} + \Delta$

$\rho_{(i+1)j} = \rho_{ij} + \delta$

Figure 9. 2D weighted pooling CNN

requires broadly tuned sensory elements may explain that topographic maps never contain narrowly tuned neurons.

The general idea of weighted pooling can be easily modeled in 2D or 3D by a CNN using the global analogy 'processor wires' (Section 2 in Reference 2c). Figure 9 shows it schematically. This 2D weighted pooling CNN can be used in diverse fields, producing hyperacuity in space, time, and frequency (depending on coding). Moreover, since the global analogy 'processor wire' is a single wire indeed, summing the currents flowing into it, the VLSI implementation is trivial. The different weights of the consecutive cells can be implemented by simple operational transconductance amplifiers as well.

7. TOPOGRAPHIC REPRESENTATION OF INFORMATION CODED NON-TOPOGRAPHICALLY BY PERIPHERAL RECEPTORS

Topographic neuronal maps often arise through the topographic projection of sensory surfaces, such as our retina or mechanosensory body surface, upon strata of higher-order neurons. One could thus argue that these maps are a 'simple' byproduct of ontogenetic self-organization. The computational 'benefits' derived from the topographic representation of continuous stimulus variables, however, become very obvious in those cases where animals construct such representations even if the original receptor input lacks this topographic order. An example is the auditory system of owls which constructs a topographic neuronal map of auditory space in the inferior colliculus by computing horizontal and vertical angles of sound locations on the basis of interaural differences in the phase and the amplitude of arriving sound waves.[6a] This auditory spatial map is then used to enhance spatial contrasts by a center-surround organization of receptive fields and to detect the motion of sound sources. And finally, a convergence with the visual map of the superior colliculus then yields a multi-modal spatial representation.[6a] Further examples of such 'computed' topographic representations are maps in the auditory system of echolocating bats that represent variables such as echo delays (i.e. target distances) and Doppler shifts (i.e. relative velocities).[16]

In the artificial implementation of the processing of non-topographic information by topographic CNN processing arrays, the first step is the conversion representation. The champion species mentioned above provide good examples. Abstract decompositions of signals in the time domain can be mapped onto the CNN code as well.[15]

8. ANALOG EVENTS AND THE COMPLEXITY OF ANALOG INFORMATION

Consider a characteristic part of a melody, a typical action of a tennis player, an expressive part of a brilliant painting, a song of a bird, the face of a loved member of our family, etc. These are information units, described in space and time as 'analog events'.[2] Their digital representation is a 'mess' of digits, their recognition is a time consuming yet mostly ambiguous process. Surprisingly, even simple animals can solve these kinds of problems with ease. Without going into the formal definition and the processing methods of analog events,[2] we want to state that their representation and processing could be made very efficient by using programmable dual CNNs.

We have discussed the representation of topographic and non-topographic information. The rotational-invariant and translation-invariant detection of features is a generic property of CNNs with isotropic and isomorphic templates. More complex templates can be used for more complicated features. The additional logic could be efficiently used for pattern recognition or detecting complex situations using different layers for different types and modalities of information. The dual CNN dynamics processes the analog event as a whole.

One way to measure the complexity of analog information is by defining the space of analog events. The whole space is restricted by the attention. For example, while looking for our children we are searching within a very restricted face-space, though each face, as a 2D analog event, could be a complex entity. The complexity of a 2D analog event can be defined by the complexity of their dual CNN algorithm. The algorithmic elements are the various templates and the elementary logic decisions.

When defining the elementary analog templates for determining the complexity of a dual CNN algorithm, it is instructive to learn about these elementary templates in biology. We may also ask whether nature has employed prototype solutions for elementary tasks and to what extent these solutions are shared across different evolutionary levels, including our own.

ACKNOWLEDGEMENT

This work has been supported by the research grant No. 2578/91 of the National Research Fund of Hungary (OTKA).

REFERENCES

1a. L. O. Chua and L. Yang, 'Cellular neural networks: theory', *IEEE Trans. Circuits and Systems*, **CAS-35**, 1257–1272 (1988).
1b. L. O. Chua and L. Yang, 'Cellular neural networks: applications', *IEEE Trans. Circuits and Systems*, **CAS-35**, 1273–1290 (1988).
2. T. Roska, 'Analog events and a dual computing structure using analog and digital circuits and operators', *Discrete Event Systems: Models and Applications* (eds. P. Varaiya and A. B. Kurzhanski), Springer-Verlag, Berlin, pp. 225–238 (1988).
3a. T, Roska and L. O. Chua, 'Cellular neural networks with nonlinear and delay-type template elements' *Proc. IEEE CNNA–90*, pp. 12–25, 1990.
3b. T. Roska and L. O. Chua, 'Cellular neural networks with nonlinear and delay-type template elements and non-uniform grids', *Int J. Circuit Theory and Applications*, **20**, 469–481 (1992).
3c. T. Roska and L. O. Chua, 'The dual CNN analog software—a programmable analog, nonlinear, dynamic, 3D computing array plus logic to form a "multi-screen treater" on silicon', Report DNS-2-1992, Dual and Neural Computing Systems Laboratory, Comp. Aut. Inst., Hungarian Academy of Sciences, Budapest, 1992
4a. W. Heiligenberg, 'The neural basis of behavior: a neuroethological view', *Ann. Rev. Neurosci.*, **14**, 247–267 (1991).
4b. W. Heiligenberg, 'Neural nets in electric fish'. *The Computational Neuroscience Series*, The MIT Press, Cambridge, Massachusetts, 1991.
4c. W. Heiligenberg, 'Central processing of sensory information in electric fish'. *J. Comp. Physiol. A*, **161**, 621–631 (1987).
4d. P. Baldi and W. Heiligenberg, 'How sensory maps could enhance resolution through ordered arrangements of broadly tuned receivers', *Biol. Cybern.*, **59**, 313–318 (1988).
5. J. Hámori, T. Pasik, P. Pasik and J.Szentágothai, 'Triadic synaptic arrangements and their possible significance in the lateral geniculate nucleus of the monkey' *Brain Research*, **80**, 379–393 (1974).
6a. M. Konishi, 'Deciphering the brain's codes', *Neural Computation*, **3**, 1–18 (1991).
6b. M. Konishi *et al.*, 'Neurophysiological and anatomical substrates of sound localization in the owl' *Auditory function* (eds. G. M. Edelman, W. E. Gall, and W. M. Cowan), Wiley, New York, pp. 721–745, (1988).
7a. J. Allman, F. Miezin, and E. McGuiness, 'Stimulus specific responses beyond the classical receptive field: neurophysiological mechanisms for local-global comparisons in visual neurons', *Ann. Rev. Neurosci.*, **8**, 407–430 (1985).
7b. G. Montgomery, 'The mind's eye', *Discover*, pp. 51–56, May (1991).
8. D. C. Van Essen, C. H. Anderson, and D. J. Felleman, 'Information processing in the primate visual system: an integrated systems perspective', *Science*, **255**, 419–423, January 24, (1992).
9. J. Lazarro and C. Mead, 'A silicon model of auditory localization', *Neural Computation*, **1**, 47–57, (1989).
10. M. A. Mahowald and C. Mead, 'The silicon retina', *Scientific American*, **264**, No. 5, 76–82, May (1991).
11. T. Roska, K. Lotz, J. Hámori, E. Labós and J. Takács, *et al.*, 'The CNN model in the visual pathway—Part I: The CNN-Retina and some direction- and length-selective mechanisms', Report DNS-10-1991, Dual and Neural Computing Systems Res. Lab., Comp. Aut. Inst., Hungarian Academy of Sciences, Budapest, 1991
12. F. Wörgötter and C. Koch, 'A detailed model of the visual pathway in the cat: Comparison of afferent excitory and intracortical inhibitory connection schemes for orientation selectivity', *J. Neuroscience*, **11**, 1959–1979 (1991).
13a. K. Halonen, V. Porra, T. Roska and L. O. Chua, 'VLSI implementation of a reconfigurable CNN containing local logic', *Proc. CNNA-90*, pp. 206–215, (1990).
13b. K. Halonen, V. Porra, T. Roska and L. O. Chua, 'Programmable analogue VLSI CNN chip with local logic', *Int. J. of Circuit Theory and Applications*, **20**, 573–582, (1992).
14. J. M. Cruz and L. O. Chua, 'A CNN chip for connected component detection', *IEEE Trans. Circuits and Systems*, **CAS-28** 812–817 (1991).
15. S. Özdemir and I. C. Göknar, 'CNN templates for Gaussian expansion of sampled signals', *Proc. ECCTD 91 (6th European Conference on Circuit Theory and Design, Copenhagen)*, **1**, pp. 451–460, (1991).
16. N. Suga 'Functional organization of the auditory cortex. Representation beyond tonotopy in the bat', *Cortical Sensory Organization*, vol. 3, *Multiple Auditory Areas*, (ed. Clinton N. Woolsey), The Humana Press Inc., Clifton, New Jersey (1982).
17. L. O. Chua and T. Roska, 'The CNN Universal Machine—Part 1: the architecture', *Proc. IEEE International Workshop on Cellular Neural Networks and their Applications, CNNA'92*, pp. 1–10, (1992).

CELLULAR NEURAL NETWORK SIMULATOR USER'S MANUAL CNNS VERSION 3.6

1. INTRODUCTION

The Cellular Neural Network Simulator (CNNS) provides a simple experimental tool for studying cellular neural networks. It offers a quick and easy way to become familiar with the basics of the CNN paradigm and allows reproduction of many results reported in the literature on one's own. Transients of CNNs with linear, nonlinear, and delay-type templates can be monitored graphically. A menu-driven user interface provides full control of the program.

This software simulator is, at the same time, an introduction to our much more powerful system called CNN Workstation[4]. As to the CNN paradigm, we refer to the original papers[1,2] and to a recent one[3] as well as the first chapter of this book.

2. INSTALLATION

2.1. System requirements

The CNNS program runs on any AT compatible machine under DOS version 3.3 or higher. It can be operated both by the mouse and the keyboard. The proper operation requires **ANSI.SYS** (or equivalent) to be installed in the **CONFIG.SYS** file. To correctly install the simulator follow the steps below.

2.2. Installation procedure

1. Insert the diskette with the simulator into drive A (or B) and switch to the appropriate drive typing

   `a: (or b:)`

2. Install the simulator files executing the command

   `install c:\my_cnns\`

   where the parameter `c:\my_cnns` specifies the path to the target directory in which the program is intended to be. (Target directory need not exist.)
3. Check the **CONFIG.SYS** file on your system and make sure that it contains a line similar to this one:

   `DEVICE=C:\DOS\ANSI.SYS`

   If not, add this line with the appropriate DOS path to the file.
4. If you wish to use a specific text editor to look at the template files, edit the file **CNNEDTEM.BAT** in the target directory and change the text `edlin` to the path and name of your favorite text editor.
5. Switch to the target directory and type `cnn` to start the simulator.

3. GETTING STARTED

The simulator starts up with a screen divided into two parts. On the left-hand side the CNN cell array will be displayed coding the state values with colors. On the right-hand side the menu, file, and parameter windows are located. At startup the main menu is displayed in the menu window. To activate a menu item move the highlight using the arrow keys or moving the mouse and press the *enter* key or click with the left mouse button when the desired item is lit. Pressing the capitalized letters shown in a menu item has the same effect.

To begin with, start the 'Transfer' submenu and choose 'Load file'. In the load submenu click on 'init. State' to set the initial state of the cell array to an image. In the next window to the right a list of image files appears. Select the file 'CHINESE' which contains a chinese character. When the selection is made the image shows up in the network window as it appears on the output of the cell array. At the same time the file window shows that the same image is loaded onto the inputs of the network as well. It happens at the first transfer only. Now load a template file by invoking the 'Template' item. A list of template files appears in the file window. Use the arrow and page up/page down keys to search among the files. Scrolling the list is also possible. Move the mouse to the bottom of the window until 'DOWN' appears on the border. Hold the left button pressed and the window starts scrolling down. Similarly, clicking on 'UP' at the top border of the window scrolls it in the opposite direction.

In every file list there is an 'Other' item. Choosing that allows you to access any file on your system.

To proceed, choose the 'CORNER' template and exit the load menu by selecting 'Exit' or pressing *escape* or clicking the right mouse button. Repeat this to get back to the main menu.

With all necessary files loaded the simulation starts by invoking 'Run'. Evolution of the transient is shown like a movie as the computation progresses. At any time the transient can be interrupted by pressing *escape* or clicking the right mouse button. To continue press 'Run' again. When the final state is reached the extracted corners appear in the output picture and the simulation stops.

With the 'teSt' command the input of the network can be displayed which still contains the original character. Any key pressed returns to the previous screen.

To redo the simulation do 'restOre'. The program prompts for saving the output file. Choose 'No' or click the right mouse button to restore the initial state. Pressing'Run' again restarts the transient. To perform a different transformation, load another template after the state is restored and run the simulation.

Now you are familiar with the basic operation of the program and can proceed to understand it in more detail. A full description is given in the following sections.

4. PROGRAM FEATURES

4.1. *Operating modes*

The CNNS simulator has two basic operating modes: *simulation* and *animation*. When simulated, the CNN's transient is computed and displayed simultaneously. Sometimes the state transition is too fast; sometimes, at larger images, the simulation runs slowly. For this reason animation of the computed transient is also possible. First, the transient is computed and displayed as usually, but all computed data is dumped into an animation file. Then, using this file the transient can be played back at the desired speed.

From a given initial state and with a specified input image, state transition of the CNN array is computed using the forward or backward Euler integration formula. All parameter and variable values are measured in normalized units. On startup, formula type, time step and other parameters are read from the file con.par. The program looks for this file first in the actual DOS directory; therefore different users may have their own parameter settings. There is of course the possibility to modify these parameters from the menu system. When exiting the program actual parameter values are saved automatically.

4.2. Images and templates

Images and templates are stored in text files. The `IPL` and `CTL` directories contain image and template files, respectively. The simulator handles images up to 44×44 pixels and templates up to size 11×11 (neighborhood size 5). Three types of templates can be used: linear, delay-type, and nonlinear templates. All are contained in `*.tem` files whose exact format is described later. Due to implementation issues there are limitations on the template element values. The absolute value of a template element should not exceed 6 and the absolute sum of all template elements should remain under 25.

Images are stored in files with extension `.cni` or `.cns`. These files can be generated by the program saving the output or state pictures or using a text editor. There is also a graphical editor with can be called from the simulator or run independently. Input and output pictures in `*.cni` files have pixels valued in the range $[-1, +1]$. Black pixels correspond to $+1$, white pixels to -1. State values of the network can also be saved to a file at any iteration step. These files, usually called `*cns`, have the same format as `*.cni` files but pixel values are not necessarily in the $[-1, +1]$ range.

Templates and pictures can be viewed or edited without exiting the program. A graphical editor called `edcni.exe` provides easy access to `*.cni` pictures. It explains the color coding of the pixel values as well. Template files can be modified using a text editor. As default DOS' `edlin` is used but it is advised you replace this by a more convenient editor as described in section 2.

4.3. Animation

Animation files containing a graphical representation of a CNN transient are stored in files having an `.ani` extension. They are usually stored in the `IPL` directory. Animation files generated and read by the simulator are binary files, so therefore no editing is possible.

4.4. Printing

Image files can be printed from within the program. Both `*.cni` and `*.cns` files can be printed as a text, while `*.cni` files can be printed graphically as well. There are two possibilities for graphical printing: black-and-white and grey-scale. In case of grey-scale printing five grey levels are used and the maximum printable image size is 40×40.

When graphical printing is initiated the printer has to be switched to graphics mode. This is performed by an escape sequence contained in a `*.set` file. The escape sequence for the Citizen 120D printer is supplied in the `citibm.set` file. This file is used as the default, but any other can be specified.

5. THE MENU SYSTEM

This section describes the menu system and briefly explains the use of menu items and some error messages.

5.1. Transfer

5.1.1. Load file

- **Load initial state** Loads initial state values from `*.cni` files. If no input was loaded the same values are loaded to the input of the network. When new state values overwrite some previous result, saving is offered.
- **Load input values** Loads an image to the inputs of the network. If no initial state was loaded the same values are loaded to the initial states of the network.
- **Load template** Loads a template file.
- **Load Bias** Allows to attach space variant biasing current values $I_{i,j}$ to the circuit. The bias values corresponding to the CNN cells are also stored in image files. The format of the file is the same as that of the input picture files. In fact, any `*.cni` file can be loaded as space variant bias.

5.1.2. Save file

Allows saving of input, output, and state variables.

- **Save output values**
- **Save state values**
- **Save input values**

5.1.3. Delete file

Allows file deletion in the default directories.

- **Delete initial state file**
- **Delete input file**
- **Delete template file**
- **Delete animation file**

5.1.4. Rename file

Allows renaming in the default directories.

- **Rename initial state file**
- **Rename input file**
- **Rename template file**
- **Rename animation file**

5.1.5. Rewind file Input files may contain a series of images to facilitate time-varying input to the CNN. This command rewinds the input file in case of simulations with time-varying input picture. In the next step the first snapshot in the file will be fed to the input of the network. Rewinding is performed automatically whenever the `Restore` function is activated.

5.1.6. Output to input Copies the network's output values onto the inputs. Convenient when a sequence of templates is used.

5.1.7. Inverse state Switches the signs of the state variables.

5.1.8. Print Allows file printing.

- **Print output values**
  - — **Print output values graphically**
  - — **print output values numerically**
- **Print state**
- **Print input value**
- **Printer mode** Specifies grey-scale or black-and-white printing mode.

5.2. Edit

Allows editing picture and template files.

5.2.1. Edit picture Graphical editing of `*.cni` files. The default picture editor `EDCNI.EXE` is specified in the `CNNEDPIC.BAT` file.

5.2.2. Edit template Text editing of template files. You are advised to use your own text editor which can be installed as described in the installation procedure.

5.3. Environment

Default file extensions and search paths defined in the environment can be modified here. This menu also allows saving and loading simulation parameters and reads the printer driver file, as well.

5.3.1. Define initial state file

5.3.2. Define input file

5.3.3. Define template file

5.3.4. Define animation file

5.3.5. Define final state file

5.3.6. Set printer

5.3.7. Read parameter file

5.3.8. Write parameter file

5.4. Parameters

This menu allows modification of simulation parameters.

5.4.1. Time step Sets the value of the time step in the range [0·1, 3].

5.4.2. Accuracy Specifies the accuracy of computation. This parameter is used when steady state is detected.

5.4.3. Running mode

- **Steady state** Simulation runs until steady state is reached.
- **Normal** Simulation stops after 20 iteration steps.
- **Step by step** One iteration step is performed every time a `Run` command is issued.

5.4.4. Formula Specifies integration formula.

- **Implicit** Simulation uses forward Euler formula.
- **Explicit** Simulation uses backward Euler formula.

5.4.5. Animation Time Specifies delay between two frames in animation mode in relative units.

5.4.6. Oversampling An input picture file may contain a series of input images instead of a single image. Oversampling rate defines how long a given snapshot is kept on the input of the network in case of time-varying input pictures. A value of k means that the simulator will get a new picture from the input file in every kth time step. The default value is $k = 1$.

5.4.7. Filter When this option is set, a nonlinear output filter with the following transfer characteristic is connected to the output of the network:

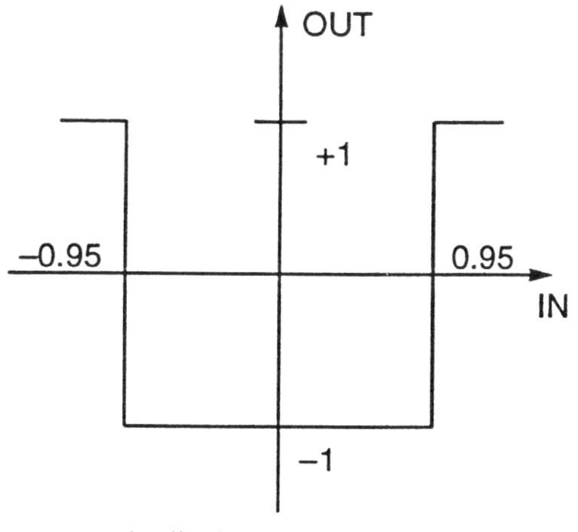

When the filter is connected, its output is displayed as the network's output.

5.5. Restore

This function reloads the last successfully loaded initial state file. If any iteration step has already been completed since the last state loading, the saving possibility is offered.

5.6. Run

Invoking `Run` starts the simulation. From the actual state time evolution of the network's state variables is computed using the specified numerical integration formula. Iteration steps are counted in the parameter window. Computation stops when the stopping condition specified by the running mode is met.

Error Message

'*No initial state file specified!*' There was no state information loaded since startup.

'*No template file specified!*' There was no template loaded since startup.

'*State and input images are of different sizes!*' State and input image sizes differ.

'*Numerical error! Decreasing time step!*' Occurs in case of implicit integration formula only. It means that in the correction step fix point iteration diverges. This is corrected by decreasing the time step, and the simulation continues. Note that in this case, although further computation is accurate, accuracy of the last few points is not guaranteed. You should repeat the simulation with smaller step size.

'*Fatal error! Fixed point iteration does not converge!*' Step size decreased under its possible lowest value (0.1). Further computation is not possible, simulation ends.

5.7. Run and dump file

This command is identical to `Run` except that it dumps the values of the state variables into a binary animation file. Before starting the computation a file name is asked. If the file exists already, confirmation is requested.

5.8. Animation

When activated, a list of animation files shows up in the file window. Animation starts by selecting a file. Time delay between two frames is controlled by the *animation time* parameter (see Parameters menu).

5.9. Quit

Returns to DOS.

6. FILE FORMATS

6.1. Parameter file

Parameter settings are stored in the `cnn.par` file. The program looks for this file in the actual DOS directory. If no parameter file is found, the following default file is generated:

```
TIME STEP:                        1.000000
RUNNING MODE:                     STEADY_STATE
FORMULA:                          EXPLICIT
ACCURACY:                         5
OVERSAMPLING:                     1
ANIMATION TIME:                   100
STATE INITIALIZATION:             ZERO
INPUT INITIALIZATION:             ZERO

INITIAL STATE FILE PATH:          .\ipl\
TEMPLATE FILE PATH:               .\ctl\
INPUT FILE PATH:                  .\ipl\
ANIMATION FILE PATH:              .\ipl\
OUTPUT VALUE FILE PATH:           .\ipl\

INITIAL STATE FILE EXTENSION:     .cni
TEMPLATE FILE EXTENSION:          .tem
INPUT FILE EXTENSION:             .cni
ANIMATION FILE EXTENSION:         .ani

OUTPUT VALUE FILE EXTENSION:      .cns

PRINTER CHARACTER SET:            CITIBM
```

Default values can be modified from the program or edited with a text editor.

6.2. Picture files

Picture files contain input, output, and state images up to 44×44 pixels. They have the following format:

```
3
5
 1.00 -0.28 -0.04 -0.15  0.24
 0.00 -0.40 -0.22 -1.00  0.25
-1.00 -0.37  0.00 -0.42 -0.37
```

where the first and second decimal integers define height and width of the image, respectively. Numbers following specify the pixel values: $+1$ (or greater) is displayed black, -1 (or smaller) displayed white, values in between are shown in different colors.

Time-varying input to the network is possible using picture files containing multiple images. Consecutive pictures have the above format and are separated by a line containing the '####' string only. A simple example of three 2×2 input frames is the following:

```
2 2
1.00 1.00
1.00 0.00
####
2 2
1.00 0.00
0.00 0.00
####
2 2
0.00 1.00
0.00 1.00
```

6.3. Template files

Template files contain keywords and numbers to specify template elements. The simulator handles templates up to size 11 x 11 (neighborhood size 5). Sintax rules for linear templates with or without delay and for nonlinear templates are described here.

6.3.1. Linear and delay-type templates Linear templates have the following format:

```
neighborhood: 2
feedback:
0 0 0 0 0
0 0 1 0 0
0 1 2 1 0
0 0 1 0 0
0 0 0 0 0
current: -1
control:
-0.1 -0.1 -0.1 -0.1 -0.1
-0.1  0.2  0.2  0.2 -0.1
-0.1  0.2 -0.3  0.2 -0.1
-0.1  0.2  0.2  0.2 -0.1
-0.1 -0.1 -0.1 -0.1 -0.1
delay: 5
delayed_feedback:
0 0  0 0 0
0 0  0 0 0
0 0 -1 0 0
0 0  0 0 0
0 0  0 0 0
delayed_control:
0.1  0.1  0.1  0.1  0.1
0.1 -0.2 -0.2 -0.2  0.1
0.1 -0.2  0.3 -0.2  0.1
0.1 -0.2 -0.2 -0.2  0.1
0.1  0.1  0.1  0.1  0.1
```

The `neighborhood` parameter specifies the neighborhood size and determines the expected number of template elements. Feedback templates denoted by `feedback` or `delayed_feedback` keywords are specified by a series of numbers. If organized in an appropriate matrix (as shown in the example) these numbers can be trivially mapped to the CNN array. Similarly, `control` and `delayed_control` operators denote feedforward templates. The constant current source is given by the `current` parameter, the delay factor is determined by the number following the `delay` keyword. In case of simple linear templates delay-type keywords should be omitted.

6.3.2. Nonlinear templates Nonlinear templates have the following format:

```
neighborhood:  1
nonlin_feedback:
      a*      0      B+
      c0      2      1
      b-      A0     C+
C 1   4    -2.5   3   -1   0  1 1   2  2.5
a 0   2    -2   2    2 2
b 1   3     0  0    1 -1    2 0
```

```
current: -3
nonlin_control:
  a-      a-     a-
  a-      0      a-
  a-      a-     a-
a03-0.180.50.18-110.5
```

Nonlinear template files obey the following rules:

- Upper and lower case letters are equivalent.
- Every item must be separated by space, tab, or newline character(s).
- Keywords must follow the above order.
- A template element can be a constant (like in the linear case) or a function identifier which is denoted by a letter instantly followed by one of the following characters:

  0: Output of the preceding function is computed for cell (i,j) using the output value of cell (k, l).
  +: The function is computed for cell (i,j) using the sum of the output values of cell (i,j) and (k,l).
  −: The function is computed using the difference of the output values: $\text{out}(k,l) - \text{out}(i,j)$.
  *: The funtion is computed using the product of the output values of cell (i,j) and (k,l).
- Function specifications can be in any order.
- A function specification consists of

  1. a function identifier (an alphabetic character),
  2. a 0 or 1 digit determining the type of interpolation: 0 specifies a step function:

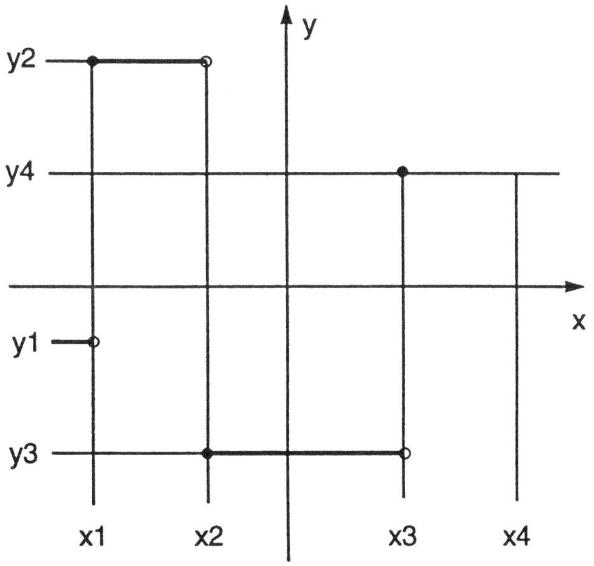

which is given by `b 0 4 x1 y1 x2 y2 x3 y3 x4 y4`

1 specifies linear interpolation:

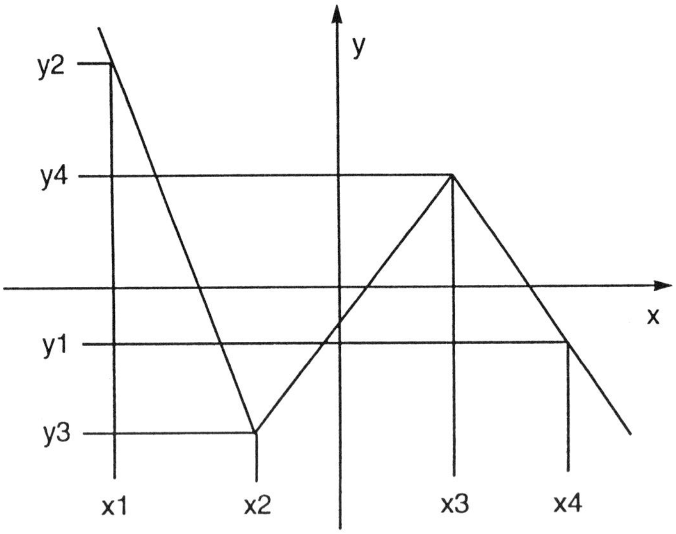

which is given by `b 1 4 x1 y1 x2 y2 x3 y3 x4 y4`

3. a decimal number specifying the number of representative points. (There must be at least two points specified.)
4. series of representative points in *increasing* order of variable x given by coordinate pairs as shown above.

- `nonlin_control` and `nonlin_feedback` templates are independent, and may contain different functions with the same identifier.

REFERENCES

1. L. O. Chua and L. Yang, 'Cellular neural networks: theory', *IEEE Trans. Circuits and Systems*, **CAS-35**, 1257–1272, (1988).
2. L. O. Chua and L. Yang, 'Cellular neural networks: application', *IEEE Trans Circuits and Systems*, **CAS-35**, 1273–1290, (1988)
3. T. Roska and L. O. Chua, 'The CNN universal machine—analogic array computer', *IEEE Trans. Circuits and Systems—Part II*, **CAS-40**, 163–173, (1993).
4. CNN Workstation, V.5.1, User's Manual, MTA-SzTAKI, Budapest (1992).

INDEX